INTERNATIONAL UNION OF PURE AND APPLIED CHEMISTRY

ANALYTICAL CHEMISTRY DIVISION
COMMISSION ON SOLUBILITY DATA

SOLUBILITY DATA SERIES

Volume 8

OXIDES OF NITROGEN

SOLUBILITY DATA SERIES

Volume 1 H. L. Clever, *Helium and Neon*

Volume 2 H. L. Clever, *Krypton, Xenon and Radon*

Volume 3 M. Salomon, *Silver Azide, Cyanide, Cyanamides, Cyanate, Selenocyanate and Thiocyanate*

Volume 4 H. L. Clever, *Argon*

Volume 5/6 C. L. Young, *Hydrogen and Deuterium*

Volume 7 R. Battino, *Oxygen and Ozone*

Volume 8 C. L. Young, *Oxides of Nitrogen*

Volume 11 B. Scrosati and C. A. Vincent, *Alkali Metal, Alkaline-Earth Metal and Ammonium Halides. Amide Solvents*

Volume 18 O. Popovych, *Tetraphenylborates*

Volumes in preparation

C. L. Young, *Oxides of Sulfur*

R. Battino, *Nitrogen*

H. L. Clever and W. Gerrard, *Hydrogen Halides in Non-Aqueous Solvents*

E. M. Woolley, *Silver Halides*

Z. Galus and C. Guminski, *Metals and Intermetallic Compounds in Mercury*

J. W. Lorimer, *Alkaline-Earth Metal Sulfates*

A. L. Horvath and F. W. Getzen, *Halogenated Benzenes, Toluenes, Xylenes and Phenols with Water*

H. L. Clever, D. M. Mason and C. L. Young, *Carbon Dioxide*

W. Gerrard and C. L. Young, *Hydrides and Methylamines*

H. L. Clever and C. L. Young, *Methane*

W. Hayduk, *Ethane*

NOTICE TO READERS

Dear Reader

If your library is not already a standing-order customer or subscriber to the Solubility Data Series, may we recommend that you place a standing order or subscription order to receive immediately upon publication all new volumes published in this valuable series. Should you find that these volumes no longer serve your needs, your order can be cancelled at any time without notice.

Robert Maxwell
Publisher at Pergamon Press

SOLUBILITY DATA SERIES

Volume 8

OXIDES OF NITROGEN

Volume Editor

COLIN L. YOUNG
University of Melbourne
Parkville, Victoria
Australia

Evaluators

RUBIN BATTINO
Wright State University
Dayton, Ohio, USA

WILLIAM GERRARD
Polytechnic of North London
Holloway, London, UK

Compilers

H. LAWRENCE CLEVER
Emory University
Atlanta, Georgia, USA

M. ELIZABETH DERRICK
Valdosta State University
Valdosta, Georgia, USA

SUSAN A. JOHNSON
Emory University
Atlanta, Georgia, USA

ALAN S. VOSPER
Portsmouth Polytechnic
Portsmouth, UK

QD543
S6629
v.8
1981

PERGAMON PRESS

OXFORD · NEW YORK · TORONTO · SYDNEY · PARIS · FRANKFURT

U.K.	Pergamon Press Ltd., Headington Hill Hall, Oxford OX3 0BW, England
U.S.A.	Pergamon Press Inc., Maxwell House, Fairview Park, Elmsford, New York 10523, U.S.A.
CANADA	Pergamon Press Canada Ltd., Suite 104, 150 Consumers Rd., Willowdale, Ontario M2J 1P9, Canada
AUSTRALIA	Pergamon Press (Aust.) Pty. Ltd., P.O. Box 544, Potts Point, N.S.W. 2011, Australia
FRANCE	Pergamon Press SARL, 24 rue des Ecoles, 75240 Paris, Cedex 05, France
FEDERAL REPUBLIC OF GERMANY	Pergamon Press GmbH, 6242 Kronberg-Taunus, Hammerweg 6, Federal Republic of Germany

---

Copyright © 1981 International Union of Pure and Applied Chemistry

All Rights Reserved. No part of this publication may be reproduced, stored in a retrieval system or transmitted in any form or by any means: electronic, electrostatic, magnetic tape, mechanical, photocopying, recording or otherwise, without permission in writing from the copyright holders.

First edition 1981

British Library Cataloguing in Publication Data

Oxides of nitrogen.—(Solubility data series, ISSN 0191-5622; v. 8)
1. Nitrogen oxides
2. Solubility
I. Young, Colin L. II. Series
546'.711'2 QD181.N1
ISBN 0-08-023924-2

In order to make this volume available as economically and as rapidly as possible the authors' typescripts have been reproduced in their original forms. This method unfortunately has its typographical limitations but it is hoped that they in no way distract the reader.

Printed in Great Britain by A. Wheaton & Co. Ltd., Exeter

CONTENTS

SOLUBILITY DATA SERIES

Editor-in-Chief

A. S. KERTES
The Hebrew University
Jerusalem, Israel

EDITORIAL BOARD

H. Akaiwa (Japan)
A. F. M. Barton (Australia)
R. Battino (USA)
A. Bylicki (Poland)
H. L. Clever (USA)
R. Cohen-Adad (France)
W. Gerrard (UK)
F. W. Getzen (USA)
L. H. Gevantman (USA)
A. L. Horvath (UK)
C. Kalidas (India)
J. W. Lorimer (Canada)
A. Maczynski (Poland)
J. D. Navratil (USA)
M. Salomon (USA)
P. W. Schindler (Switzerland)
D. G. Shaw (USA)
A. Szafranski (USA)
E. Wilhelm (Austria)
B. A. Wolf (FRG)
E. M. Woolley (USA)
S. H. Yalkowsky (USA)
C. L. Young (Australia)

INTERNATIONAL UNION OF PURE AND APPLIED CHEMISTRY

IUPAC Secretariat: Bank Court Chambers, 2-3 Pound Way,
Cowley Centre, Oxford OX4 3YF, UK

FOREWORD

If the knowledge is undigested or simply wrong, more is not better.

How to communicate and disseminate numerical data effectively in chemical science and technology has been a problem of serious and growing concern to IUPAC, the International Union of Pure and Applied Chemistry, for the last two decades. The steadily expanding volume of numerical information, the formulation of new interdisciplinary areas in which chemistry is a partner, and the links between these and existing traditional subdisciplines in chemistry, along with an increasing number of users, have been considered as urgent aspects of the information problem in general, and of the numerical data problem in particular.

Among the several numerical data projects initiated and operated by various IUPAC commissions, the *Solubility Data Project* is probably one of the most ambitious ones. It is concerned with preparing a comprehensive critical compilation of data on solubilities in all physical systems, of gases, liquids and solids. Both the basic and applied branches of almost all scientific disciplines require a knowledge of solubilities as a function of solvent, temperature and pressure. Solubility data are basic to the fundamental understanding of processes relevant to agronomy, biology, chemistry, geology and oceanography, medicine and pharmacology, and metallurgy and materials science. Knowledge of solubility is very frequently of great importance to such diverse practical applications as drug dosage and drug solubility in biological fluids, anesthesiology, corrosion by dissolution of metals, properties of glasses, ceramics, concretes and coatings, phase relations in the formation of minerals and alloys, the deposits of minerals and radioactive fission products from ocean waters, the composition of ground waters, and the requirements of oxygen and other gases in life support systems.

The widespread relevance of solubility data to many branches and disciplines of science, medicine, technology and engineering, and the difficulty of recovering solubility data from the literature, lead to the proliferation of published data in an ever increasing number of scientific and technical primary sources. The sheer volume of data has overcome the capacity of the classical secondary and tertiary services to respond effectively.

While the proportion of secondary services of the review article type is generally increasing due to the rapid growth of all forms of primary literature, the review articles become more limited in scope, more specialized. The disturbing phenomenon is that in some disciplines, certainly in chemistry, authors are reluctant to treat even those limited-in-scope reviews exhaustively. There is a trend to preselect the literature, sometimes under the pretext of reducing it to manageable size. The crucial problem with such preselection - as far as numerical data are concerned - is that there is no indication as to whether the material was excluded by design or by a less than thorough literature search. We are equally concerned that most current secondary sources, critical in character as they may be, give scant attention to numerical data.

On the other hand, tertiary sources - handbooks, reference books, and other tabulated and graphical compilations - as they exist today, are comprehensive but, as a rule, uncritical. They usually attempt to cover whole disciplines, thus obviously are superficial in treatment. Since they command a wide market, we believe that their service to advancement of science is at least questionable. Additionally, the change which is taking place in the generation of new and diversified numerical data, and the rate at which this is done, is not reflected in an increased third-level service. The emergence of new tertiary literature sources does not parallel the shift that has occurred in the primary literature.

With the status of current secondary and tertiary services being as briefly stated above, the innovative approach of the *Solubility Data Project* is that its compilation and critical evaluation work involve consolidation and reprocessing services when both activities are based on intellectual and scholarly reworking of information from primary sources. It comprises compact compilation, rationalization and simplification, and the fitting of isolated numerical data into a critically evaluated general framework.

The *Solubility Data Project* has developed a mechanism which involves a number of innovations in exploiting the literature fully, and which contains new elements of a more imaginative approach for transfer of reliable information from primary to secondary/tertiary sources. *The fundamental trend of the Solubility Data Project is toward integration of secondary and tertiary services with the objective of producing in-depth critical analysis and evaluation which are characteristic to secondary services, in a scope as broad as conventional tertiary services.*

Fundamental to the philosophy of the project is the recognition that the basic element of strength is the active participation of career scientists in it. Consolidating primary data, producing a truly critically-evaluated set of numerical data, and synthesizing data in a meaningful relationship are demands considered worthy of the efforts of top scientists. Career scientists, who themselves contribute to science by their involvement in active scientific research, are the backbone of the project. The scholarly work is commissioned to recognized authorities, involving a process of careful selection in the best tradition of IUPAC. This selection in turn is the key to the quality of the output. These top experts are expected to view their specific topics dispassionately, paying equal attention to their own contributions and to those of their peers. They digest literature data into a coherent story by weeding out what is wrong from what is believed to be right. To fulfill this task, the evaluator must cover *all* relevant open literature. No reference is excluded by design and every effort is made to detect every bit of relevant primary source. Poor quality or wrong data are mentioned and explicitly disqualified as such. In fact, it is only when the reliable data are presented alongside the unreliable data that proper justice can be done. The user is bound to have incomparably more confidence in a succinct evaluative commentary and a comprehensive review with a complete bibliography to both good and poor data.

It is the standard practice that any given solute-solvent system consists of two essential parts: I. Critical Evaluation and Recommended Values, and II. Compiled Data Sheets.

The Critical Evaluation part gives the following information:

(i) a verbal text of evaluation which discusses the numerical solubility information appearing in the primary sources located in the literature. The evaluation text concerns primarily the quality of data after consideration of the purity of the materials and their characterization, the experimental method employed and the uncertainties in control of physical parameters, the reproducibility of the data, the agreement of the worker's results on accepted test systems with standard values, and finally, the fitting of data, with suitable statistical tests, to mathematical functions;

(ii) a set of recommended numerical data. Whenever possible, the set of recommended data includes weighted average and standard deviations, and a set of smoothing equations derived from the experimental data endorsed by the evaluator;

(iii) a graphical plot of recommended data.

The compilation part consists of data sheets of the best experimental data in the primary literature. Generally speaking, such independent data sheets are given only to the best and endorsed data covering the known range of experimental parameters. Data sheets based on primary sources where the data are of a lower precision are given only when no better data are available. Experimental data with a precision poorer than considered acceptable are reproduced in the form of data sheets when they are the only known data for a particular system. Such data are considered to be still suitable for some applications, and their presence in the compilation should alert researchers to areas that need more work.

The typical data sheet carries the following information:

(i) components - definition of the system - their names, formulas and Chemical Abstracts registry numbers;

(ii) reference to the primary source where the numerical information is reported. In cases when the primary source is a less common periodical or a report document, published though of limited availability, abstract references are also given;

(iii) experimental variables;

(iv) identification of the compiler;

(v) experimental values as they appear in the primary source. Whenever available, the data may be given both in tabular and graphical form. If auxiliary information is available, the experimental data are converted also to SI units by the compiler.

Under the general heading of Auxiliary Information, the essential experimental details are summarized:

(vi) experimental method used for the generation of data;

(vii) type of apparatus and procedure employed;

(viii) source and purity of materials;

(ix) estimated error;

(x) references relevant to the generation of experimental data as cited in the primary source.

This new approach to numerical data presentation, developed during our four years of existence, has been strongly influenced by the diversity of background of those whom we are supposed to serve. We thus deemed it right to preface the evaluation/compilation sheets in each volume with a detailed discussion of the principles of the accurate determination of relevant solubility data and related thermodynamic information.

Finally, the role of education is more than corollary to the efforts we are seeking. The scientific standards advocated here are necessary to strengthen science and technology, and should be regarded as a major effort in the training and formation of the next generation of scientists and engineers. Specifically, we believe that there is going to be an impact of our project on scientific-communication practices. The quality of consolidation adopted by this program offers down-to-earth guidelines, concrete examples which are bound to make primary publication services more responsive than ever before to the needs of users. The self-regulatory message to scientists of 15 years ago to refrain from unnecessary publication has not achieved much. The literature is still, in 1981, cluttered with poor-quality articles. The Weinberg report (in "Reader in Science Information", Eds. J. Sherrod and A. Hodina, Microcard Editions Books, Indian Head, Inc., 1973, p.292) states that "admonition to authors to restrain themselves from premature, unnecessary publication can have little effect unless the climate of the entire technical and scholarly community encourages restraint..." We think that projects of this kind translate the climate into operational terms by exerting pressure on authors to avoid submitting low-grade material. The type of our output, we hope, will encourage attention to quality as authors will increasingly realize that their work will not be suited for permanent retrievability unless it meets the standards adopted in this project. It should help to dispel confusion in the minds of many authors of what represents a permanently useful bit of information of an archival value, and what does not.

If we succeed in that aim, even partially, we have then done our share in protecting the scientific community from unwanted and irrelevant, wrong numerical information.

A. S. Kertes

PREFACE

Users of this volume will find (1) the experimental solubility data of gaseous nitrous oxide and nitric oxide in liquids as reported in the scientific literature (2) tables of smoothed mole fraction solubility values for a limited number of systems which have been studied over a temperature range (3) critical evaluations of the experimental data and a limited number of tables of either tentative or recommended solubility data when two or more laboratories reported data over the same range of temperature and pressure for a system. The number of smoothed data values given is considerably less than in previous volumes on gas solubility (*i.e.*, Volumes 1, 2 and 4) since the number of systems for which two or more groups of workers have reported solubility data is very limited.

The present volume covers the literature up the end of 1979. Some 1980 papers are cited.

Some words of explanation are required with respect to units, corrections, smoothing equations, auxiliary data and data sources, nomenclature and other points. The experimental data are presented in the units found in the original paper. In addition, the original data are often converted to other units, especially mole fraction. Temperatures have been converted to Kelvin. In evaluations of solubility data, SI units are used where practical.

Only in the past 10 to 15 years have experimental methods for the determination of the solubility of gases in liquids developed to the point where 0.5 per cent or better accuracy is attained. Only a small fraction of the literature's gas solubility data is accurate to 0.5 per cent. The corrections for non-ideal gas behaviour and for expansion of the liquid phase on dissolution of the gas are small and usually well within the normal experimental error. Thus such corrections were not made for gas solubility data of the oxides of nitrogen at low pressure unless stated otherwise. In general, measurements of the solubility of the oxides of nitrogen are not of very high accuracy. Many of the measurements were undertaken over forty years ago and by present day standards are of low accuracy.

Indeed, the compilation, evaluation and editing of this volume was made particularly difficult by the fact that although fairly extensive data are available for the nitrous oxide and nitric oxide solubilities particularly in aqueous solution, the data have been presented in a manner which makes it difficult (a) to compare with other workers and (b) in some cases to be sure what the author intended. It must be appreciated that the compilers and to some slightly lesser extent the evaluators, are almost completely limited by the authors' presentation.

Most gas solubility measurements carried out near atmospheric conditions are measured at a total pressure near one atmosphere with the gas saturated with the solvent vapor. Usually the actual partial pressure of the gas is not known. In such experiments the Ostwald coefficient is the directly measured unit. The Bunsen coefficient and the mole fraction gas solubility at one atmosphere gas partial pressure are calculated from the Ostwald coefficient assuming that the Ostwald coefficient is independent of pressure.

Solvent density data were often required in making solubility unit conversions. The density data were not directly referenced. The main sources of density data are

Circular 461 of the U.S. National Bureau of Standards
American Petroleum Research Project 44 Publications
The International Critical Tables, Volume III (E. W. Washburn, Editor), McGraw-Hill Co., 1931
Smow Table, *Pure and Applied Chemistry*, 1976, 45, 1-9
Thermodynamic Properties of Aliphatic Alcohols, R. C. Wilhoit and B. J. Zwolinski, *J. Phys. Chem. Ref. Data*, 1973, 2, Supplement No. 1
Organic Solvents, J. A. Riddick and W. B. Bunger (Technique of Chemistry, Volume II, A. Weissberger, Editor), Wiley-Interscience, New York, 1970, 3rd Ed.

Chemical Abstracts registry numbers were used throughout. Common names are cross referenced to Chemical Abstract recommended names in the Index. There is a Registry Number Index.

The Editor would appreciate users calling errors and omissions to his attention.

The Editor gratefully acknowledges the advice and comments of the IUPAC Commission on Solubility Data, the cooperation and hard work of the Evaluators and Compilers, the work of typists Joy Wall and Lesley Flanagan and the help of Kerri Hubbard in obtaining photocopies of papers. Acknowledgement is also made to the University of Melbourne for a travel grant which facilitated consultation with the evaluators.

Colin L. Young

Melbourne, Victoria
January 1981

THE SOLUBILITY OF GASES IN LIQUIDS

R. Battino, H. L. Clever and C. L. Young

INTRODUCTION

The Solubility Data Project aims to make a comprehensive search of the literature for data on the solubility of gases, liquids and solids in liquids. Data of suitable accuracy are compiled into data sheets set out in a uniform format. The data for each system are evaluated and where data of sufficient accuracy are available values recommended and in some cases a smoothing equation suggested to represent the variation of solubility with pressure and/or temperature. A text giving an evaluation and recommended values and the compiled data sheets are published on consecutive pages.

DEFINITION OF GAS SOLUBILITY

The distinction between vapor-liquid equilibria and the solubility of gases in liquids is arbitrary. It is generally accepted that the equilibrium set up at 300K between a typical gas such as argon and a liquid such as water is gas liquid solubility whereas the equilibrium set up between hexane and cyclohexane at 350K is an example of vapor-liquid equilibrium. However, the distinction between gas-liquid solubility and vapor-liquid equilibrium is often not so clear. The equilibria set up between methane and propane above the critical temperature of methane and below the critical temperature of propane may be classed as vapor-liquid equilibrium or as gas-liquid solubility depending on the particular range of pressure considered and the particular worker concerned.

The difficulty partly stems from our inability to rigorously distinguish between a gas, a vapor, and a liquid, which has been discussed in numerous textbooks. We have taken a fairly liberal view in these volumes and have included systems which may be regarded, by some workers, as vapor-liquid equilibria.

UNITS AND QUANTITIES

The solubility of gases in liquids is of interest to a wide range of scientific and technological disciplines and not solely to chemistry. Therefore a variety of ways for reporting gas solubility have been used in the primary literature and inevitably sometimes, because of insufficient available information, it has been necessary to use several quantities in the compiled tables. Where possible, the gas solubility has been quoted as a mole fraction of the gaseous component in the liquid phase. The units of pressure used are bar, pascal, millimeters of mercury and atmosphere. Temperatures are reported in Kelvin.

EVALUATION AND COMPILATION

The solubility of comparatively few systems is known with sufficient accuracy to enable a set of recommended values to be presented. This is true both of the measurement near atmospheric pressure and at high pressures. Although a considerable number of systems have been studied by at least two workers, the range of pressures and/or temperatures is often sufficiently different to make meaningful comparison impossible.

Occasionally, it is not clear why two groups of workers obtained very different sets of results at the same temperature and pressure, although both sets of results were obtained by reliable methods and are internally consistent. In such cases, sometimes an incorrect assessment has been given. There are several examples where two or more sets of data have been classified as tentative although the sets are mutually inconsistent.

Many high pressure solubility data have been published in a smoothed form. Such data are particularly difficult to evaluate, and unless specifically discussed by the authors, the estimated error on such values can only be regarded as an "informed guess".

Many of the high pressure solubility data have been obtained in a more general study of high pressure vapor-liquid equilibrium. In such cases a note is included to indicate that additional vapor-liquid equilibrium data are given in the source. Since the evaluation is for the compiled data, it is possible that the solubility data are given a classification which is better than that which would be given for the complete vapor-liquid data (or vice versa). For example, it is difficult to determine coexisting liquid and vapor compositions near the critical point of a mixture using some widely used experimental techniques which yield accurate high pressure solubility data. For example, conventional methods of analysis may give results with an expected error which would be regarded as sufficiently small for vapor-liquid equilibrium data but an order of magnitude too large for acceptable high pressure gas-liquid solubility.

It is occasionally possible to evaluate data on mixtures of a given substance with a member of a homologous series by considering all the available data for the given substance with other members of the homologous series. In this study the use of such a technique has been very limited.

The estimated error is often omitted in the original article and sometimes the errors quoted do not cover all the variables. In order to increase the usefulness of the compiled tables estimated errors have been included even when absent from the original article. If the error on *any* variable has been inserted by the compiler this has been noted.

PURITY OF MATERIALS

The purity of materials has been quoted in the compiled tables where given in the original publication. The solubility is usually more sensitive to impurities in the gaseous component than to liquid impurities in the liquid component. However, the most important impurities are traces of a gas dissolved in the liquid. Inadequate degassing of the absorbing liquid is probably the most often overlooked serious source of error in gas solubility measurements.

APPARATUS AND PROCEDURES

In the compiled tables brief mention is made of the apparatus and procedure. There are several reviews on experimental methods of determining gas solubilities and these are given in References 1-7.

METHODS OF EXPRESSING GAS SOLUBILITIES

Because gas solubilities are important for many different scientific and engineering problems, they have been expressed in a great many ways:

The Mole Fraction, x (g)

The mole fraction solubility for a binary system is given by:

$$x(\mathrm{g}) = \frac{n(\mathrm{g})}{n(\mathrm{g}) + n(\mathrm{l})}$$

$$= \frac{W(\mathrm{g})/M(\mathrm{g})}{[W(\mathrm{g})/M(\mathrm{g})] + [W(\mathrm{l})/M(\mathrm{l})]}$$

here n is the number of moles of a substance (an *amount* of substance), W is the mass of a substance, and M is the molecular mass. To be unambiguous, the partial pressure of the gas (or the total pressure) and the temperature of measurement must be specified.

The Weight Per Cent Solubility, wt%

For a binary system this is given by

$$\mathrm{wt\%} = 100\, W(\mathrm{g})/[W(\mathrm{g}) + W(\mathrm{l})]$$

where W is the weight of substance. As in the case of mole fraction, the pressure (partial or total) and the temperature must be specified. The weight per cent solubility is related to the mole fraction solubility by

$$x(g) = \frac{[wt\%/M(g)]}{[wt\%/M(g)] + [(100 - wt\%)/M(1)]}$$

The Weight Solubility, C_W

The weight solubility is the number of moles of dissolved gas per gram of solvent when the partial pressure of gas is 1 atmosphere. The weight solubility is related to the mole fraction solubility at one atmosphere partial pressure by

$$x(g) \text{ (partial pressure 1 atm)} = \frac{C_W M(1)}{1 + C_W M(1)}$$

where M(1) is the molecular weight of the solvent.

The Moles Per Unit Volume Solubility, n

Often for multicomponent systems the density of the liquid mixture is not known and the solubility is quoted as moles of gas per unit volume of liquid mixture. This is related to the mole fraction solubility by

$$x = \frac{n\, v^o(1)}{1 + n\, v^o(1)}$$

where $v^o(1)$ is the molar volume of the liquid component.

The Bunsen Coefficient, α

The Bunsen coefficient is defined as the volume of gas reduced to 273.15K and 1 atmosphere pressure which is absorbed by unit volume of solvent (at the temperature of measurement) under a partial pressure of 1 atmosphere. If ideal gas behavior and Henry's law is assumed to be obeyed,

$$\alpha = \frac{V(g)}{V(1)} \frac{273.15}{T}$$

where V(g) is the volume of gas absorbed and V(1) is the original (starting) volume of absorbing solvent. The mole fraction solubility x is related to the Bunsen coefficient by

$$x(1 \text{ atm}) = \frac{\alpha}{\alpha + \frac{273.15}{T} \frac{v^o(g)}{v^o(1)}}$$

where $v^o(g)$ and $v^o(1)$ are the molar volumes of gas and solvent at a pressure of one atmosphere. If the gas is ideal,

$$x = \frac{\alpha}{\alpha + \frac{273.15R}{v^o(1)}}$$

Real gases do not follow the ideal gas law and it is important to establish the real gas law used for calculating α in the original publication and to make the necessary adjustments when calculating the mole fraction solubility.

The Kuenen Coefficient, S

This is the volume of gas, reduced to 273.15K and 1 atmosphere pressure, dissolved at a partial pressure of gas of 1 atmosphere by 1 gram of solvent.

The Ostwald Coefficient, L

The Ostwald coefficient, L, is defined at the ratio of the volume of gas absorbed to the volume of the absorbing liquid, all measured at the same temperature:

$$L = \frac{V(g)}{V(1)}$$

If the gas is ideal and Henry's Law is applicable, the Ostwald coefficient is independent of the partial pressure of the gas. It is necessary, in practice, to state the temperature and total pressure for which the Ostwald coefficient is measured. The mole fraction solubility, x , is related to the Ostwald coefficient by

$$x = \left[\frac{RT}{P(g)\ L\ v^{o}(1)} + 1\right]^{-1}$$

where P is the partial pressure of gas. The mole fraction solubility will be at a partial pressure of P(g).

The Absorption Coefficient, β

There are several "absorption coefficients", the most commonly used one being defined as the volume of gas, reduced to 273.15K and 1 atmosphere, absorbed per unit volume of liquid when the total pressure is 1 atmosphere. β is related to the Bunsen coefficient by

$$\beta = \alpha(1-P(1))$$

where P(1) is the partial pressure of the liquid in atmosphere.

The Henry's Law Contant

A generally used formulation of Henry's Law may be expressed as

$$P(g) = K_H x$$

where K_H is the Henry's Law constant and x the mole fraction solubility. Other formulations are

$$P(g) = K_2 C(1)$$

or

$$C(g) = K_c C(1)$$

where K_2 and K_c are constants, C the concentration, and (1) and (g) refer to the liquid and gas phases. Unfortunately, K_H, K_2 and K_c are all sometimes referred to as Henry's Law constants. Henry's Law is a limiting law but can sometimes be used for converting solubility data from the experimental pressure to a partial gas pressure of 1 atmosphere, provided the mole fraction of the gas in the liquid is small, and that the difference in pressures is small. Great caution must be exercised in using Henry's Law.

The Mole Ratio, N

The mole ratio, N, is defined by

$$N = n(g)/n(1)$$

Table 1 contains a presentation of the most commonly used inter-conversions not already discussed.

For gas solubilities greater than about 0.01 mole fraction at a partial pressure of 1 atmosphere there are several additional factors which must be taken into account to unambiguously report gas solubilities. Solution densities or the partial molar volume of gases must be known. Corrections should be made for the possible non-ideality of the gas or the non-applicability of Henry's Law.

TABLE 1 Interconversion of parameters used for reporting solubility

$$L = \alpha(T/273.15)$$

$$C_w = \alpha/v_o\rho$$

$$K_H = \frac{17.033 \times 10^6 \rho(\text{soln})}{M(1)} - 760$$

$$L = C_w\, v_{t,gas}\, \rho$$

where v_o is the molal volume of the gas in $cm^3 mol^{-1}$ at 0^oC, ρ the density of the solvent at the temperature of the measurement, ρ(soln) the density of the solution at the temperature of the measurement, and $v_{t,gas}$ the molal volume of the gas ($cm^3 mol^{-1}$) at the temperature of the measurement.

SALT EFFECTS

The effect of a dissolved salt in the solvent on the solubility of a gas is often studied. The activity coefficient of a dissolved gas is a function of the concentration of all solute species (see ref. 8). At a given temperature and pressure the logarithm of the dissolved gas activity coefficient can be represented by a power series in C_s, the electrolyte concentration, and C_i, the nonelectrolyte solute gas concentration

$$\log f_i = \sum_{m,n} k_{mn} C_s^{\,n} C_i^{\,m}$$

It is usually assumed that only the linear terms are important for low C_s and C_i values when there is negligible chemical interaction between solute species.

$$\log f_i = k_s C_s + k_i C_i$$

where k_s is the salt effect parameter and k_i is the solute-solute gas interaction parameter. The dissolved gas activity is the same in the pure solvent and a salt solution in that solvent for a given partial pressure and temperature

$$a_i = f_i S_i = f_i^{\,o} S_i^{\,o} \text{ and } f_i = f_i^{\,o} \frac{S_i^{\,o}}{S_i}$$

where S_i and $S_i^{\,o}$ are the gas solubility in the salt solution and in the pure solvent, respectively, and the f's are the corresponding activity coefficients. If follows that $\log f_i/f_o = \log S_i^{\,o}/S_i = k_s C_s + k_i(S_i - S_i^{\,o})$. When the quantity $(S_i - S_i^{\,o})$ is small the second term is negligible even though k_s and k_i may be of similar magnitude. This is generally the case for gas solubilities and the equation reduces to

$$\log \frac{f_i}{f_i^{\,o}} = \log \frac{S_i^{\,o}}{S_i} = k_s C_s$$

which is the form of the empirical Setschenow equation in use since the 1880's. A salt that increases the activity coefficient of the dissolved gas is said to salt-out and a salt that decreases the activity coefficient of the dissolved gas is said to salt-in.

Although salt effect studies have been carried out for many years, there appears to be no common agreement of the units for either the gas solubility or the salt concentration. Both molar ($mol\ dm^{-3}$) and molal ($mol\ kg^{-1}$) are used for the salt concentration. The gas solubility ratio $S_i^{\,o}/S_i$ is given as Bunsen coefficient ratio and Ostwald coefficient ratio, which would be the same as a molar ratio; Kueunen coefficient ratio, volume dissolved in 1 g or 1 kg of solvent which would be a molal ratio; and mole fraction ratio. Recent theoretical treatments use salt concentration in mol dm^{-3} and S_i^0/S_i ratio as mole fraction ratio with each salt ion acting as a mole. Evaluations which compare the results of several workers are made in the units most compatible with present theory.

TEMPERATURE DEPENDENCE OF GAS SOLUBILITY

In a few cases it has been found possible to fit the mole fraction solubility at various temperatures using an equation of the form.

$$\ln x = A + B / (T/100K) + C \ln (T/100K)$$

It is then possible to write the thermodynamic functions $\Delta\bar{G}_1^0, \Delta\bar{H}_1^0, \Delta\bar{S}_1^0$ and $\Delta\bar{C}^0_{p_1}$ for the transfer of the gas from the vapor phase at 101.325 Pa partial pressure to the (hypothetical) solution phase of unit mole fraction as :

$$\Delta\bar{G}_1^0 = -RAT - 100\ RB - RCT \ln (T/100)$$

$$\Delta\bar{S}_1^0 = RA + RC \ln (T/100) + RC$$

$$\Delta\bar{H}_1^0 = -100\ RB + RCT$$

$$\Delta\bar{C}^0_{p_1} = RC$$

In cases where there are solubilities at only a few temperatures it is convenient to use the simpler equations.

$$\Delta\bar{G}_1^0 = -RT \ln x = A + BT$$

in which case $A = \Delta\bar{H}_1^0$ and $-B = \Delta\bar{S}_1^0$.

REFERENCES

1. Battino, R.; Clever, H.L. *Chem. Rev.* 1966, *66*, 395.
2. Clever, H.L.; Battino, R. in *Solutions and Solubilities*, Ed. M.R.J. Dack, J. Wiley & Sons, New York, 1975, Chapter 7.
3. Hildebrand, J. H.; Prausnitz, J.M.; Scott, R.L. *Regular and Related Solutions*, Van Nostrand Reinhold, New York, 1970, Chapter 8.
4. Markham, A.E.; Kobe, K.A. *Chem. Rev.* 1941, *63*,449.
5. Wilhelm, E.; Battino, R. *Chem. Rev.* 1973, *73*, 1.
6. Wilhelm, E.; Battino, R.; Wilcock, R.J. *Chem. Rev.* 1977, *77*, 219.
7. Kertes, A.S.; Levy, O.; Markovits, G.Y. in *Experimental Thermochemistry* Vol. II, Ed. B. Vodar and B. LeNaindre, Butterworth, London, 1974, Chapter 15.
8. Long, F.A.; McDevit, W.F. *Chem. Rev.* 1952, *51*, 119.

COMPONENTS:	EVALUATOR:
1. Nitrous oxide; N_2O; [10024-97-2] 2. Water; H_2O; [7732-18-5]	Rubin Battino, Department of Chemistry, Wright State University, Dayton, Ohio 45431, U.S.A.

CRITICAL EVALUATION:

The data used by eleven workers was considered to be of sufficient accuracy to use for the smoothing equation. In fitting the data, those points which were about two standard deviations greater than the smoothed data were rejected. In the data sheets which follow the points which were used are marked with asterisks. Twenty-three data points were used as follows (reference - number of data points used from that reference): 1-5, 2-1, 3-1, 4-1, 5-8, 6-1, 7-2, 8-1, 9-1, 10-1, 11-1. The fitting equation used was

$$\ln x_1 = A + B/(T/100K) + C \ln (T/100K) \quad (1)$$

Using T/100K as the variable gave coefficients of comparable magnitude. The best fit for the 23 points was

$$\ln x_1 = -60.7467 + 88.8280/(T/100K) + 21.2531 \ln (T/100K) \quad (2)$$

where x_1 is the mole fraction solubility of nitrous oxide at 101.325 kPa (1 atm) partial pressure of gas. The smoothing equation gave a fit in the mole fraction (one standard deviation) of 1.2%. Table 1 gives smoothed values of the mole fraction solubility at 101.325 kPa and the Ostwald coefficient at 5K intervals.

Table 1 also gives values of the thermodynamics functions ΔG_1°, ΔH_1°, ΔS_1°, and $\Delta C_{p_1}^\circ$ for the transfer of the gas from the vapor phase at 101.325 kPa partial gas pressure to the (hypothetical) solution phase of unit mole fraction. These were calculated from the smoothing equation according to the following equations:

$$\Delta G_1^\circ = -RAT - 100RB - RCT \ln (T/100) \quad (3)$$

$$\Delta S_1^\circ = RA + RC \ln (T/100) + RC \quad (4)$$

$$\Delta H_1^\circ = -100RB + RCT \quad (5)$$

$$\Delta C_{p_1}^\circ = RC \quad (6)$$

The heat capacity turns out to be independent of temperature since the three-constant fit was all that was needed for these data.

Several sets of data were rejected for purposes of fitting the smoothing equation. Roth's older data were consistently high by 6 to 8% (12). The chromatographically determined values of Jay *et al.*, (13) were 7% high. The two values determined by the Van Slyke method by Christoforides and Hedley-White (14) were 4 to 6% high. Hikita *et al.*'s single value was about 3% low (15). The Orcutt and Seevers value was 10% high (16).

Schwab and Berninger (17) reported data at 10K intervals from 293 to 353K, but their results for this gas and others are highly erratic and are rejected. Schröder (18) determined the solubility from 20 to 178 °C at a pressure of 40 atm. The results are given in graphical form and the author gives the equation

$$\ln \alpha = 3.925 - 5251/(T/K) + 1.109 \times 10^6/(T/K)$$

where α is the Bunsen coefficient.

<u>NOTE added by editor</u>: Additional data are given on pages 12-14, 21, 22. The recent data of Weiss and Price (19) are believed to be accurate and are classified as tentative.

COMPONENTS:

1. Nitrous oxide; N_2O; [10024-97-2]

2. Water; H_2O; [7732-18-5]

EVALUATOR:

Rubin Battino,
Department of Chemistry,
Wright State University,
Dayton, Ohio 45431,
U.S.A.

CRITICAL EVALUATION:

References

1. Geffcken, G. *Z. Physik. Chem.* 1904, *49*, 257-302.
2. Knopp, W. *Z. Physik. Chem.* 1904, *48*, 97-108.
3. Findlay, A.; Creighton, H. J. M. *J. Chem. Soc.* 1910, *97*, 536-61.
4. Findlay, A.; Howell, O. R. *J. Chem. Soc.* 1914, *105*, 291-8.
5. Kunerth, W. *Phys. Rev.* 1922, *19*, 512-24.
6. Manchot, W.; Jahrstorfer, M.; Zepter, H. *Z. Anorg. Allgem. Chem.* 1924, *141*, 45-81.
7. Markham, A. E.; Kobe, K. A. *J. Amer. Chem. Soc.* 1941, *63*, 449-54.
8. Joosten, G. E. H.; Danckwerts, P. V. *J. Chem. Eng. Data* 1972, *17*, 452-4.
9. Sada, E.; Kito, S.; Ito, Y. *J. Chem. Eng. Japan* 1974, *7*, 57-9.
10. Sada, E.; Kito, S.; Ito, Y. *Ind. Eng. Chem. Fundam.* 1975, *14*, 232-7.
11. Sada, E.; Kumazawa, H.; Butt, M. A. *J. Chem. Eng. Data* 1977, *22*, 277-9.
12. Roth, W. *Z. Physik. Chem.* 1897, *24*, 114-51.
13. Jay, B. E.; Wilson, R. H.; Doty, V.; Pingree, H.; Hargis, B. *Anal. Chem.* 1962, *34*, 414-8.
14. Christoforides, C.; Hedley-White, J. *Federation Proceedings* 1970, *29*:A330.
15. Hikita, H.; Asai, S.; Ishikawa, H.; Esaka, N. *J. Chem. Eng. Data* 1974, *19*, 89-92.
16. Orcutt, F. S.; Seevers, M. H. *J. Biol. Chem.* 1937, *117*, 501-7.
17. Schwab, G. M.; Berninger, E. *Z. Physik. Chem.* 1928, *A138*, 55.
18. Schröder, W. *Chem. Ing. Tech.* 1973, *45*, 603.
19. Weiss, R. F.; Price, B. A. *Marine Chem.* 1980, *8*, 347.

TABLE 1. Smoothed values of nitrous oxide solubility in water and thermodynamic functions using equation [a] (2) at 101.325 kPa (1 atm) partial pressure of gas.

T/K	$x_1 \times 10^4$ [b]	L [c]	$\Delta\bar{G}_1^\circ$ [d]	$\Delta\bar{H}_1^\circ$ [d]	$\Delta\bar{S}_1^\circ$ [e]
273.15	10.378	1.292	15.60	-25.59	-150.8
278.15	8.505	1.078	16.35	-24.70	-147.6
283.15	7.067	0.912	17.08	-23.80	-144.4
288.15	5.948	0.780	17.79	-22.94	-141.4
293.15	5.068	0.676	18.49	-22.05	-138.3
298.15	4.367	0.592	19.18	-21.17	-135.3
303.15	3.805	0.523	19.85	-20.29	-132.4
308.15	3.348	0.467	20.50	-19.40	-129.5
313.15	2.975	0.421	21.14	-18.52	-126.7

a. $\bar{C}_{p_1}^\circ$ was independent of temperature and has the value 178 J K^{-1} mol^{-1}.

b. Mole fraction solubility at 101.325 kPa partial pressure of gas.

c. Ostwald coefficient.

d. Units are K J mol^{-1}, cal_{th} = 4.184 J.

e. Units are J K^{-1} mol^{-1}.

COMPONENTS:

1. Nitrous oxide; N_2O; [10024-97-2]
2. Water; H_2O; [7732-18-5]

ORIGINAL MEASUREMENTS:

Roth, W.

Z. Physik. Chem., 1897, *24*, 114-51.

VARIABLES:

T/K: 278-298

PREPARED BY:

R. Battino

EXPERIMENTAL VALUES:

T/K [a]	$x_1 \times 10^4$ [b]	L [c]	α [d]
278.15	9.157	1.1612	1.1403
283.15	7.615	0.9826	0.9479
288.15	6.348	0.8330	0.7896
293.15	5.355	0.7141	0.6654
298.15	4.635	0.6279	0.5752

a. Temperature reported to 0.01^0C.

b. Mole fraction solubility at 101,325 Pa (1 atm) partial pressure of gas. Calculated by compiler.

c. Ostwald coefficient. Calculated by compiler.

d. Bunsen coefficient. Smoothed best values given by author calculated from $\alpha = 1.3668 - 0.04870\ t + 0.00068145\ t^2$ from measurements in the temperature range 3.5 to 24.7^0C.

AUXILIARY INFORMATION

METHOD/APPARATUS/PROCEDURE:

The Ostwald apparatus as described by Timofejew (1) was used.

SOURCE AND PURITY OF MATERIALS:

1. Nitrous oxide - prepared from pure ammonium nitrate.
2. Water - no comment by author.

ESTIMATED ERROR:

$\delta\alpha/\alpha = 0.01$ (compiler's estimate)

REFERENCES:

1. Timofejew, W., *Z. Physik. chem.*, 1890, *6*, 141.

COMPONENTS:

1. Nitrous oxide; N_2O; [10024-97-2]

2. Water; H_2O; [7732-18-5]

ORIGINAL MEASUREMENTS:

Geffcken, G.

Z. Physik. Chem., 1904, *49*, 257-302.

VARIABLES:

T/K: 278-298

PREPARED BY:

R. Battino

EXPERIMENTAL VALUES:

	T/K [a]	$x_1 \times 10^4$ [b]	L [c]	No. Detns.	σ
*	278.15	8.415	1.067	3	.0027
*	283.15	7.054	0.9101	2	.0002
*	288.15	5.932	0.7784	4	.0008
*	293.15	5.066	0.6756	5	.0008
*	298.15	4.386	0.5942	6	.0013

a. Temperature reported to 1^0C, but values are not interpolated, i.e., measured at even temperatures - 5, 10, 15, 20, 25^0C.

b. Mole fraction solubility at 101.325 kPa (1 atm) partial pressure of gas. Calculated by compiler.

c. Ostwald coefficient. These are averages of the number of determinations listed in column 4 with standard deviations listed in column 5.

AUXILIARY INFORMATION

METHOD/APPARATUS/PROCEDURE:

Used the basic Ostwald method. Solvent is degassed by boiling. Details and a drawing given in the paper.

SOURCE AND PURITY OF MATERIALS:

1. Nitrous oxide - prepared from pure ammonium nitrate. Details in paper.

2. Water - distilled.

ESTIMATED ERROR:

$\delta L/L = 0.01$ (compiler's estimate)

REFERENCES:

COMPONENTS:

1. Nitrous oxide; N_2O; [10024-97-2]
2. Water; H_2O; [7732-18-5]

ORIGINAL MEASUREMENTS:

Knopp, W.

Z. Physik. Chem., 1904, *48*, 97-108.

VARIABLES:

PREPARED BY:

R. Battino

EXPERIMENTAL VALUES:

	T/K [a]	$x_1 \times 10^4$ [b]	L [c]	α [d]
*	293.15	5.046	0.6729	0.6270

a. Temperature reported to 1^{0}C.

b. Mole fraction solubility of 101,325 kPa (1 atm) partial pressure of gas. Calculated by compiler.

c. Ostwald coefficient. Calculated by compiler.

d. Bunsen coefficient.

AUXILIARY INFORMATION

METHOD/APPARATUS/PROCEDURE:

Used the Ostwald apparatus as modified by Braun (1).

SOURCE AND PURITY OF MATERIALS:

1. Nitrous oxide - prepared from pure ammonium nitrate.
2. Water - no comment by author.

ESTIMATED ERROR:

$\delta\alpha/\alpha$ = 0.01 (compiler's estimate)

REFERENCES:

1. Braun, L., *A. Physick. Chem.*, 1900, *33*, 721.

COMPONENTS:

1. Nitrous oxide; N_2O; [10024-97-2]
2. Water; H_2O; [7732-18-5]

ORIGINAL MEASUREMENTS:

Findlay, A.; Creighton, H.J.M.,

J. Chem. Soc., 1910, *97*, 536-61

VARIABLES:

PREPARED BY:

R. Battino

EXPERIMENTAL VALUES:

P/mmHg [a]	L [b]	P/mmHg	L	P/mmHg	L
758	0.592	758	0.592	758	0.591
842	0.593	831	0.593	888	0.592
967	0.592	997	0.592	971	0.591
1041	0.593	1082	0.593	1091	0.592
1185	0.592	1214	0.594	1190	0.593
1362	0.592	1351	0.592	1281	0.593

T/K [c]	$x_1 \times 10^4$ [d]	L [e]
* 298.15	4.370	0.592

a. Partial pressure of nitrous oxide. All measurements at 298.15 K.
b. Ostwald coefficient.
c. Temperatures reported to 0.1 °C.
d. Mole fraction solubility at 101.325 kPa (1 atm) partial pressure of gas. Calculated by compiler.
e. Ostwald coefficient - mean of values at varying pressures cited above.

AUXILIARY INFORMATION

METHOD/APPARATUS/PROCEDURE:

Used the Geffcken (1) apparatus. Liquids degassed by boiling.

SOURCE AND PURITY OF MATERIALS:

1. Nitrous oxide - prepared by heating pure ammonium nitrate.
2. Water - distilled.

ESTIMATED ERROR:

$\delta L/L = 0.01$ (compiler's estimate) (Authors' estimate is 0.0025).

REFERENCES:

1. Geffcken, G., *Z. Physik. Chem.*, 1904, *49*, 257-302.

COMPONENTS:

1. Nitrous oxide; N_2O; [10024-97-2]

2. Water; H_2O; [7732-18-5]

ORIGINAL MEASUREMENTS:

Findlay, A.; Howell, O.R.,

J. Chem. Soc., 1914, *105*, 291-8.

VARIABLES:

PREPARED BY:

R. Battino

EXPERIMENTAL VALUES:

P/mmHg [a]	L [b]	P/mmHg	L
282.5	0.585	272.8	0.585
396.1	0.585	393.2	0.585
562.9	0.584	548.6	0.585
664.5	0.585	652.4	0.585
789.3	0.585	751.0	0.585
1027.5	0.585	1021.7	0.586

a. Partial pressure of nitrous oxide. All measurements at 298.15 K.

b. Ostwald coefficient.

T/K [a]	$x_1 \times 10^4$ [b]	L [c]
* 298.15	4.319	0.585

a. Temperatures reported to 0.1°C.

b. Mole fraction solubility at 101.325 kPa (1 atm) partial pressure of gas. Calculated by compiler.

c. Ostwald coefficient. Average of 12 values in previous table.

AUXILIARY INFORMATION

METHOD/APPARATUS/PROCEDURE:

The apparatus used was that described earlier (1).

SOURCE AND PURITY OF MATERIALS:

1. Nitrous oxide - prepared by heating carefully purified ammonium nitrate.

2. Water - distilled.

ESTIMATED ERROR:

$\delta L/L = 0.01$ (compiler's estimate)

$= 0.0025$ (author's estimate)

REFERENCES:

1. Findlay, A; Williams, T. *J. Chem. Soc.*, 1913, *103*, 636.

COMPONENTS:

1. Nitrous oxide; N_2O; [10024-97-2]
2. Water; H_2O; [7732-18-5]

ORIGINAL MEASUREMENTS:

Kunerth, W.

Phys. Rev., 1922, *19*, 512-24.

VARIABLES:

T/K: 291-309

PREPARED BY:

R. Battino

EXPERIMENTAL VALUES:

T/K [a]	$x_1 \times 10^4$ [b]	L [c]
* 291.15	5.306	0.703
* 293.15	5.062	0.675
* 295.15	4.754	0.638
* 297.15	4.569	0.617
* 299.15	4.320	0.587
301.15	4.104	0.561
* 303.15	3.854	0.530
305.15	3.708	0.513
* 307.15	3.435	0.478
* 309.15	3.208	0.449

a. Temperatures reported to 0.1°C.

b. Mole fraction solubility at 101,325 Pa (1 atm) partial pressure of gas. Calculated by compiler.

c. Ostwald coefficient.

AUXILIARY INFORMATION

METHOD/APPARATUS/PROCEDURE:

Apparatus is similar to that of McDaniel (1). It uses a 120 cm^3 thermostatted gas buret. The absorption pipet is 31.3 cm^3 and connected to the gas buret via a glass capillary. The solvent is degassed by boiling under vacuum. When gas is allowed into the absorption pipet the entire apparatus is mechanically shaken. Details and a drawing in the original paper.

SOURCE AND PURITY OF MATERIALS:

1. Nitrous oxide - from S.S. White Dental Company. 99.7% pure.
2. Water - no comment by author.

ESTIMATED ERROR:

$\delta L/L = 0.01$ (compiler's estimate)

REFERENCES:

1. McDaniel, A.S., *J. Phys. Chem.*, 1911, *15*, 587.

COMPONENTS:

1. Nitrous oxide; N_2O; [10024-97-2]

2. Water; H_2O; [7732-18-5]

ORIGINAL MEASUREMENTS:

Manchot, W.; Jahrstorter, M.; Zepter, H.

Z. Anorg. Allgem. Chem., 1924, *141*, 45-81.

VARIABLES:

PREPARED BY:

R. Battino

EXPERIMENTAL VALUES:

	T/K [a]	$x_1 \times 10^4$ [b]	L [c]	S [d]
*	298.15	4.281	0.5800	53.3

a. Temperature reported to $\pm 1^0$C.

b. Mole fraction solubility at 101,325 Pa (1 atm) partial pressure of gas. Calculated by compiler.

c. Ostwald coefficient calculated by compiler.

d. Solubility in units of cm^3 at 273.15 K and 1 atm dissolved in 100 g of water.

AUXILIARY INFORMATION

METHOD APPARATUS/PROCEDURE:

Determined in a gasometer. Details given in reference (1).

SOURCE AND PURITY OF MATERIALS:

1. Nitrous oxide - chemically prepared and purified.

2. Water - no comment by authors.

ESTIMATED ERROR:

$\delta S/S = 0.03$ (estimated by compiler)

REFERENCES:

1. Manchot, W., *Z. Anorg. Allgem. Chem.*, 1924, *141*, 38-44.

COMPONENTS:	ORIGINAL MEASUREMENTS:
1. Nitrous oxide; N_2O; [10024-97-2] 2. Water; H_2O; [7732-18-5]	Markham, A.E.; Kobe, K.A., *J. Amer. Chem. Soc.*, 1941, *63*, 449-54.

VARIABLES:	PREPARED BY:
T/K: 273-313	R. Battino

EXPERIMENTAL VALUES:

	T/K [a]	$x_1 \times 10^4$ [b]	L [c]	α [d]	No. Detns.	Avg. Devn.
*	273.35	10.415	1.2980	1.2970	3	0.0001
*	298.15	4.345	0.5886	0.5392	11	0.0012
	313.15	2.898	0.4103	0.3579	7	0.0010

a. Temperature reported to ±0.1°C.

b. Mole fraction solubility at 101, 325 Pa (1 atm) partial pressure of gas. Calculated by compiler.

c. Ostwald coefficient. Calculated by compiler.

d. Bunsen coefficient. Average of number of determinations listed in column 5. The average deviation is listed in column 6.

AUXILIARY INFORMATION

METHOD/APPARATUS/PROCEDURE:

Used the Ostwald method with modifications. Absorption flask consists of two bulbs connected at top and bottom via three-way stopcocks. One bulb is double the volume of the other. Critical parts including the manometer and gas buret are thermostatted in a water bath. The gas is saturated before exposure to the degassed liquid. The bulks are mechanically agitated. Volumes are calibrated with mercury weighings. Details and a drawing are given in the original paper.

SOURCE AND PURITY OF MATERIALS:

1. Nitrous oxide - from commercial cylinders. 99.7% pure.

2. Water - distilled.

ESTIMATED ERROR:

$\delta\alpha/\alpha = 0.01$ (compiler's estimate)

REFERENCES:

COMPONENTS:	ORIGINAL MEASUREMENTS:
1. Nitrous oxide; N_2O; [10024-97-2] 2. Water; H_2O; [7732-18-5]	Jay, B.E.; Wilson, R.H.; Doty, V.; Pingree, H.; Hargis, B. *Anal. Chem.*, 1962, *34*, 414-8.
VARIABLES:	PREPARED BY: R. Battino

EXPERIMENTAL VALUES:

T/K[a]	$x_1 \times 10^4$ [b]	L[c]
309.15	3.036	0.425

a. Temperature reported to $1^{0}C$.

b. Mole fraction solubility at 101.325 kPa (1 atm) partial pressure of gas. Calculated by compiler.

c. Ostwald coefficient. Average of 11 values with a standard deviation of 1.1%.

AUXILIARY INFORMATION

METHOD/APPARATUS/PROCEDURE:

A gas chromatographic method for determining solubility is described in the paper with illustrations.

SOURCE AND PURITY OF MATERIALS:

1. Nitrous oxide - from Matheson Gas Co. Stated to be "100% N_2O" - but not as a measure of purity (compiler).

2. Water - no comment by authors.

ESTIMATED ERROR:

$\delta L/L = 0.02$ (estimated by compiler)

REFERENCES:

COMPONENTS:	ORIGINAL MEASUREMENTS:
1. Nitrous oxide; N_2O; [10024-97-2] 2. Water; H_2O; [7732-18-5]	Sy, W.P.; Hasbrouck, J.D. *Anesthesiology*, 1964, *25*, 59.
VARIABLES:	**PREPARED BY:** C.L. Young

EXPERIMENTAL VALUES:

T/K	Partial pressure[+] of nitrous oxide P/mmHg	Ostwald coefficient, L
37	699.8	0.367
	699.2	0.370
	699.2	0.367
	694.7	0.372
	694.9	0.365
	695.2	0.366
	695.5	0.369
	695.8	0.361
	695.8	0.367

Mean Ostwald coefficient = 0.367
Standard deviation = 0.003

\+ Calculated by subtracting vapor pressure of water from total pressure.

AUXILIARY INFORMATION

METHOD APPARATUS/PROCEDURE:

Nitrous oxide bubbled through water - allowed to stand for 15 minutes. Samples analysed in Van Slyke - Neill apparatus (1).

SOURCE AND PURITY OF MATERIALS:

Water distilled, no other details given.

ESTIMATED ERROR:

$\delta T/K = \pm 0.1$; $\delta P/\text{mmHg} = \pm 0.1$

REFERENCES:

1. Van Slyke, D.D.; Neill, J.M. *J. Biol. Chem.* 1924, *61*, 523.

COMPONENTS:

1. Nitrous Oxide; N_2O; [10024-97-2]
2. Water; H_2O; [7732-18-5]

ORIGINAL MEASUREMENTS:

Borgstedt, H.H.; Gillies, A.J.

Anesthesiology, 1965, *26*, 675-8

VARIABLES:

Temperature, pressure

PREPARED BY:

C.L. Young

EXPERIMENTAL VALUES:

T/K	Bunsen coefficient, α	Number of Observations	10^4 Mole fraction[+] $10^4 x_{N_2O}$
303	0.485 ± 0.0026	20	3.91
305.5	0.469 ± 0.0024	20	3.78
310	0.429 ± 0.0028	40	3.46
313	0.403 ± 0.0025	20	3.25
318	0.365 ± 0.0017	20	2.94
323	0.338 ± 0.0026	20	2.73

\+ calculated by compiler

AUXILIARY INFORMATION

METHOD/APPARATUS/PROCEDURE:

A gas chromatographic method in which a sample of water saturated with gas was injected directly into the chromatograph. Water in sample was absorbed by molecular sieve type 3A and nitrous oxide detected with a thermal conductivity detector. Detector response compared with that of samples of known amount. Details in source.

SOURCE AND PURITY OF MATERIALS:

Water stated to be pure. No other details given.

ESTIMATED ERROR:

$\delta T/K = \pm 0.1$

REFERENCES:

COMPONENTS:	ORIGINAL MEASUREMENTS:
1. Nitrous oxide; N_2O; [10024-97-2] 2. Water; H_2O; [7732-18-5]	Saidman, L.J.; Eger, E.I.; Munson, E.S.; Severinghaus, J.W. *Anesthesiology*, 1966, *27*, 180-184.
VARIABLES:	PREPARED BY:
Temperature	C.L. Young

EXPERIMENTAL VALUES:

T/°C	T/K	No of samples	Ostwald coefficient, *L* Mean	Standard deviation
37	310.2	6	0.444	0.010
25	298.2	6	0.594	0.008
20	293.2	6	0.657	0.017

AUXILIARY INFORMATION

METHOD APPARATUS/PROCEDURE:

Modified Scholander apparatus used. Known amount of water equilibrated with a known volume of gas and change in volume used to estimate Ostwald coefficient. Details in source and ref. (1).

SOURCE AND PURITY OF MATERIALS:

1. No details given.
2. Degassed

ESTIMATED ERROR:

$\delta T/K = \pm 0.1$ (estimated by compiler).

REFERENCES:

1. Douglas, E.
J. Phys. Chem. 1964, *68*, 169.

COMPONENTS:	ORIGINAL MEASUREMENTS:
1. Nitrous oxide; N_2O; [10024-97-2] 2. Water; H_2O; [7732-18-5]	Christoforides, C.; Hedley-White, J. *J. Federation Proceedings*, 1970, *29*: A330.
VARIABLES:	PREPARED BY:
T/K: 298-310	R. Battino

EXPERIMENTAL VALUES:

T/K [a]	$x_1 \times 10^4$ [b]	L [c]	α [d]
298.15	4.520	0.6123	0.561
310.15	3.405	0.4780	0.421

a. Temperatures reported as 25^0 and 37^{0}C.

b. Mole fraction solubility at 101,325 kPa (1 atm) partial pressure of gas. Calculated by compiler.

c. Ostwald coefficient. Calculated by compiler.

d. Bunsen coefficient.

AUXILIARY INFORMATION

METHOD/APPARATUS/PROCEDURE:	SOURCE AND PURITY OF MATERIALS:
Determined by Van Slyke manometry.	1. Nitrous oxide - no comment by authors. 2. Water - no comment by authors.
	ESTIMATED ERROR: $\delta\alpha/\alpha$ = 0.02 (compiler's estimate)
	REFERENCES:

COMPONENTS:

1. Nitrous oxide; N_2O; [10024-97-2]
2. Water; H_2O; [7732-18-5]

ORIGINAL MEASUREMENTS:

Joosten, G.E.H.; Danckwerts, P.V.,

J. Chem. Eng. Data, 1972, *17*, 452-4.

VARIABLES:

PREPARED BY:

R. Battino

EXPERIMENTAL VALUES:

	T/K [a]	$x_1 \times 10^4$ [b]	L [c]	$1/K_H \times 10^5$ [d]
*	298.15	4.409	0.5973	2.44

a. Temperature reported as 25°C.

b. Mole fraction solubility at 101,325 kPa (1 atm) partial pressure of gas. Calculated by compiler.

c. Ostwald coefficient. Calculated by compiler.

d. Henry's law constant in units of cm^3 atm mol^{-1}.

AUXILIARY INFORMATION

METHOD/APPARATUS/PROCEDURE:

Used the apparatus and procedures of Markham and Kobe (1).

SOURCE AND PURITY OF MATERIALS:

1. Nitrous oxide - no comment by authors.
2. Water - no comment by authors.

ESTIMATED ERROR:

$\delta K_H/K_H = 0.01$ (compiler's estimate)

REFERENCES:

1. Markham, A.E.; Kobe, K.A., *J. Amer. Chem. Soc.*, 1941, *63*, 449-54.

COMPONENTS:	ORIGINAL MEASUREMENTS:
1. Nitrous oxide; N_2O; [10024-97-2] 2. Water; H_2O; [7732-18-5]	Hikita, H.; Asai, S.; Ishikawa, H.; Esaka, N., *J. Chem. Eng. Data*, 1974, *19*, 89-92.
VARIABLES:	PREPARED BY: R. Battino

EXPERIMENTAL VALUES:

T/K [a]	$x_1 \times 10^4$ [b]	L [c]	M [d]
298.15	4.255	0.5764	0.02356

a. Temperature reported as 25°C.

b. Mole fraction solubility at 101,325 kPa (1 atm) partial pressure of gas. Calculated by compiler.

c. Ostwald coefficient. Calculated by compiler.

d. Molarity in mol l^{-1} at 1 atm partial gas pressure.

AUXILIARY INFORMATION

METHOD APPARATUS/PROCEDURE:

Used the gas volumetric method of Markham and Kobe (1) in an apparatus similar to that used by Onda *et al.* (2).

SOURCE AND PURITY OF MATERIALS:

1. Nitrous oxide - from a commercial cylinder. 99.8% purity.
2. Water - distilled.

ESTIMATED ERROR:

$\delta M/M = 0.01$ (compiler's estimate)

REFERENCES:

1. Markham, A.E.; Kobe, K.A., *J. Amer. Chem. Soc.*, 1941, *63*, 449.
2. Onda, K.; Sada, E.; Kobayashi, T.; Kito, S.; Ito, K., *J. chem. Eng. Japan*, 1970, *3*, 18.

COMPONENTS:	ORIGINAL MEASUREMENTS:
1. Nitrous oxide; N_2O; [10024-97-2] 2. Water; H_2O; [7732-18-5]	Sada, E.; Kito, S.; Ito, Y., *J. Chem. Eng. Japan*, 1974, *7*, 57-9.
VARIABLES:	PREPARED BY: R. Battino

EXPERIMENTAL VALUES:

	T/K [a]	$x_1 \times 10^4$ [b]	L [c]	α [d]
*	298.15	4.441	0.6016	0.5512

a. Temperature reported as 25^{0}C.

b. Mole fraction solubility at 101.325 kPa (1 atm) partial pressure of gas. Calculated by compiler.

c. Ostwald coefficient. Calculated by compiler.

d. Bunsen coefficient.

AUXILIARY INFORMATION

METHOD/APPARATUS/PROCEDURE:

Details are given in (1). A measured volume of gas is brought into contact with a quantity of degassed liquid. Agitation is via a magnetic stirrer. The gas volumes are determined on a gas buret. All critical components are thermostatted in a water bath to $\pm 0.01^0$C.

SOURCE AND PURITY OF MATERIALS:

1. Nitrous oxide - Showa Denko Co. Ltd., Tokyo. 99.8% pure.
2. Water - distilled.

ESTIMATED ERROR:

$\delta\alpha/\alpha = 0.01$ (compiler's estimate)

REFERENCES:

1. Onda, K.; Sada, E.; Kobayashi, T.; Kito, S.; Ito, K., *J. Chem. Eng. Japan*, 1970, *3*, 18-24.

COMPONENTS:

1. Nitrous oxide; N_2O; [10024-97-2]

2. Water; H_2O; [7732-18-5]

ORIGINAL MEASUREMENTS:

Sada, E.; Kito, S.; Ito, Y.,

Ind. Eng. Chem. Fundam., 1975, *14*, 232-7.

VARIABLES:

PREPARED BY:

R. Battino

EXPERIMENTAL VALUES:

	T/K [a]	$x_1 \times 10^4$ [b]	L [c]	H [d]
*	298.15	4.310	0.5839	2320.1

a. Temperature reported to $\pm 0.01^0$C.

b. Mole fraction solubility at 101,325 Pa (1 atm) partial pressure of gas. Calculated by compiler.

c. Ostwald coefficient calculated by compiler.

d. Henry's law coefficient in atm.

AUXILIARY INFORMATION

METHOD/APPARATUS/PROCEDURE:

Details are given in reference (1) with a drawing. A measured volume of gas is brought into contact with a quantity of degassed liquid. Agitation is via a magnetic stirrer. The gas volumes are determined with a gas buret. All critical components are in a water bath controlled to $\pm 0.01^0$C.

SOURCE AND PURITY OF MATERIALS:

1. Nitrous oxide - prepared by Showa Denka Co. Ltd. (Tokyo) and 99.8% pure.

2. Water - "carefully distilled".

ESTIMATED ERROR:

$\delta H/H = 0.01$ (compiler's estimate)

REFERENCES:

1. Onda, K.; Sada, E.; Kobayashi, T.; Kito, S.; Ito, K., *J. Chem. Eng. Japan*, 1970, *3*, 18-24.

COMPONENTS:	ORIGINAL MEASUREMENTS:
1. Nitrous oxide; N_2O; [10024-97-2] 2. Water; H_2O; [7732-18-5]	Sada, E.; Kumazawa, H.; Butt, M.A., *J. Chem. Eng. Data*, 1977, *22*, 277-9.
VARIABLES:	PREPARED BY: R. Battino

EXPERIMENTAL VALUES:

	T/K [a]	$x_1 \times 10^4$ [b]	L [c]	α [d]
*	298.15	4.441	0.6016	0.5512

a. Temperature reported to $\pm 0.01^0$C.

b. Mole fraction solubility of gas at 101,325 Pa (1 atm) partial pressure of gas. Calculated by compiler.

c. Ostwald coefficient calculated by compiler.

d. Bunsen coefficient.

AUXILIARY INFORMATION

METHOD/APPARATUS/PROCEDURE:

Details are given in reference (1). A measured volume of gas is brought into contact with a quantity of degassed liquid. Agitation is via a magnetic stirrer. The gas volumes are determined on a gas buret. All critical components are thermostatted in a water bath to $\pm 0.01^0$C.

SOURCE AND PURITY OF MATERIALS:

1. Nitrous oxide - from commercial cylinder with minimum purity of 99.8%.

2. Water - "carefully distilled".

ESTIMATED ERROR:

$\delta\alpha/\alpha$ = 0.01 (compiler's estimate)

REFERENCES:

1. Onda, K.; Sada, E.; Kobayashi, T.; Kito, S.; Ito, K., *J. Chem. Eng. Japan*, 1970, *3*, 18-24.

COMPONENTS:

1. Nitrous oxide; N_2O; [10024-97-2]
2. Water; H_2O; [7732-18-5]

ORIGINAL MEASUREMENTS:

Weiss, R.F.; Price, B.A.

Marine Chemistry, 1980, *8*, 347-359.

VARIABLES:

Temperature

PREPARED BY:

C.L. Young

EXPERIMENTAL VALUES:

T/K	Solubility,+ K_0 /mol dm^{-3} atm^{-1}	Bunsen coefficient, α
273.44	0.05870	1.3054
273.44	0.05858	1.3028
273.44	0.05870	1.3054
273.44	0.05870	1.3054
273.44	0.05858	1.3028
283.16	0.04016	0.8931
283.15	0.04017	0.8933
283.15	0.04007	0.8911
283.15	0.04010	0.8918
283.16	0.04013	0.8925
283.15	0.04019	0.8938
293.11	0.02882	0.6409
293.13	0.02880	0.6405
293.13	0.02873	0.6389
293.12	0.02878	0.6400
293.12	0.02879	0.6403
293.13	0.02875	0.6394
303.35	0.02156	0.4795
303.34	0.02153	0.4788
303.36	0.02155	0.4793
303.36	0.02149	0.4779

AUXILIARY INFORMATION

METHOD/APPARATUS/PROCEDURE:

The Scholander microgasometric technique as adapted by Douglas (1), (2) was used. The equilibrium chamber was enlarged to contain approximately 10 ml of solvent. The proceduresfor degassing the water and transferring the gas were checked for air contamination by gas chromatography. All volumes were read on a micrometer which displaced mercury.

SOURCE AND PURITY OF MATERIALS:

1. Matheson Ultra High purity sample, purity better than 99.99 mole per cent.
2. Distilled.

ESTIMATED ERROR:

$\delta T/K = \pm 0.01$; $\delta K_0 = \pm 0.3\%$

REFERENCES:

1. Douglas, E. *J. Phys. Chem.* 1964, *68*, 169 and 1965, *69*, 2608.
2. Weiss, R.F. *Marine Chem.* 1974, *2*, 203.

COMPONENTS:	ORIGINAL MEASUREMENTS:
1. Nitrous oxide; N_2O; [10024-97-2] 2. Water; H_2O; [7732-18-5]	Weiss, R.F.; Price, B.A. *Marine, Chemistry*, 1980, *8*, 347-359.

EXPERIMENTAL VALUES:

T/K	Solubility,$^+$ K_o /mol dm^{-3} atm^{-1}	Bunsen coefficient, α
303.37	0.02151	0.4784
313.23	0.01693	0.3765
313.23	0.01692	0.3763
313.24	0.01691	0.3761
313.23	0.01696	0.3772
313.24	0.01694	0.3767

\+ defined as α/V where V is the molar volume at 273.15K and 101.325kPa (22239 cm^{3}mol^{-1})

COMPONENTS:	EVALUATOR:
1. Nitrous oxide; N_2O; [10024-97-2] 2. Seawater	Colin L. Young, School of Chemistry, University of Melbourne, Parkville, Victoria 3052, Australia. August 1980

CRITICAL EVALUATION:

The only measurements of the solubility of nitrous oxide in sea-water are those of Weiss and Price (1), who used the microgasometric technique as refined by Weiss (2), (3) in previous work. The method is of high precision and these workers made corrections for the deviations from ideality of nitrous oxide. The work of Weiss and coworkers on other gases, e.g., argon (4) has been compared in detail (5) with that of other workers and is thought to be reliable. Therefore in the absence of any evidence to the contrary, their data are classified as tentative. Weiss and Price reported various smoothing equations including one giving the volumetric solubility coefficient K_o/mol dm^{-3} atm^{-1} as a function of temperature and salinity, S, in parts per thousand ‰. K_o is equal to the Bunsen coefficient/(molar volume of nitrous oxide at 273.15 K and 101.325 kPa).

$$\ln (K_o/\text{mol dm}^{-3}\ \text{atm}^{-1}) = -62.7062 + 97.3066\ (100\ T/K) + 24.1406 \ln (T/100\ K) + S[-0.058420 + 0.033193\ (T/100\ K) - 0.0051313\ (T/100\ K)^2]$$

Values calculated from this equation are given below.

The volumetric solubility coefficient (10^2K_o/mol dm^{-3} atm^{-1}) for nitrous oxide at various temperatures and salinities.

T/K	Salinity (‰) 0	10	20	30	35	38	40
272.15	---	---	5.480	5.156	5.001	4.911	4.851
273.15	5.933	5.585	5.258	4.950	4.803	4.717	4.660
274.15	5.691	5.360	5.048	4.755	4.615	4.533	4.479
275.15	5.461	5.147	4.850	4.570	4.437	4.358	4.307
276.15	5.245	4.945	4.662	4.396	4.268	4.193	4.144
277.15	5.040	4.754	4.484	4.230	4.108	4.037	3.990
278.15	4.846	4.573	4.315	4.073	3.956	3.888	3.843
279.15	4.662	4.401	4.156	3.923	3.812	3.747	3.704
281.15	4.322	4.084	3.860	3.647	3.546	3.486	3.447
283.15	4.016	3.799	3.593	3.398	3.305	3.250	3.214
285.15	3.741	3.541	3.352	3.172	3.086	3.036	3.003
287.15	3.492	3.307	3.133	2.968	2.889	2.842	2.811
289.15	3.266	3.096	2.935	2.782	2.709	2.666	2.637
291.15	3.061	2.904	2.754	2.613	2.545	2.505	2.479
293.15	2.875	2.729	2.590	2.459	2.395	2.358	2.334
295.15	2.705	2.569	2.440	2.318	2.259	2.224	2.201
296.15	2.551	2.424	2.303	2.189	2.134	2.102	2.080
298.15	2.409	2.291	2.178	2.071	2.019	1.989	1.969
301.15	2.280	2.169	2.063	1.963	1.914	1.886	1.867
303.15	2.161	2.057	1.958	1.863	1.818	1.791	1.773
305.15	2.053	1.954	1.861	1.771	1.728	1.703	1.686
307.15	1.953	1.860	1.771	1.687	1.646	1.622	1.607
309.15	1.861	1.773	1.689	1.609	1.570	1.548	1.533
311.15	1.776	1.692	1.613	1.537	1.500	1.479	1.465
313.15	1.698	1.618	1.542	1.470	1.435	1.415	1.401

(cont.)

COMPONENTS:	EVALUATOR:
1. Nitrous oxide; N_2O; [10024-97-2] 2. Seawater	Colin L. Young, School of Chemistry, University of Melbourne, Parkville, Victoria 3052, Australia. August 1980

CRITICAL EVALUATION:

References:

1. Weiss, R. F.; Price, B. A. *Marine Chem.* 1980, *8*, 347.

2. Weiss, R. F. *J. Chem. Eng. Data* 1971, *16*, 235.

3. Weiss, R. F. *Marine Chem.* 1974, *2*, 203.

4. Weiss, R. F. *Deep-Sea Res.* 1971, *18*, 225.

5. Chen, C. T. in "Argon" *Solubility Data Series* Vol. 4, p.27. H. L. Clever, Ed. Pergamon 1980.

COMPONENTS:

1. Nitrous oxide; N_2O; [10024-97-2]

2. Seawater

ORIGINAL MEASUREMENTS:

Weiss, R.F.; Price, B.A.

Marine Chemistry, 1980, *8*, 347-359

VARIABLES:

Temperature

PREPARED BY:

C.L. Young

EXPERIMENTAL VALUES:

T/K	Salinity/‰	Solubility,[+] K_o / mol dm^{-3} atm^{-1}	Bunsen coefficient, α
283.24	18.060	0.03629	0.8071
283.24		0.03610	0.8028
283.25		0.03614	0.8037
283.25		0.03615	0.8039
283.25		0.03612	0.8033
283.25		0.03606	0.8019
303.00		0.01985	0.4414
303.00		0.01987	0.4419
303.00		0.01982	0.4408
303.00		0.01984	0.4412
303.01		0.01983	0.4410
273.45	36.130	0.04716	1.0488
273.44		0.04714	1.0483
273.44		0.04728	1.0515
273.45		0.04706	1.0466
273.44		0.04716	1.0488
278.08		0.03943	0.8769
278.08		0.03938	0.8758
278.08		0.03940	0.8762
278.09		0.03947	0.8778
278.08		0.03941	0.8764
287.19		0.03273	0.7279
287.19		0.03282	0.7299
287.19		0.03280	0.7294

AUXILIARY INFORMATION

METHOD/APPARATUS/PROCEDURE:

The Scholander microgasometric technique as adapted by Douglas (1), (2) was used. The equilibrium chamber was enlarged to contain approximately 10 ml of solvent. The procedures for degassing the water and transferring the gas were checked for air contamination by gas chromatography. All volumes were read on a micrometer which displaced mercury.

SOURCE AND PURITY OF MATERIALS:

1. Matheson Ultra high purity sample, purity better than 99.99 mole per cent.

2. Surface seawater, poisoned with $HgCl_2$, filtered.

ESTIMATED ERROR:

$\delta T/K = \pm 0.01$; $\delta K_o = \pm 0.3\%$
δ salinity $= \pm 0.004$.

REFERENCES:

1. Douglas, E.
J. Phys. Chem. 1964, *68*, 169 and 1965, *69*, 2608.

2. Weiss, R.F. *Marine Chem.* 1974, *2*, 203.

COMPONENTS:	ORIGINAL MEASUREMENTS:
1. Nitrous oxide; N_2O; [10024-97-2] 2. Seawater.	Weiss, R.F.; Price, B.A. *Marine Chemistry*, 1980, *8*, 347-359

EXPERIMENTAL VALUES:

T/K	Salinity/‰	Solubility,+ K_0 /mol $dm^{-3}atm^{-1}$	Bunsen coefficient, α
287.19	36.130	0.03281	0.7297
287.19		0.03276	0.7285
293.30		0.02372	0.5275
293.30		0.02372	0.5275
293.29		0.02369	0.5268
293.30		0.02366	0.5262
293.30		0.02373	0.5277
293.29		0.02374	0.5280
303.10		0.01812	0.4030
303.11		0.01813	0.4032
303.11		0.01810	0.4025
303.11		0.01811	0.4027
303.10		0.01810	0.4025
313.23		0.01428	0.3176
313.24		0.01425	0.3169
313.24		0.01421	0.3160
313.24		0.01422	0.3162
313.24		0.01425	0.3169

\+ defined as α/V where V is the molar volume at 273.15K and 101.325 kPa (22239 cm^3 mol^{-1}).

COMPONENTS:	EVALUATOR:
1. Nitrous oxide; N_2O; [10024-97-2] 2. Water; H_2O; [7732-18-5] 3. Electrolyte	Colin L. Young, School of Chemistry, University of Melbourne, Parkville, Victoria 3052, Australia. February 1981

CRITICAL EVALUATION:

An Evaluation of the Solubility of Nitrous Oxide in Aqueous Electrolyte Solutions at a Nitrous oxide Partial Pressure of 101.325 kPa

There are few measurements of the solubility of nitrous oxide in any one aqueous electrolyte system over common ranges of concentration and temperature. Consequently it is not possible to recommend solubility values. Most of the available data are classified as tentative.

In order to have a common basis for comparison, the solubility data have been converted to Sechenow salt effect parameters in the form

$$k_{sc\alpha}/\mathrm{dm^3\ mol^{-1}} = (1/(c/\mathrm{mol\ dm^{-3}}))\ \log(\alpha°/\alpha)$$

where c is the electrolyte concentration in units of mol dm^{-3} and $\alpha°$ and α are the Bunsen coefficients in pure water and electrolyte solution, respectively. The Ostwald coefficient ratio, $L°/L$, will give the same value, but the salt effect parameter is symbolized, k_{scL}. Both ratios are equivalent to a molar gas solubility ratio.

Another form of the salt effect parameter which will be found in the subsequent discussion is

$$k_{sm\alpha}/\mathrm{kg\ mol^{-1}} = (1/(m/\mathrm{mol\ kg^{-1}}))\ \log\ (\alpha°/\alpha)$$

where m is the electrolyte molality.

The salt effect parameter, $k_{sc\alpha}$, is often assumed to be independent of electrolyte concentration. This is only approximately true for most solutions. There are indications throughout the literature that the salt effect parameter values are larger in dilute solutions than at higher concentrations but there are no definitive studies on this point.

The effect of the electrolyte concentration on the salt effect parameter was checked by one of two ways. A graph was prepared either of log $(\alpha°/\alpha)$ *vs.* c or of $k_{sc\alpha}$ *vs.* c. A linear log $(\alpha°/\alpha)$ *vs.* c plot shows no concentration dependence to the salt effect parameter and the slope is $k_{sc\alpha}$. A linear $k_{sc\alpha}$ *vs.* c plot of zero slope shows no concentration dependence of the salt effect parameter.

The solubility of nitrous oxide in water has been taken from the paper under evaluation for the calculation of the salt effect parameter. The reason for using the water solubility of the author, instead of the recommended solubility of nitrous oxide in water, is that systematic errors in a given author's work may cancel in the ratio $\alpha°/\alpha$. The use of a salt effect parameter of the Sechenow type should not be taken to mean that it is necessarily the best way to represent salt effect results. It is used here as a convenient parameter for the comparison of data from several authors.

The largest group of electrolyte solutions has been studied by Manchot *et al.* (1). In many cases only two concentrations of electrolyte were studied and the work was restricted to 298.15 K. Seidell attributes some work to Manchot *et al.* (1) which is not in numerical form in the original. The data are evaluated separately as the status of these results is not clear.

Nitrous oxide measurements have been reported for over forty aqueous electrolyte systems. Each system is discussed briefly on the following pages. Unless otherwise stated, the salt effect parameters are

COMPONENTS:	EVALUATOR:
1. Nitrous oxide; N_2O; [10024-97-2] 2. Water; H_2O; [7732-18-5] 3. Electrolyte	Colin L. Young, School of Chemistry, University of Melbourne, Parkville, Victoria 3052, Australia. February 1981

CRITICAL EVALUATION:

for a concentration of one mol dm^{-3} solution at a nitrous oxide partial pressure of 0.101325 MPa. Tentative values of the salt effect parameters are given as calculated from the various papers. The systems are given in the order of the standard arrangement for electrolytes used in U.S. National Bureau of Standards publications.

1. Nitrous oxide + water + Hydrochloric Acid [7647-01-0]

This system has been studied by Geffcken (2) at 288.16 K and 298.16 K and the salt effect parameters, k_{scL}, are

$$k_{scL}/dm^3\ mol^{-1} = 0.0281 - 0.0052\ (c/mol\ dm^{-3})$$

$$k_{scL}/dm^3\ mol^{-1} = 0.0272 - 0.0060\ (c/mol\ dm^{-3}),$$

respectively. See also, the evaluation of Seidell's values following this evaluation.

2. Nitrous oxide + water + Sulfuric Acid [7664-93-9]

This system has also been studied by Geffcken (2) at 288.16 K and 298.16 K. The salt effect parameters, k_{scL} are not linear with concentration and are as given below.

T/K = 288.16

$k_{scL}/dm^3\ mol^{-1}$	0.1003	0.0970	0.0882	0.0897	0.0806
conc/mol dm^{-3}	0.2615	0.2630	0.5250	0.5252	1.0210
$k_{scL}/dm^3\ mol^{-1}$	0.0812	0.0749	0.0749	0.0715	0.0704
conc/mol dm^{-3}	1.0235	1.4855	1.4815	1.9485	1.9865

T/K = 298.16

$k_{scL}/dm^3\ mol^{-1}$	0.0843	0.0812	0.0715	0.0762	0.0664
conc/mol dm^{-3}	0.2615	0.2630	0.5250	0.5252	1.0210
$k_{scL}/dm^3\ mol^{-1}$	0.0659	0.0612	0.0614	0.0586	0.0571
conc/mol dm^{-3}	1.0235	1.4855	1.4815	1.9485	1.9865

See also the evaluation of Seidell's values following this evaluation.

3. Nitrous oxide + water + Nitric acid [7697-37-2]

Geffcken (2) studied this system at 288.16 K and 298.16 K at six concentrations. The salt effect parameters, k_{scL}, are small and the average values are

$$k_{scL}/dm^3\ mol^{-1} = 0.001$$

$$k_{scL}/dm^3\ mol^{-1} = -0.005$$

at 288.16 K and 298.16 K, respectively. See also the evaluation of Seidell's values following this evaluation.

COMPONENTS:	EVALUATOR:
1. Nitrous oxide; N_2O; [10024-97-2] 2. Water; H_2O; [7732-18-5] 3. Electrolyte	Colin L. Young, School of Chemistry, University of Melbourne, Parkville, Victoria 3052, Australia. February 1981

CRITICAL EVALUATION:

4. Nitrous oxide + water + Phosphoric acid [7664-38-2]

This system was studied by Roth (3) at twelve different concentrations over a range of temperature from 277 K to 298 K. The salt effect parameters show very large scatter and the data of Roth (3) are classified as doubtful.

5. Nitrous oxide + water + Ammonium chloride [12125-02-9]

This system has been studied by Geffcken (2) at 288.15 K and 298.15 K and by Manchot *et al.* (1) at 298.15 K. There is reasonable agreement between the salt effect parameters calculated from these data, Geffcken's data give

$$k_{scL} = 0.052 \text{ dm}^3 \text{ mol}^{-1} \text{ at } 288.15 \text{ K}$$

$$k_{scL} = 0.050 \text{ dm}^3 \text{ mol}^{-1} \text{ at } 298.15 \text{ K}$$

at a salt concentration of one mol dm^{-3} and the value decreases with increasing concentration, whereas the data of Manchot *et al.* (1) at 298.15 K give

$$k_{scL} = 0.053 \text{ dm}^3 \text{ mol}^{-1}$$

at a salt concentration of one mol dm^{-3} and again the value decreases with increasing concentration.

6. Nitrous oxide + water + Ammonium bromide [12124-97-9]

This system has been investigated by Manchot *et al.* (1) at three concentrations at 298.15 K. The value of k_{scL} decreases with increasing concentration and k_{scL} = 0.0048 dm^3 mol^{-1} at a concentration of one mol dm^{-3}.

7. Nitrous oxide + water + Ammonium sulfate [7783-20-2]

This system has been investigated at 298.15 K at two concentrations by Manchot *et al.* (1). The values of k_{scL} at concentrations of 1.346 and 2.18 mol dm^{-3} are 0.217 dm^3 mol^{-1} and 0.221 dm^3 mol^{-1}, respectively.

8. Nitrous oxide + water + Ammonium nitrate [6484-52-2]

This system has been investigated by Manchot *et al.* (1) and by Sada *et al.* (4) at 298.15 K. There is a slight variation in $k_{sc\alpha}$ (or k_{scL}) between these two groups of workers. The average value of $k_{sc\alpha}$ (or k_{scL}) is 0.032 ± 0.001 dm^3 mol^{-1}.

9. Nitrous oxide + water + Zinc nitrate [7779-88-6]
Cadmium nitrate [10325-94-7]
Copper(II) nitrate [3251-23-8]
Nickel(II) sulfate [7786-81-4]
Cobalt(II) sulfate [10124-43-3]
Iron(II) sulfate [7720-78-7]
Iron(III) sulfate [10028-22-5]
Manganese(II) sulfate [7785-87-7]
Chromium(III) sulfate [10101-53-8]
Aluminium sulfate [10043-01-3]
Aluminium nitrate [13473-90-0]

COMPONENTS:	EVALUATOR:
1. Nitrous oxide; N_2O; [10024-97-2] 2. Water; H_2O; [7732-18-5] 3. Electrolyte	Colin L. Young, School of Chemistry, University of Melbourne, Parkville, Victoria 3052, Australia. February 1981

CRITICAL EVALUATION:

These systems have all been investigated at two concentrations at 298.15 K by Manchot *et al.* (1). Values of k_{scL}/dm^3 mol^{-1} are given below. The numbers in parentheses are concentrations in mol dm^{-3}.

Zinc nitrate	0.151 (0.84)	0.156 (1.68)
Cadmium nitrate	0.232 (0.781)	0.188 (1.562)
Copper(II) nitrate	0.252 (0.69)	0.204 (1.38)
Nickel(II) sulfate	0.357 (0.937)	0.312 (1.874)
Cobalt(II) sulfate	0.363 (0.788)	0.312 (1.576)
Iron(II) sulfate	0.269 (0.72)	0.272 (1.438)
Iron(III) sulfate	0.473 (0.66)	0.461 (1.32)
Manganese(II) sulfate	0.255 (0.94)	0.256 (1.93)
Chromium(III) sulfate	0.391 (0.57)	0.408 (1.14)
Aluminium sulfate	0.726 (0.5166)	0.735 (0.8141)
Aluminium nitrate	0.350 (0.4795)	0.270 (0.959).

10. Nitrous oxide + water + Magnesium chloride [7786-30-3]

This system has been studied by Sada *et al.* (4) at four concentrations. The salt effect parameters scatter considerably but appear to decrease with increasing concentration and have an approximate value of 0.19 dm^3 mol^{-1} at a salt concentration of one mol dm^{-3}.

11. Nitrous oxide + water + Magnesium sulfate [7487-88-9]

This system has been investigated by three groups. The two concentrations studied by Manchot *et al.* (1) lead to an average value of k_{scL} of 0.289 dm^3 mol^{-1} at 298.15 K. The data of Markham and Kobe (5) give salt effect parameters which show no definite concentration dependence. The average values are:

T/K	273.15	298.15	313.15
$k_{sm\alpha}$/kg mol^{-1}	0.336	0.289	0.271
$k_{sc\alpha}$/dm^3 mol^{-1}	0.339	0.292	0.274

Gordon (6) did not give values of the Bunsen coefficient for pure water at the same temperatures for which he studied the aqueous salt solutions but reported smoothed values for both water and the solutions at 278.15, 283.15, 288.15 and 293.15 K. The values of the salt effect parameters show some scatter but, in the case of magnesium sulfate solutions, exhibit no definite concentration dependence. The average values of $k_{sc\alpha}$ are given below.

T/K	278.15	283.15	288.15	293.15
$k_{sc\alpha}$/dm^3 mol^{-1}	0.285	0.277	0.281	0.314

The accuracy of Gordon values is probably no better than 0.02 and Markham and Kobe (5) and Manchot *et al.* (1) values are to be preferred and are classified as tentative.

12. Nitrous oxide + water + Magnesium nitrate [10377-60-3]

This system has been investigated by Markham and Kobe (5) at 273.15, 298.15 and 313.15 K. The salt effect parameter decreases with increasing concentration and increasing temperature. The salt effect parameters are given by:

COMPONENTS:	EVALUATOR:
1. Nitrous oxide; N_2O; [10024-97-2] 2. Water; H_2O; [7732-18-5] 3. Electrolyte	Colin L. Young, School of Chemistry, University of Melbourne, Parkville, Victoria 3052, Australia. February 1981

CRITICAL EVALUATION:

$$k_{sm\alpha}/\text{kg mol}^{-1} = 0.224 - 0.013\ c_3/\text{mol kg}^{-1} \text{ at } 273.15 \text{ K}$$

$$k_{sm\alpha}/\text{kg mol}^{-1} = 0.158 - 0.0075\ c_3/\text{mol kg}^{-1} \text{ at } 298.15 \text{ K}$$

$$k_{sm\alpha}/\text{kg mol}^{-1} = 0.1405 - 0.0065\ c_3/\text{mol kg}^{-1} \text{ at } 313.15 \text{ K}$$

where c_3 is the concentration of salt. When values of $k_{sc\alpha}$ are calculated from the data of Markham and Kobe (5), it is found that the values are more or less independent of concentrations, the average values being 0.216, 0.157 and 0.140 $dm^3\ mol^{-1}$ at 273.15, 298.15 and 313.15 K, respectively.

The salt effect parameters, k_{scL}, calculated from Manchot *et al.* (1) data at 298.15 K are 0.136 and 0.140 $dm^3\ mol^{-1}$ at concentrations of 0.97 and 1.93 mol dm^{-3}, respectively. The data of Markham and Kobe (5) are classified as tentative and are to be preferred to those of Manchot *et al.* (1).

13. Nitrous oxide + water + Calcium chloride [10043-52-4]

This system has been investigated by Manchot *et al.* (1) at 298.15 K, by Gordon (6) at temperatures between 281.26 K to 295.46 K and by Sada *et al.* (4) at 298.15 K. The data of Manchot *et al.* (1) lead to a value of k_{scL} of 0.210 $dm^3\ mol^{-1}$ whereas the data of Sada *et al.* (4) lead to a value of $k_{sc\alpha}$ of 0.208 $dm^3\ mol^{-1}$ at the same temperature.

Gordon (6) did not give values of the Bunsen coefficient for pure water at the same temperatures as he studied the aqueous salt solutions but reported "smoothed" values for both water and the solutions at 278.15, 283.15, 288.15 and 293.15 K. The values of the salt effect parameters show some scatter ($\sim \pm 0.01$-02 $dm^3\ mol^{-1}$) but, in the case of calcium chloride solutions, exhibit no definite concentration dependence. The average values of $k_{sc\alpha}$ are given below.

T/K	278.15	283.15	288.15	293.15
$k_{sc\alpha}/dm^3\ mol^{-1}$	0.228	0.216	0.210	0.230

The data of Manchot *et al.* (1) and Sada *et al.* (4) are classified as tentative and are to be preferred to those of Gordon (6).

14. Nitrous oxide + water + Calcium nitrate [10124-37-5]

This system has been investigated at 298.15 K by Manchot *et al.* (1) and the salt effect parameter

$$k_{scL} = 0.160\ dm^3\ mol^{-1}$$

does not show significant concentration dependence.

15. Nitrous oxide + water + Strontium chloride [10476-85-4]

This system has been studied by Gordon (6) who gave "smoothed" values of the Bunsen coefficient of water and aqueous solutions at 278.15, 283.15, 288.15 and 293.15 K. The salt effect parameters appear to decrease with increasing concentration. Values calculated from Gordon's (6) "smoothed" values are given below.

COMPONENTS:	EVALUATOR:
1. Nitrous oxide; N_2O; [10024-97-2] 2. Water; H_2O; [7732-18-5] 3. Electrolyte	Colin L. Young, School of Chemistry, University of Melbourne, Parkville, Victoria 3052, Australia. February 1981

CRITICAL EVALUATION:

Values of $k_{sc\alpha}/dm^3\ mol^{-1}$

T/K	278.15	283.15	288.15	293.15
Concentration /wt-%				
3.309 (0.22)*	0.334	0.313	0.300	0.357
5.732 (0.38)	0.291	0.297	0.279	0.259
13.239 (0.94)	0.245	0.240	0.240	0.263

* approximate concentration/mol dm^{-3}

16. Nitrous oxide + water + Barium chloride [10361-37-2]

This system has also been investigated at 298.15 K by Manchot *et al.* and the value of $k_{scL}/dm^3\ mol^{-1}$ at a concentration of one mol dm^{-3} is 0.24. The value of $k_{scL}/dm^3\ mol^{-1}$ appears to decrease with increasing concentration.

17. Nitrous oxide + water + Lithium chloride [7447-41-8]

This system has been studied by Gordon (6) and by Geffcken (2). The results of Geffcken (2) give salt effect parameters, $k_{scL}/dm^3\ mol^{-1}$ of 0.096 and 0.091 at 288.16 and 298.16 K, respectively. Gordon (6) gave "smoothed" values of the Bunsen coefficient of water and aqueous solutions at four temperatures. The salt effect parameters appear to decrease with increasing concentration. Values of $k_{sc\alpha}/dm^3\ mol^{-1}$ are given below.

T/K	278.15	283.15	288.15	293.15
Concentration /wt-%				
1.346 (0.32)*	0.142	0.139	0.145	0.203
3.853 (0.93)	0.103	0.100	0.100	0.118
11.476 (2.9)	0.089	0.088	0.087	0.127

* approximate concentration/mol dm^{-3}

18. Nitrous oxide + water + Lithium sulfate [10377-48-7]

This system has been investigated by Gordon (6) who gave "smoothed" values of the Bunsen coefficient of water and aqueous solutions at four temperatures. The salt effect parameters appear to decrease with increasing concentration. Values of $k_{sc\alpha}/dm^3\ mol^{-1}$ are given below.

T/K	278.15	283.15	288.15	293.15
Concentration /wt-%				
2.369 (0.22)*	0.314	0.297	0.297	0.381
5.463 (0.52)	0.266	0.270	0.278	0.313
8.560 (0.84)	0.273	0.262	0.254	0.264

* approximate concentration/mol dm^{-3}

COMPONENTS:	EVALUATOR:
1. Nitrous oxide; N_2O; [10024-97-2] 2. Water; H_2O; [7732-18-5] 3. Electrolyte	Colin L. Young, School of Chemistry, University of Melbourne, Parkville, Victoria 3052, Australia. February 1981

CRITICAL EVALUATION:

19. Nitrous oxide + water + Sodium chloride [7647-14-5]

This system has been investigated by four groups. The least satisfactory measurements are those of Roth (3) which give rise to salt effect parameters which show considerable scatter. His data are classified as doubtful. The data of Manchot *et al.* (1) at 298.15 K give a salt effect parameter, $k_{scL}/dm^3\ mol^{-1}$ of 0.117 which decreases to 0.113 at the highest concentration of 4.32 mol dm^{-3}. The data of Markham and Kobe (5) lead to salt effect parameters as given below. The values of $k_{sm\alpha}$ decrease with increasing concentration and increasing temperature.

T/K	273.15	298.15	313.15
Conc./mol kg^{-1}			
1.0	0.150	0.127	0.119
2.0	0.144	0.116	0.108
3.0	0.136	0.111	0.104

When values of $k_{sc\alpha}$ are calculated from the data of Markham and Kobe (5) it is found that the salt effect parameters are less dependent on salt concentration and are given by

$$k_{sc\alpha}/dm^3\ mol^{-1} = 0.157 - 0.0046\ c_3/mol\ dm^{-3} \text{ at } 273.15\ K$$

$$k_{sc\alpha}/dm^3\ mol^{-1} = 0.134 - 0.0054\ c_3/mol\ dm^{-3} \text{ at } 298.15\ K$$

and $$k_{sc\alpha}/dm^3\ mol^{-1} = 0.121 - 0.0036\ c_3/mol\ dm^{-3} \text{ at } 313.15\ K.$$

This system has also been studied by Gordon (6) who gave "smoothed" values of the Bunsen coefficient of water and aqueous solutions at four temperatures. The salt effect parameters appear to decrease with increasing concentration. Values of $k_{sc\alpha}/dm^3\ mol^{-1}$ are given below.

T/K	278.15	283.15	288.15	293.15
Conc./wt-%				
6.20 (1.1)*	0.123	0.117	0.112	0.119
8.88 (1.6)	0.115	0.113	0.115	0.125
12.78 (2.4)	0.099	0.099	0.100	0.106

* approximate concentration/mol dm^{-3}

The concentration of the salt effect parameters is greater for values calculated from Gordon's (6) data than for values from Markham and Kobe's (5) data. The data of Markham and Kobe (5) are probably the most reliable and are classified as tentative.

20. Nitrous oxide + water + Sodium bromide [7647-15-6]

This system has been investigated by Manchot *et al.* (1). The salt effect parameter, k_{scL}, is 0.109 $dm^3\ mol^{-1}$ at a salt concentration of one mol dm^{-3} and 298.15 K. The parameter decreases only marginally with increasing concentration to a value of 0.107 $dm^3\ mol^{-1}$ at a salt concentration of 4 mol dm^{-3}.

21. Nitrous oxide + water + Sodium sulfate [7757-82-6]

This system has been investigated by three groups. Gordon (6) gave

COMPONENTS:	EVALUATOR:
1. Nitrous oxide; N_2O; [10024-97-2] 2. Water; H_2O; [7732-18-5] 3. Electrolyte	Colin L. Young, School of Chemistry, University of Melbourne, Parkville, Victoria 3052, Australia. February 1981

CRITICAL EVALUATION:

"smoothed" values of the Bunsen coefficient of water and aqueous solutions at four temperatures. Although the scatter is fairly great the salt effect parameters appear to decrease with increasing concentration. Values of $k_{sc\alpha}/dm^3\ mol^{-1}$ are given below.

T/K	278.15	283.15	288.15	293.15
Conc./wt-%				
5.77 (0.43)*	0.307	0.311	0.293	0.339
8.53 (0.65)	0.306	0.315	0.322	0.341
12.44 (0.97)	0.299	0.285	0.278	0.297

* approximate concentration/mol dm^{-3}

The data of Manchot *et al.* (1) lead to a value of $k_{scL}/dm^3\ mol^{-1}$ at 298.15 K of 0.338 at a concentration of one mol dm^{-3} and the parameter appears to decrease with increasing concentration.

The salt effect parameters calculated from the data of Markham and Kobe (5) are given below.

Values of $k_{sm\alpha}/kg\ mol^{-1}$

T/K	298.15	313.15
Conc./mol kg^{-1}		
0.5	0.359	0.338
1.0	0.338	0.318
1.5	0.331	0.310

While these salt effect parameters decrease with increasing concentration the relationship between parameter and concentration is not linear.

When values of $k_{sc\alpha}/dm^3\ mol^{-1}$ are calculated from Markham and Kobe's data (5) the salt effect parameters are less concentration-dependent than values of $k_{sm\alpha}/dm^3\ mol^{-1}$ and are given below.

Values of $k_{sc\alpha}/dm\ \ mol^{-1}$

T/K	298.15	313.15
Conc./mol kg^{-1}		
0.5	0.364	0.345
1.0	0.348	0.331
1.5	0.346	0.328

22. Nitrous oxide + water + Sodium sulfate [7757-82-6] + Sulfuric acid [7664-93-9]

Kobe and Kenton (8) made measurements of the solubility of nitrous oxide in a mixture that contained 1.76 mol kg^{-1} of Na_2SO_4 and 0.90 mol kg^{-1} of H_2SO_4. Using the sum of the component molalities gives the salt effect parameter, $k_{sm\alpha}$ = 0.215. This value is slightly greater than would be expected if

$$k_{sm\alpha}\ m(\text{overall}) = \sum (k_{sm\alpha})_i\ m_i .$$

The value appears reasonable and is classified as tentative.

COMPONENTS:	EVALUATOR:
1. Nitrous oxide; N_2O; [10024-97-2] 2. Water; H_2O; [7732-18-5] 3. Electrolyte	Colin L. Young, School of Chemistry, University of Melbourne, Parkville, Victoria 3052, Australia. February 1981

CRITICAL EVALUATION:

23. Nitrous oxide + water + Sodium nitrate [7631-99-4]

This system has been investigated by Knopp (7) at 293.15 K at four concentrations up to 1.12 mol dm^{-3}. The salt effect parameter does not show significant concentration dependence and the average value of $k_{sc\alpha}/dm^3\ mol^{-1}$ is 0.094. Similarly the salt effect parameter calculated from the data of Manchot *et al.* (1) at 298.15 K show no discernible concentration dependence and has a value of $k_{scL}/dm^3\ mol^{-1}$ of 0.093.

24. Nitrous oxide + water + Sodium phosphate [7601-54-9] + Sodium hydrogen phosphate [7558-79-4]

These two systems have only been studied at one concentration at 298.15 K by Manchot *et al.* (1). The salt effect parameters, $k_{scL}/dm^3\ mol^{-1}$ are 0.53 and 0.37 for sodium phosphate and sodium hydrogen phosphate, respectively.

25. Nitrous oxide + water + Sodium carbonate [497-19-8] + Sodium bicarbonate [144-55-8]

In order to evaluate the data for this system it is convenient to write

$$\log (\alpha°/\alpha) = K_s \sum_i I_i$$

where I_i is the ionic strength of ions of species i and K_s is the overall salt effect parameter. Values of K_s vary with R, the ratio of the concentration of bicarbonate ion to that of the carbonate and are given below.

T/K = 298.15

R	0	0.2	1.0	2.0	5.0
K_s/dm^3 g-ion^{-1}	0.118	0.118	0.129	0.149	0.181

The data of Hikita *et al.* (9) appear to be reliable and were determined with an apparatus of proven design for moderate accuracy and are therefore classified as tentative.

26. Nitrous oxide + water + Potassium hydroxide [1310-58-3]

This system has been studied by Geffcken (2) at 288.16 and 298.15 K. The salt effect parameters increase with increasing concentration and are given by

$$k_{scL}/dm^3\ mol^{-1} = 0.120 + 0.024\ (c/mol\ dm^{-3})\ \text{at } 288.16\ K$$

and $$k_{scL}/dm^3\ mol^{-1} = 0.112 + 0.0206\ (c/mol\ dm^{-3})\ \text{at } 298.15\ K.$$

27. Nitrous oxide + water + Potassium chloride [7447-40-7]

This system has been investigated by four groups of workers. Gordon (6) gave "smoothed" values of the Bunsen coefficient of water and aqueous solutions at four temperatures. The salt effect parameters appear to decrease with increasing concentration but the scatter is too great to allow a meaningful equation to be derived. Values of $k_{sc\alpha}/dm^3\ mol^{-1}$ are given below.

COMPONENTS:	EVALUATOR:
1. Nitrous oxide; N_2O; [10024-97-2] 2. Water; H_2O; [7732-18-5] 3. Electrolyte	Colin L. Young, School of Chemistry, University of Melbourne, Parkville, Victoria 3052, Australia. February 1981

CRITICAL EVALUATION:

T/K	278.15	283.15	288.15	293.15
Conc./wt-%				
4.90 (0.68)*	0.139	0.129	0.122	0.140
7.64 (1.04)	0.126	0.114	0.111	0.140
14.58 (2.15)	0.104	0.095	0.090	0.096
22.08 (3.41)	0.089	0.088	0.088	0.091

* approximate concentration/mol dm^{-3}

Geffcken's data lead to salt effect parameters of $k_{scL}/dm^3\ mol^{-1}$ of 0.102 and 0.098 at 288.16 K and 298.16 K, respectively, at a concentration of one mol dm^{-3}. These parameters decrease with increasing concentration but the data are too limited to allow evaluation of the concentration-dependence of k_{scL}.

Markham and Kobe's data (5) yield values of $k_{sm\alpha}$ which decrease with increasing concentration and values of $k_{sm\alpha}/kg\ mol^{-1}$ are given by

$$k_{sm\alpha}/kg\ mol^{-1} = 0.125 - 0.0075c_3/mol\ kg^{-1} \text{ at } 273.15\ K$$

$$k_{sm\alpha}/kg\ mol^{-1} = 0.101 - 0.006c_3/mol\ kg^{-1} \text{ at } 298.15\ K$$

$$k_{sm\alpha}/kg\ mol^{-1} = 0.101 - 0.0077c_3/mol\ kg^{-1} \text{ at } 313.15\ K$$

where c_3 is the concentration of salt solution.

When values of $k_{sc\alpha}/dm^3\ mol^{-1}$ are calculated from Markham and Kobe's data (5), the salt effect parameters are less concentration-dependent than values of $k_{sm\alpha}/kg\ mol^{-1}$ and are given by

$$k_{sc\alpha}/dm^3\ mol^{-1} = 0.1253 - 0.0045c_3/mol\ kg^{-1} \text{ at } 273.15\ K$$

$$k_{sc\alpha}/dm^3\ mol^{-1} = 0.1018 - 0.0035c_3/mol\ kg^{-1} \text{ at } 298.15\ K$$

and $$k_{sc\alpha}/dm^3\ mol^{-1} = 0.1033 - 0.0055c_3/mol\ kg^{-1} \text{ at } 313.15\ K.$$

The data of Manchot *et al.* (1) lead to values of the salt effect parameter as given by

$$k_{sc\alpha}/dm^3\ mol^{-1} = 0.092 - 0.002c_3/mol\ dm^{-3}.$$

The salt effect parameters from the work of Geffcken (2), Manchot *et al.* (1) and Markham and Kobe (5) agree fairly well at 298.15 K and at a concentration of one mol dm^{-3}. However, the data of Gordon (6) are considerably greater and are classified as doubtful.

28. Nitrous oxide + water + Potassium bromide [7758-02-3]

This system has been studied by Geffcken (2) at 288.16 and 298.16 K. The salt effect parameter appears to decrease with increasing concentration but the scatter is too great for an accurate estimation of the concentration dependence. The values of $k_{scL}/dm^3\ mol^{-1}$ at a concentration of one mol dm^{-3} are 0.093 and 0.087 at 288.16 and 298.16 K, respectively. The parameters calculated from the data of Manchot *et al.* (1) also appear to decrease with increasing concentration but the scatter in this case is also too large for an accurate estimation of the concentration dependence. The value of $k_{scL}/dm^3\ mol^{-1}$ at 298.15 K is 0.083.

COMPONENTS:	EVALUATOR:
1. Nitrous oxide; N_2O; [10024-97-2] 2. Water; H_2O; [7732-18-5] 3. Electrolyte	Colin L. Young, School of Chemistry, University of Melbourne, Parkville, Victoria 3052, Australia. February 1981

CRITICAL EVALUATION:

29. Nitrous oxide + water + Potassium iodide [7681-11-0]

The salt effect parameter calculated from the data of Sada *et al.* (4) are given below. They appear to decrease with increasing concentration and then increase at the highest concentration.

$k_{sc\alpha}/dm^3\ mol^{-1}$	0.0846	0.0814	0.0787	0.0782	0.0807
conc./mol dm^{-3}	0.644	1.093	1.661	2.196	2.859

The salt effect parameters calculated from Geffcken's data (2) give average values of $k_{scL}/dm^3\ mol^{-1}$ of 0.091 and 0.082 at 288.16 and 298.16 K, respectively.

30. Nitrous oxide + water + Potassium sulfate [7778-80-5]

This system has been studied by two groups but only to a very limited extent. Manchot *et al.* (1) studied only one concentration at 298.15 K which gives a salt effect parameter of 0.29 $dm^3\ mol^{-1}$. Gordon (6) studied two low concentrations and values of $k_{sc\alpha}/dm^3\ mol^{-1}$ calculated from his smoothed data are given below.

T/K	278.15	283.15	288.15	293.15
Conc./wt-%				
2.623 (0.15)*	0.295	0.287	0.299	0.370
4.784 (0.29)	0.268	0.285	0.305	0.368

* approximate concentration/mol dm^{-3}

Both sets of data are classified as doubtful.

31. Nitrous oxide + water + Potassium nitrate [7757-79-1]

Values of the salt effect parameters from the four studies of this system are given below. There is approximate agreement between the values of the parameters from the results of Manchot *et al.* (1), Sada *et al.* (4) and Knopp (7) near concentrations of one mol dm^{-3} but the data of Markham and Kobe (5) give smaller values. The concentration dependence of $k_{sc\alpha}$ (or k_{scL}) is moderately large for values calculated from the data of Markham and Kobe (5) and Manchot *et al.* (1) but insignificant when calculated from the data of Sada *et al.* (4) The data of Knopp (7) are only over a small range of concentration and the values scatter too greatly to enable any statement to be made of the concentration dependence of $k_{sc\alpha}$.

Author	T/K	Conc./mol dm^{-3}	$k_{sc\alpha}/dm^3\ mol^{-1}$	$k_{sm\alpha}/kg\ mol^{-1}$
Knopp	293.15	0.1061	0.064	-
		0.2764	0.070	-
		0.5630	0.072	-
		1.1683	0.070	-
Manchot *et al.*	298.15	1.02	0.073	-
		2.15	0.066	-
Sada *et al.*	298.15	1.381	0.069	-
		2.645	0.069	-

(cont.)

COMPONENTS:	EVALUATOR:
1. Nitrous oxide; N_2O; [10024-97-2] 2. Water; H_2O; [7732-18-5] 3. Electrolyte	Colin L. Young, School of Chemistry, University of Melbourne, Parkville, Victoria 3052, Australia. February 1981

CRITICAL EVALUATION:

Author	T/K	Conc./mol dm^{-3}	$k_{sc\alpha}/dm^3\ mol^{-1}$	$k_{sm\alpha}/kg\ mol^{-1}$
Markham and Kobe	273.35	0.2*	0.122	0.121
		0.5*	0.102	0.100
		1.0*	0.093	0.090
	298.15	1.0*	0.059	0.056
		2.0*	0.054	0.050
		3.0*	0.050	0.045
	323.15	1.0*	0.047	0.045
		2.0*	0.042	0.039
		3.0*	0.041	0.036

* concentration/mol kg^{-1}

32. Nitrous oxide + water + Potassium carbonate [584-08-7] + Potassium bicarbonate [298-14-6]

This system has been studied by Joosten and Danckwerts (10). This author analysed their data according to an equation which may be written

$$\log \alpha_o/\alpha = KI$$

where K is the overall salt effect parameter and I is the ionic strength of the solution. The value of K was 0.105 ± 0.001 dm^3 g-ion^{-1} for equal concentrations of carbonate and bicarbonate.

33. Nitrous oxide + water + Potassium periodate [7790-21-8]

This system has been studied at one concentration (0.0164 mol dm^{-3}) by Manchot *et al.* (1). The small concentration and consequent slight change in Bunsen coefficient make an accurate estimation of the salt effect parameter impossible.

34. Nitrous oxide + water + Rubidium chloride [7791-11-9]

This system has been studied by Geffcken (2) at 288.16 and 298.16 K. The values of k_{scL} decrease with increase in concentration. The value of k_{scL} at a concentration of approximately one mol dm^{-3} are 0.094 dm^3 mol^{-1} and 0.089 dm^3 mol^{-1} at 288.16 and 298.16 K, respectively.

35. Nitrous oxide + water + Caesium chloride [7647-17-8]

This system has been studied by Geffcken (2) at two concentrations near 0.5 mol dm^{-3}. The values of the salt effect parameter $k_{scL}/dm^3\ mol^{-1}$ were 0.081 and 0.076 at 288.16 and 298.16 K, respectively. The data are classified as tentative.

References

1. Manchot, von W.; Jahrstorfer, M.; Zepter, H. *Z. Anorg. Allg. Chem.* 1924, *141*, 45.

2. Geffcken, G. *Z. Phys. Chem.* 1904, *49*, 257.

COMPONENTS:	EVALUATOR:
1. Nitrous oxide; N_2O; [10024-97-2] 2. Water; H_2O; [7732-18-5] 3. Electrolyte.	Colin L. Young. School of Chemistry University of Melbourne, Parkville, Victoria 3052. AUSTRALIA: February, 1981

CRITICAL EVALUATION

3. Roth, W. *Z. Phys. Chem.* 1897, *24*, 114.

4. Sada, E.; Ando, N.; Kito, S. *J. Appl. Chem. Biotechnol.* 1972, *22*, 1185.

5. Markham, A.E.; Kobe, K.A. *J. Amer. Chem. Soc.* 1941, *63*, 449.

6. Gordon, V. *Z. Phys. Chem.* 1895, *18*, 1.

7. Knopp, W. *Z. Phys. Chem.* 1904, *48*, 97.

8. Kobe, K.A.; Kenton, F.H. *Ind. Eng. Chem.* 1938, *10*, 76

9. Hikita, H.; Asai, S.; Ishikawa, H.; Esaka, H. *J. Chem. Engng. Data.* 1974, *19*, 89.

10. Joosten, G.E.H.; Danckwerts, P.V. *J. Chem. Engng. Data.* 1972, *17*, 452.

NOTE:

Kreitus and co-workers (11,12) have recently investigated the solubility of nitrous oxides in concentrated salt solutions. The concentration dependence of the salt effect parameter for potassium chloride is in fair agreement with that of Markham and Kobe (5) whereas the salt effect parameters for lithium chloride is in marked disagreement with those obtained from the data of Gordon (6) and also considerably smaller than that obtained from Geffcken's data (2). The salt effect parameter for cesium chloride is in reasonable agreement with that obtained from the data of Geffcken (2).

The presentation of the data of Kreitus and co-workers (11,12) is such that less approximation is required to calculate $k_{sm\alpha}$ rather than $k_{sc\alpha}$. Values of $k_{sm\alpha}$ are given below.

LiCl		KCl		KF		CsCl	
Conc./mol kg^{-1} (water)	$k_{sm\alpha}$	Conc./mol kg^{-1} (water)	$k_{sm\alpha}$	Conc./mol kg^{-1} (water)	$k_{sm\alpha}$	Conc./mol kg^{-1} (water)	$k_{sm\alpha}$
0.97	0.070	0.51	0.098	1.21	0.136	0.50	0.084
2.66	0.069	1.03	0.094	5.84	0.131	0.93	0.106
5.44	0.077	2.13	0.081	7.78	0.107	2.45	0.093
9.10	0.073	2.70	0.082	12.33	0.118	4.85	0.076
12.2	0.071	3.61	0.077			7.25	0.081
15.5	0.072	4.63	0.076				
18.7	0.068						
19.8	0.067						

These measurements are classified as tentative.

11. Kreitus, I.; Gorbovitskalya, T.I. *Latv. PRS. Zinat. Akad. Vestis. Khim. Ser.* 1979, 664.

12. Kreitus, I.; Abramenkov, A. *Latv. PRS. Zinat. Akad. Vestis. Khim. Ser.* 1980, 238

COMPONENTS:	EVALUATOR:
Nitrous oxide; N_2O; [10024-97-2] Water; H_2O; [7732-18-5] Acids	Colin Young School of Chemistry University of Melbourne Parkville, AUSTRALIA March, 1981

In addition to data discussed in the preceding evaluation Seidell (1) quotes values of Bunsen coefficients for the solubility of nitrous oxide in aqueous solutions of acids at 25°C (298.2K). It is not clear whether the exact source was a private communication or a thesis but the numerical data do not appear in the reference given i.e. "Manchot Jahrstorfer and Zepter 1924" (2) although graphs in the original indicate the existence of experimental result for these systems. The data are reported below and Setschenow salt effect parameters have been calculated as in the preceding evaluation.

Conc. of acid $/g\ dm^{-3}$	Specific gravity before adsorption d_4^{25}	10^2 x Bunsen coefficient, $10^2\ \alpha$	$k_{sc\alpha}$ $/dm^3\ mol^{-1}$
	Nitric acid, HNO_3; [7697-37-2]		
69.32	1.0351	54.1	-0.007
142.42	1.0731	55.1	-0.007
229.38	1.1191	56.2	-0.007
	Hydrochloric acid; HCl; [7647-01-0]		
39.387	1.0168	51.2	0.018
77.318	1.0335	50.1	0.013
167.03	1.0741	49.9	0.007
237.06	1.1050	52.1	0.001
	Periodic acid; HIO_4; [13444-71-8]		
204.21	1.1740	38.7	0.132
470.23	1.4066	23.8	0.143
	Sulfuric acid; H_2SO_4; [7664-93-9]		
110.84	1.0680	44.2	0.073
269.74	1.1630	38.7	0.051
454.15	1.2687	38.2	0.031
588.53	1.3363	39.9	0.021
	Phosphoric acid; H_3PO_4; [7664-38-2]		
114.73	1.0593	46.4	0.053
186.31	1.0964	43.8	0.046
495.20	1.2557	35.3	0.036

The values of $k_{sc\alpha}$ from the present data agree with those of Geffcken (2) for nitric acid. However values of the salt effect parameter for sulfuric acid and hydrochloric acid do not agree with the values obtained from Geffcken's results. Although the values for both sets of results for both acid show a decrease for an increase in concentration.

REFERENCES:

1. Seidell, A. *Solubilities of Inorganic and Metal Organic Compounds*, 3rd. edn. Vol. 1, 1953, p.1136.

2. Geffcken, G. *Z. Phys. Chem.* 1904, *49*, 271.

COMPONENTS:

1. Nitrous oxide; N_2O; [10024-97-2]
2. Water; H_2O; [7732-18-5];
3. Hydrogen chloride; HCl; [7647-01-0]

ORIGINAL MEASUREMENTS:

Geffcken, G. *Z. Phys. Chem.* **1904**, *49*, 257-302.

VARIABLES:

Temperature, concentration

PREPARED BY:

W. Gerrard/C. L. Young

EXPERIMENTAL VALUES:

T/K	Conc. of acid/mol dm^{-3} (soln.)	Ostwald coefficient, L
288.16	0.549	0.7550
	0.550	0.7528
	1.089	0.7360
	1.093	0.7347
	2.300	0.7103
	2.340	0.7122
298.16	0.549	0.5775
	0.550	0.5759
	1.089	0.5670
	1.093	0.5657
	2.300	0.5546
	2.340	0.5564

AUXILIARY INFORMATION

METHOD /APPARATUS/PROCEDURE:

Measurement of volume of N_2O absorbed by the aqueous solution. Detailed description and diagram given in source.

SOURCE AND PURITY OF MATERIALS:

1. Nitrous oxide self prepared and attested.

ESTIMATED ERROR:

REFERENCES:

COMPONENTS:	ORIGINAL MEASUREMENTS:
1. Nitrous oxide; N_2O; [10024-97-2] 2. Water; H_2O; [7732-18-5] 3. Sulfuric acid; H_2SO_4; [7664-93-9]	Geffcken, G. *Z. Phys. Chem.* 1904, *49*, 257-302.
VARIABLES:	PREPARED BY:
Temperature, concentration	W. Gerrard

EXPERIMENTAL VALUES:

T/K	Conc of acid / mol dm^{3}	Ostwald coefficient, L
288.16	0.2615	0.7328
	0.2630	0.7340
	0.5250	0.6997
	0.5252	0.6984
	1.0210	0.6440
	1.0235	0.6428
	1.4855	0.6024
	1.4815	0.6030
	1.9485	0.5648
	1.9865	0.5640
298.16	0.2615	0.5648
	0.2630	0.5657
	0.5250	0.5426
	0.5252	0.5419
	1.0210	0.5083
	1.0235	0.5087
	1.4855	0.4819
	1.4815	0.4820
	1.9485	0.4569
	1.9865	0.4577

AUXILIARY INFORMATION

METHOD/APPARATUS/PROCEDURE:	SOURCE AND PURITY OF MATERIALS:
Measurement of volume of gas absorbed by the aqueous solution. Detailed description and diagram given in source.	1. Self prepared and attested.
	ESTIMATED ERROR:
	REFERENCES:

COMPONENTS:	ORIGINAL MEASUREMENTS:
1. Nitrous oxide; N_2O; [10024-97-2] 2. Water; H_2O; [7732-18-5]; 3. Nitric acid; HNO_3; [7697-37-2]	Geffcken, G. *Z. Phys. Chem.* 1904, *49*, 257-302.
VARIABLES:	PREPARED BY:
Temperature, concentration	W. Gerrard/C. L. Young

EXPERIMENTAL VALUES:

T/K	Conc. of acid/mol dm^{-3} (soln.)	Ostwald coefficient, L
288.16	0.610	0.7770
	0.614	0.7766
	1.253	0.7767
	1.254	0.7767
	2.405	0.7735
	2.435	0.7737
298.16	0.610	0.5969
	0.614	0.5980
	1.253	0.6045
	1.254	0.6061
	2.405	0.6156
	2.435	0.6149

AUXILIARY INFORMATION

METHOD /APPARATUS/PROCEDURE:

Measurement of volume of N_2O absorbed by the aqueous solution. Detailed description and diagram given in source.

SOURCE AND PURITY OF MATERIALS:

1. Nitrous oxide self prepared and attested.

ESTIMATED ERROR:

REFERENCES:

COMPONENTS:	ORIGINAL MEASUREMENTS:
1. Nitrous oxide; N_2O; [10024-97-2] 2. Water; H_2O; [7732-18-5]; 3. Phosphoric acid; H_3PO_4; [7664-38-2]	Roth, W. *Z. Phys. Chem.* 1897, *24*, 114-151.
VARIABLES:	PREPARED BY:
Temperature, concentration	W. Gerrard

EXPERIMENTAL VALUES: t = T/K - 273.16; density = *d*; α = Bunsen coefficient.

Concn. of phosphoric acid, %: 3.098
Change of *d* with t: $d = 1.01779 - 0.000027696t - 0.0000051374t^2$
Change of α with t: $\alpha = 1.2652 - 0.045826t + 0.00067862t^2$

T/K	298.24	293.16	288.13	283.14	278.51
α	0.5428	0.6188	0.7313	0.8770	1.0395

Concn. of phosphoric acid, %: 3.659
Change of *d* with t: $d = 1.02099 - 0.000028273t - 0.0000052916t^2$
Change of α with t: $\alpha = 1.2587 - 0.042209t + 0.0005399t^2$

T/K	298.27	293.31	288.21	283.72	277.80
α	0.5392	0.6304	0.7457	0.8725	1.0744

Concn. of phosphoric acid, %: 4.465
Change of *d* with t: $d = 1.02570 - 0.000042813t - 0.0000050554t^2$
Change of α with t: $\alpha = 1.2123 - 0.041104t + 0.00055483t^2$

T/K	298.04	293.13	288.25	283.60	278.19
α	0.5331	0.6139	0.7184	0.8501	1.0196

Concn. of phosphoric acid, %: 4.569
Change of *d* with t: $d = 1.02630 - 0.000044265t - 0.0000050466t^2$
Change of α with t: $\alpha = 1.2668 - 0.04376t + 0.00058643t^2$

T/K	298.31	292.97	288.30	283.17	277.90
α	0.5372	0.6294	0.7387	0.8723	1.0726

(continued)

AUXILIARY INFORMATION

METHOD/APPARATUS/PROCEDURE:

Ostwald method, using gas buret and pipet. Measurement of volume of gas before and after absorption. Specific gravity of solution was determined by a Sprengel pyknometer. Vapour pressure of water, adjusted by assuming Raoult's law, was allowed for.

SOURCE AND PURITY OF MATERIALS:

1. N_2O was self prepared and purified.

3. Phosphoric acid was analysed by an appropriate method.

ESTIMATED ERROR:

REFERENCES:

COMPONENTS:

1. Nitrous oxide; N_2O; [10024-97-2]
2. Water; H_2O; [7732-18-5];
3. Phosphoric acid; H_3PO_4; [7664-38-2]

ORIGINAL MEASUREMENTS:

Roth, W.

Z. Phys. Chem. 1897, *24*, 114-151.

EXPERIMENTAL VALUES: t = T/K - 273.16; density = *d*; α = Bunsen coefficient.

Concn. of phosphoric acid, %: 5.137
Change of *d* with t: $d = 1.02971 - 0.000048262t - 0.000005142t^2$
Change of α with t: $\alpha = 1.2285 - 0.04359t + 0.00062256t^2$

T/K	298.32	293.54	288.08	283.39	277.80
α	0.5259	0.6003	0.7168	0.8417	1.0397

Concn. of phosphoric acid, %: 8.702
Change of *d* with t: $d = 1.05048 - 0.00009855t - 0.000004480t^2$
Change of α with t: $\alpha = 1.1765 = 0.039698t + 0.00053963t^2$

T/K	298.06	293.15	288.14	283.20	278.22
α	0.5226	0.6012	0.7029	0.8279	0.9894

Concn. of phosphoric acid, %: 8.855
Change of *d* with t: $d = 1.05139 - 0.00010293t - 0.000004388t^2$
Change of α with t: $\alpha = 1.1836 - 0.040624t + 0.0005469t^2$

T/K	298.05	293.14	288.31	282.90	278.21
α	0.5113	0.5952	0.6937	0.8386	0.9924

Concn. of phosphoric acid, %: 8.963
Change of *d* with t: $d = 1.05192 - 0.00010204t - 0.0000044433t^2$
Change of α with t: $\alpha = 1.1614 - 0.039037t + 0.0005219t^2$

T/K	298.05	292.97	288.00	283.29	277.61
α	0.5129	0.5961	0.6970	0.8117	0.9980

Concn. of phosphoric acid, %: 9.775
Change of *d* with t: $d = 1.05707 - 0.00012724t - 0.000003938t^2$
Change of α with t: $\alpha = 1.1601 - 0.039735t + 0.00054612t^2$

T/K	298.12	293.06	288.17	283.42	277.99
α	0.5085	0.5885	0.6867	0.7994	0.9809

Concn. of phosphoric acid, %: 10.001
Change of *d* with t: $d = 1.05836 - 0.00012525t - 0.0000040003t^2$
Change of α with t: $\alpha = 1.1256 - 0.037196t + 0.0004913t^2$

T/K	298.14	293.14	288.17	283.28	278.11
α	0.5031	0.5780	0.6780	0.7998	0.9535

Concn. of phosphoric acid, %: 13.260
Change of *d* with t: $d = 1.7852 - 0.00015737t - 0.0000039225t^2$
Change of α with t: $\alpha = 1.0921 - 0.03790t + 0.00054374t^2$

T/K	297.84	293.23	288.03	283.30	278.83
α	0.4879	0.5555	0.6487	0.7647	0.8946

Concn. of phosphoric acid, %: 13.438
Change of *d* with t: $d = 1.07950 - 0.00016617t - 0.000003869t^2$
Change of α with t: $\alpha = 1.0852 - 0.035824t + 0.00047667t^2$

T/K	298.08	293.06	288.10	283.10	277.97
α	0.4885	0.5632	0.6564	0.7756	0.9239

COMPONENTS:

1. Nitrous oxide; N_2O; [10024-97-2]
2. Water; H_2O; [7732-18-5];
3. Ammonium chloride; NH_4Cl; [12125-02-9]

ORIGINAL MEASUREMENTS:

Geffcken, G. *Z. Phys. Chem.* 1904, *49*, 257-302.

VARIABLES:

Temperature, concentration

PREPARED BY:

W. Gerrard/C. L. Young

EXPERIMENTAL VALUES:

T/K	Conc. of salt/mol dm^{-3} (soln.)	Ostwald coefficient, L
288.15	0.598	0.7203
	0.600	0.7185
	1.158	0.6800
	1.166	0.6775
298.15	0.598	0.5532
	0.600	0.5504
	1.158	0.5223
	1.166	0.5200

AUXILIARY INFORMATION

METHOD /APPARATUS/PROCEDURE:

Measurement of volume of N_2O absorbed by the aqueous solution. Detailed description and diagram given in source.

SOURCE AND PURITY OF MATERIALS:

1. Nitrous oxide self prepared and attested.

ESTIMATED ERROR:

REFERENCES:

COMPONENTS:	ORIGINAL MEASUREMENTS:
1. Nitrous oxide; N_2O; [10024-97-2] 2. Water; H_2O; [7732-18-5] 3. Ammonium chloride; NH_4Cl; [12125-02-9]	Manchot, von W.; Jahrstorfer, M.; Zepter, M. *Z. Anorg. Allg. Chem.* 1924, *141*, 45-81.
VARIABLES:	PREPARED BY:
Concentration	C. L. Young

EXPERIMENTAL VALUES:

T/K	Density, d_4^{25} of salt soln.	Conc. of salt /10^3 mol m^{-3} (soln.)	Conc. of salt /mol kg^{-1} (water)	S_1 /cm^3	S_2 /cm^3
298.15	1.0146	1.07	1.1177	46.6	45.9
	1.0312	2.25	2.4703	41.1	39.9
	1.0594	4.30	5.1848	34.7	32.7

S_1 - volume of nitrous oxide absorbed per 100 cm^3 of salt solution

S_2 - volume of nitrous oxide absorbed per 100 g of salt solution

Both S_1 and S_2 were reduced to conditions of 273.15 K and 101.3 kPa

AUXILIARY INFORMATION

METHOD/APPARATUS/PROCEDURE:

Measurement of volume of gas absorbed by means of gas buret and pipet. Volume absorbed appears to be taken as independent of pressure. (1). Density of the aqueous solution determined by Sprengel pyknometer.

SOURCE AND PURITY OF MATERIALS:

1. Prepared by heating ammonium nitrate, frozen in liquid air and then distilled.
2. Recrystallized.

ESTIMATED ERROR:

REFERENCES:

1. Manchot, W. *Z.Anorg.Chem.* 1924, *141*, 38.

COMPONENTS:

1. Nitrous oxide; N_2O; [10024-97-2]
2. Water; H_2O; [7732-18-5]
3. Ammonium bromide; NH_4Br; [12124-97-9]

ORIGINAL MEASUREMENTS:

Manchot, von W.; Jahrstorfer, M.; Zepter, H.

Z. Anorg. Allg. Chem. 1924, *141*, 45-81.

VARIABLES:

Concentration

PREPARED BY:

C. L. Young

EXPERIMENTAL VALUES:

T/K	Density, d_4^{25} of salt soln.	Conc. of salt $/10^3$ mol m^{-3} (soln.)	Conc. of salt /mol kg^{-1} (water)	S_1 $/cm^3$	S_2 $/cm^3$
298.15	1.0535	1.04	1.1598	47.4	45.0
	1.1088	2.11	2.3387	42.3	38.1
	1.2122	4.15	4.9922	35.8	29.6

S_1 - volume of nitrous oxide absorbed per 100 cm of salt solution

S_2 - volume of nitrous oxide absorbed per 100 g of salt solution

Both S_1 and S_2 were reduced to conditions of 273.15 K and 101.3 kPa

AUXILIARY INFORMATION

METHOD/APPARATUS/PROCEDURE:

Measurement of volume of gas absorbed by means of gas buret and pipet. Volume absorbed appears to be taken as independent of pressure. (1). Density of the aqueous solution determined by Sprengel pyknometer.

SOURCE AND PURITY OF MATERIALS:

1. Prepared by heating ammonium nitrate, frozen in liquid air and then distilled.

2. Recrystallized.

ESTIMATED ERROR:

REFERENCES:

1. Manchot, W. *Z.Anorg.Chem.* 1924, *141*, 38.

COMPONENTS:	ORIGINAL MEASUREMENTS:
1. Nitrous oxide; N_2O; [10024-97-2] 2. Water; H_2O; [7732-18-5] 3. Ammonium sulfate; $(NH_4)_2SO_4$; [7783-20-2]	Manchot, von W.; Jahrstorfer, M.; Zepter, H. *Z. Anorg. Allg. Chem.* 1924, *141*, 45-81.
VARIABLES:	PREPARED BY:
Concentration	C. L. Young

EXPERIMENTAL VALUES:

T/K	Density, d_4^{25} of salt soln.	Conc. of salt /10^3 mol m^{-3} (soln.)	Conc. of salt /mol kg^{-1} (water)	S_1 /cm^3	S_2 /cm^3
298.15	1.0896	1.346	1.4763	27.1	24.9
	1.1393	2.18	2.5610	17.5	15.4

S_1 - volume of nitrous oxide absorbed per 100 cm^3 of salt solution

S_2 - volume of nitrous oxide absorbed per 100 g of salt solution

Both S_1 and S_2 were reduced to conditions of 273.15 K and 101.3 kPa

AUXILIARY INFORMATION

METHOD/APPARATUS/PROCEDURE:

Measurement of volume of gas absorbed by means of gas buret and pipet. Volume absorbed appears to be taken as independent of pressure. (1). Density of the aqueous solution determined by Sprengel pyknometer.

SOURCE AND PURITY OF MATERIALS:

1. Prepared by heating ammonium nitrate, frozen in liquid air and then distilled.

2. Recrystallized.

ESTIMATED ERROR:

REFERENCES:

1. Manchot, W. *Z.Anorg.Chem.* 1924, *141*, 38.

COMPONENTS:

1. Nitrous oxide; N_2O; [10024-97-2]
2. Water; H_2O; [7732-18-5]
3. Ammonium nitrate; NH_4NO_3; [6484-52-2]

ORIGINAL MEASUREMENTS:

Manchot, von W.; Jahrstorfer, M.; Zepter, H.

Z. Anorg. Allg. Chem. **1924**, *141*, 45-81.

VARIABLES:

Concentration

PREPARED BY:

C. L. Young

EXPERIMENTAL VALUES:

T/K	Density, d_4^{25} of salt soln.	Conc. of salt $/10^3$ mol m^{-3} (soln.)	Conc. of salt /mol kg^{-1} (water)	S_1 $/cm^3$	S_2 $/cm^3$
298.15	1.0249	0.89	0.9333	49.8	48.6
	1.0527	1.85	2.0451	46.4	44.1
	1.1040	3.63	4.4627	40.9	37.1
	1.2116	7.40	11.950	30.8	25.4

S_1 - volume of nitrous oxide absorbed per 100 cm^3 of salt solution

S_2 - volume of nitrous oxide absorbed per 100 g of salt solution

Both S_1 and S_2 were reduced to conditions of 273.15 K and 101.3 kPa

AUXILIARY INFORMATION

METHOD/APPARATUS/PROCEDURE:

Measurement of volume of gas absorbed by means of gas buret and pipet. Volume absorbed appears to be taken as independent of pressure (1). Density of the aqueous solution determined by Sprengel pyknometer.

SOURCE AND PURITY OF MATERIALS:

1. Prepared by heating ammonium nitrate, frozen in liquid air and then distilled.

2. Recrystallized.

ESTIMATED ERROR:

REFERENCES:

1. Manchot, W. *Z.Anorg.Chem.* **1924**, *141*, 38.

COMPONENTS:

1. Nitrous oxide; N_2O; [10024-97-2]
2. Water; H_2O; [7732-18-5];
3. Ammonium nitrate; NH_4NO_3; [6484-52-2]

ORIGINAL MEASUREMENTS:

Sada, E.; Ando, N.; Kito, S.

J. Appl. Chem. Biotechnol. 1972, *22*, 1185-1193.

VARIABLES:

T/K: 298.15 *P*/kPa: 101.325 (1 atm)

Molarity of salt: mol l^{-1}

PREPARED BY:

W. Gerrard/C. L. Young

EXPERIMENTAL VALUES:

T/K	Concn. of salt/mol l^{-1}	Bunsen coefficient, α
298.15	0	0.5512 (pure water)
	1.322	0.5009
	3.049	0.4381
	4.484	0.3098*

* This appears to be a typographical error, it appears from the the graph given in the original paper that the value should be 0.3908

AUXILIARY INFORMATION

METHOD/APPARATUS/PROCEDURE:

Equilibrium established between a measured volume of gas and a measured amount of gas-free liquid in a cell fitted with a magnetic stirrer. Details in source and ref. 1.

SOURCE AND PURITY OF MATERIALS:

1. From commercial cylinder; 99.8 per cent, as attested by gas chromatography.
2. Distilled water was used.
3. Salt was of reagent grade.

ESTIMATED ERROR:

$\delta T/K = \pm 0.2$; $\delta\alpha = \pm 2\%$ (estimated by compiler).

REFERENCES:

1. Onda, K.; Sada, E.; Kobayashi, T.; Kito, S.; Ito, K. *J. Chem. Engng. Japan* 1970, *3*, 18; 137.

COMPONENTS:

1. Nitrous oxide; N_2O; [10024-97-2]
2. Water; H_2O; [7732-18-5]
3. Zinc sulfate; $ZnSO_4$; [7733-02-0]

ORIGINAL MEASUREMENTS:

Manchot, von W.; Jahrstorfer, M.; Zepter, H.

Z. Anorg. Allg. Chem. 1924, *141*, 45-81.

VARIABLES:

Concentration

PREPARED BY:

C. L. Young

EXPERIMENTAL VALUES:

T/K	Density, d_4^{25} of salt soln.	Conc. of salt $/10^3$ mol m^{-3} (soln.)	Conc. of salt /mol kg^{-1} (water)	S_1 /cm^3	S_2 /cm^3
298.15	1.1403	0.95	0.9626	29.9	26.2
	1.2699	1.84	1.8914	16.9	13.3

S_1 - volume of nitrous oxide absorbed per 100 cm^3 of salt solution

S_2 - volume of nitrous oxide absorbed per 100 g of salt solution

Both S_1 and S_2 were reduced to conditions of 273.15 K and 101.3 kPa

AUXILIARY INFORMATION

METHOD/APPARATUS/PROCEDURE:

Measurement of volume of gas absorbed by means of gas buret and pipet. Volume absorbed appears to be taken as independent of pressure. (1) Density of the aqueous solution determined by Sprengel pyknometer.

SOURCE AND PURITY OF MATERIALS:

1. Prepared by heating ammonium nitrate, frozen in liquid air and then distilled.

2. Recrystallized.

ESTIMATED ERROR:

REFERENCES:

1. Manchot, W. *Z.Anorg.Chem.* 1924, *141*, 38.

COMPONENTS:	ORIGINAL MEASUREMENTS:
1. Nitrous oxide; N_2O; [10024-97-2] 2. Water; H_2O; [7732-18-5] 3. Zinc nitrate; $Zn(NO_3)_2$; [7779-88-6]	Manchot, von W.; Jahrstorfer, M.; Zepter, H. *Z. Anorg. Allg. Chem.* 1924, *141*, 45-81.
VARIABLES:	PREPARED BY:
Concentration	C. L. Young

EXPERIMENTAL VALUES:

T/K	Density, d_4^{25} of salt soln.	Conc. of salt $/10^3$ mol m^{-3} (soln.)	Conc. of salt /mol kg^{-1} (water)	S_1 $/cm^3$	S_2 $/cm^3$
298.15	1.1223	0.84	0.8721	39.7	35.4
	1.2433	1.68	1.8160	29.1	23.4

S_1 - volume of nitrous oxide absorbed per 100 cm^3 of salt solution

S_2 - volume of nitrous oxide absorbed per 100 g of salt solution

Both S_1 and S_2 were reduced to conditions of 273.15 K and 101.3 kPa

AUXILIARY INFORMATION

METHOD/APPARATUS/PROCEDURE:

Measurement of volume of gas absorbed by means of gas buret and pipet. Volume absorbed appears to be taken as independent of pressure. (1). Density of the aqueous solution determined by Sprengel pyknometer.

SOURCE AND PURITY OF MATERIALS:

1. Prepared by heating ammonium nitrate, frozen in liquid air and then distilled.

2. Recrystallized.

ESTIMATED ERROR:

REFERENCES:

1. Manchot, W. *Z. Anorg.Chem.* 1924, *141*, 38.

COMPONENTS:	ORIGINAL MEASUREMENTS:
1. Nitrous oxide; N_2O; [10024-97-2] 2. Water; H_2O; [7732-18-5] 3. Cadmium nitrate; $Cd(NO_3)_2$; [10325-94-7]	Manchot, von W.; Jahrstorfer, M.; Zepter, H. *Z. Anorg. Allg. Chem.* 1924, *141*, 45-81.
VARIABLES: Concentration	PREPARED BY: C. L. Young

EXPERIMENTAL VALUES:

T/K	Density, d_4^{25} of salt soln.	Conc. of salt /10^3 mol m^{-3} (soln.)	Conc. of salt /mol kg^{-1} (water)	S_1 /cm^3	S_2 /cm^3
298.15	1.1435	0.781	0.8145	35.0	30.6
	1.2874	1.562	1.7014	27.0	21.0

S_1 - volume of nitrous oxide absorbed per 100 cm^3 of salt solution

S_2 - volume of nitrous oxide absorbed per 100 g of salt solution

Both S_1 and S_2 were reduced to conditions of 273.15 K and 101.3 kPa

AUXILIARY INFORMATION

METHOD/APPARATUS/PROCEDURE:

Measurement of volume of gas absorbed by means of gas buret and pipet. Volume absorbed appears to be taken as independent of pressure. (1). Density of the aqueous solution determined by Sprengel pyknometer.

SOURCE AND PURITY OF MATERIALS:

1. Prepared by heating ammonium nitrate, frozen in liquid air and then distilled.

2. Recrystallized.

ESTIMATED ERROR:

REFERENCES:

1. Manchot, W. *Z.Anorg.Chem.* 1924, *141* ,38.

COMPONENTS:

1. Nitrous oxide; N_2O; [10024-97-2]
2. Water; H_2O; [7732-18-5]
3. Copper(II) nitrate; $Cu(NO_3)_2$; [3251-23-8]

ORIGINAL MEASUREMENTS:

Manchot, von W.; Jahrstorfer, M.; Zepter, H.

Z. Anorg. Allg. Chem. 1924, *141*, 45-81.

VARIABLES:

Concentration

PREPARED BY:

C. L. Young

EXPERIMENTAL VALUES:

T/K	Density, d_4^{25} of salt soln.	Conc. of salt $/10^3$ mol m^{-3} (soln.)	Conc. of salt /mol kg^{-1} (water)	S_1 $/cm^3$	S_2 $/cm^3$
298.15	1.1028	0.69	0.7089	35.6	32.3
	1.2049	1.38	1.4587	27.8	23.1

S_1 - volume of nitrous oxide absorbed per 100 cm^3 of salt solution

S_2 - volume of nitrous oxide absorbed per 100 g of salt solution

Both S_1 and S_2 were reduced to conditions of 273.15 K and 101.3 kPa

AUXILIARY INFORMATION

METHOD/APPARATUS/PROCEDURE:

Measurement of volume of gas absorbed by means of gas buret and pipet. Volume absorbed appears to be taken as independent of pressure. (1). Density of the aqueous solution determined by Sprengel pyknometer.

SOURCE AND PURITY OF MATERIALS:

1. Prepared by heating ammonium nitrate, frozen in liquid air and then distilled.

2. Recrystallized.

ESTIMATED ERROR:

REFERENCES:

1. Manchot, W. *Z.Anorg.Chem.* 1924, *141*, 38.

COMPONENTS:

1. Nitrous oxide; N_2O; [10024-97-2]
2. Water; H_2O; [7732-18-5]
3. Nickel(II) sulfate; $NiSO_4$; [7786-81-4]

ORIGINAL MEASUREMENTS:

Manchot, von W.; Jahrstorfer, M.; Zepter, H.

Z. Anorg. Allg. Chem. 1924, *141*, 45-81.

VARIABLES:

Concentration

PREPARED BY:

C. L. Young

EXPERIMENTAL VALUES:

T/K	Density, d_4^{25} of salt soln.	Conc. of salt /10^3 mol m^{-3} (soln.)	Conc. of salt /mol kg^{-1} (water)	S_1 /cm^3	S_2 /cm^3
298.15	1.1355	0.937	0.946	24.6	21.7
	1.2642	1.874	1.9236	13.8	10.9

S_1 - volume of nitrous oxide absorbed per 100 cm^3 of salt solution

S_2 - volume of nitrous oxide absorbed per 100 g of salt solution

Both S_1 and S_2 were reduced to conditions of 273.15 K and 101.3 kPa

AUXILIARY INFORMATION

METHOD/APPARATUS/PROCEDURE:

Measurement of volume of gas absorbed by means of gas buret and pipet. Volume absorbed appears to be taken as independent of pressure. (1). Density of the aqueous solution determined by Sprengel pyknometer.

SOURCE AND PURITY OF MATERIALS:

1. Prepared by heating ammonium nitrate, frozen in liquid air and then distilled.

2. Recrystallized.

ESTIMATED ERROR:

REFERENCES:

1. Manchot, W. *Z.Anorg.Chem.* 1924, *141*, 38.

COMPONENTS:

1. Nitrous oxide; N_2O; [10024-97-2]
2. Water; H_2O; [7732-18-5]
3. Cobalt(II) sulfate; $CoSO_4$; [10124-43-3]

ORIGINAL MEASUREMENTS:

Manchot, von W.; Jahrstorfer, M.; Zepter, H.

Z. Anorg. Allg. Chem. 1924, *141*, 45-81.

VARIABLES:

Concentration

PREPARED BY:

C. L. Young

EXPERIMENTAL VALUES:

T/K	Density, d_4^{25} of salt soln.	Conc. of salt $/10^3$ mol m^{-3} (soln.)	Conc. of salt /mol kg^{-1} (water)	S_1 $/cm^3$	S_2 $/cm^3$
298.15	1.1131	0.788	0.7952	27.5	24.7
	1.2218	1.576	1.6123	17.1	14.0

S_1 - volume of nitrous oxide absorbed per 100 cm^3 of salt solution

S_2 - volume of nitrous oxide absorbed per 100 g of salt solution

Both S_1 and S_2 were reduced to conditions of 273.15 K and 101.3 kPa

AUXILIARY INFORMATION

METHOD/APPARATUS/PROCEDURE:

Measurement of volume of gas absorbed by means of gas buret and pipet. Volume absorbed appears to be taken as independent of pressure. (1). Density of the aqueous solution determined by Sprengel pyknometer.

SOURCE AND PURITY OF MATERIALS:

1. Prepared by heating ammonium nitrate, frozen in liquid air and then distilled

2. Recrystallized.

ESTIMATED ERROR:

REFERENCES:

1. Manchot, W. *Z.Anorg.Chem.* 1924, *141*, 38.

COMPONENTS:

1. Nitrous oxide; N_2O; [10024-97-2]
2. Water; H_2O; [7732-18-5]
3. Iron(II) sulfate; $FeSO_4$; [7720-78-7]

ORIGINAL MEASUREMENTS:

Manchot, von W.; Jahrstorfer, M.; Zepter, H.

Z. Anorg. Allg. Chem. 1924, *141*, 45-81.

VARIABLES:

Concentration

PREPARED BY:

C. L. Young

EXPERIMENTAL VALUES:

T/K	Density, d_4^{25} of salt soln.	Conc. of salt $/10^3$ mol m^{-3} (soln.)	Conc. of salt /mol kg^{-1} (water)	S_1 $/cm^3$	S_2 $/cm^3$
298.15	1.1017	0.72	0.7256	34.0	30.9
	1.2011	1.438	1.4634	21.6	18.0

S_1 - volume of nitrous oxide absorbed in 100 cm^3 of salt solution

S_2 - volume of nitrous oxide absorbed in 100 g of salt solution

Both S_1 and S_2 were reduced to conditions of 273.15 K and 101.3 kPa

AUXILIARY INFORMATION

METHOD/APPARATUS/PROCEDURE:

Measurement of volume of gas absorbed by means of gas buret and pipet. Volume absorbed appears to be taken as independent of pressure. (1). Density of the aqueous solution determined by Sprengel pyknometer.

SOURCE AND PURITY OF MATERIALS:

1. Prepared by heating ammonium nitrate, frozen in liquid air and then distilled.

2. Recrystallized.

ESTIMATED ERROR:

REFERENCES:

1. Manchot, W. *Z.Anorg.Chem.* 1924, *141*, 38.

COMPONENTS:	ORIGINAL MEASUREMENTS:
1. Nitrous oxide; N_2O; [10024-97-2] 2. Water; H_2O; [7732-18-5] 3. Iron(III) sulfate; $Fe_2(SO_4)_3$; [10028-22-5]	Manchot, von W.; Jahrstorfer, M.; Zepter, H. *Z. Anorg. Allg. Chem.* 1924, *141*, 45-81.
VARIABLES:	PREPARED BY:
Concentration	C. L. Young

EXPERIMENTAL VALUES:

T/K	Density, d_4^{25} of salt soln.	Conc. of salt $/10^3$ mol m^{-3} (soln.)	Conc. of salt /mol kg^{-1} (water)	S_1 $/cm^3$	S_2 $/cm^3$
298.15	1.2240	0.66	0.6874	25.9	21.2
	1.4319	1.32	1.460	13.1	9.2

S_1 - volume of nitrous oxide absorbed per 100 cm^3 of salt solution

S_2 - volume of nitrous oxide absorbed per 100 g of salt solution

Both S_1 and S_2 were reduced to conditions of 273.15 K and 101.3 kPa

AUXILIARY INFORMATION

METHOD/APPARATUS/PROCEDURE:

Measurement of volume of gas absorbed by means of gas buret and pipet. Volume absorbed appears to be taken as independent of pressure.(1). Density of the aqueous solution determined by Sprengel pyknometer.

SOURCE AND PURITY OF MATERIALS:

1. Prepared by heating ammonium nitrate, frozen in liquid air and then distilled.

2. Recrystallized.

ESTIMATED ERROR:

REFERENCES:

1. Manchot, W. *Z.Anorg, Chem.* 1924,*141*,38.

COMPONENTS:	ORIGINAL MEASUREMENTS:
1. Nitrous oxide; N_2O; [10024-97-2] 2. Water; H_2O; [7732-18-5] 3. Manganese(II) sulfate; $MnSO_4$; [7785-87-7]	Manchot, von W.; Jahrstorfer, M.; Zepter, H. *Z. Anorg. Allg. Chem.* 1924, *141*, 45-81.
VARIABLES:	PREPARED BY:
Concentration	C. L. Young

EXPERIMENTAL VALUES:

T/K	Density, d_4^{25} of salt soln.	Conc. of salt $/10^3$ mol m^{-3} (soln.)	Conc. of salt /mol kg^{-1} (water)	S_1 $/cm^3$	S_2 $/cm^3$
298.15	1.1226	0.94	0.9585	30.6	27.3
	1.2460	1.93	2.0219	17.0	13.6

S_1 - volume of nitrous oxide absorbed in 100 cm^3 of salt solution

S_2 - volume of nitrous oxide absorbed in 100 g of salt solution

Both S_1 and S_2 were reduced to conditions of 273.15 K and 101.3 kPa

AUXILIARY INFORMATION

METHOD/APPARATUS/PROCEDURE:	SOURCE AND PURITY OF MATERIALS:
Measurement of volume of gas absorbed by means of gas buret and pipet. Volume absorbed appears to be taken as independent of pressure. (1). Density of the aqueous solution determined by Sprengel pyknometer.	1. Prepared by heating ammonium nitrate, frozen in liquid air and then distilled. 2. Recrystallized.
	ESTIMATED ERROR:
	REFERENCES: 1. Manchot, W. *Z. Anorg. Chem.* 1924, *141*, 38.

COMPONENTS:	ORIGINAL MEASUREMENTS:
1. Nitrous oxide; N_2O; [10024-97-2] 2. Water; H_2O; [7732-18-5] 3. Chromium(III) sulfate; $Cr_2(SO_4)_3$; [10101-53-8]	Manchot, von W.; Jahrstorfer, M.; Zepter, H. *Z. Anorg. Allg. Chem.* 1924, *141*, 45-81.
VARIABLES:	PREPARED BY:
Concentration	C. L. Young

EXPERIMENTAL VALUES:

T/K	Density, d_4^{25} of salt soln.	Conc. of salt $/10^3$ mol m^{-3} (soln.)	Conc. of salt /mol kg^{-1} (water)	S_1 $/cm^3$	S_2 $/cm^3$
298.15	1.1657	0.57	0.605	31.8	27.2
	1.3280	1.14	1.294	18.2	13.7

S_1 - volume of nitrous oxide absorbed per 100 cm^3 of salt solution

S_2 - volume of nitrous oxide absorbed per 100 g of salt solution

Both S_1 and S_2 were reduced to conditions of 273.15 K and 101.3 kPa

AUXILIARY INFORMATION

METHOD/APPARATUS/PROCEDURE:

Measurement of volume of gas absorbed by means of gas buret and pipet. Volume absorbed appears to be taken as independent of pressure. (1). Density of the aqueous solution determined by Sprengel pyknometer.

SOURCE AND PURITY OF MATERIALS:

1. Prepared by heating ammonium nitrate, frozen in liquid air and then distilled.

2. Recrystallized.

ESTIMATED ERROR:

REFERENCES:

1. Manchot, W. *Z.Anorg.Chem.* 1924, *141*, 38.

COMPONENTS:

1. Nitrous oxide; N_2O; [10024-97-2]
2. Water; H_2O; [7732-18-5]
3. Aluminium sulfate; $Al_2(SO_4)_3$; [10043-01-3]

ORIGINAL MEASUREMENTS:

Manchot, von W.; Jahrstorfer, M.; Zepter, H.

Z. Anorg. Allg. Chem. 1924, *141*, 45-81.

VARIABLES:

Concentration

PREPARED BY:

C. L. Young

EXPERIMENTAL VALUES:

T/K	Density, d_4^{25} of salt soln.	Conc. of salt $/10^3$ mol m^{-3} (soln.)	Conc. of salt /mol kg^{-1} (water)	S_1 /cm^3	S_2 /cm^3
298.15	1.1558	0.5166	0.5277	22.4	19.4
	1.2381	0.8141	0.8486	13.4	10.8

S_1 - volume of nitrous oxide absorbed per 100 cm^3 of salt solution

S_2 - volume of nitrous oxide absorbed per 100 g of salt solution

Both S_1 and S_2 were reduced to conditions of 273.15 K and 101.3 kPa

AUXILIARY INFORMATION

METHOD/APPARATUS/PROCEDURE:

Measurement of volume of gas absorbed by means of gas buret and pipet. Volume absorbed appears to be taken as independent of pressure. (1). Density of the aqueous solution determined by Sprengel pyknometer.

SOURCE AND PURITY OF MATERIALS:

1. Prepared by heating ammonium nitrate, frozen in liquid air and then distilled.

2. Recrystallized.

ESTIMATED ERROR:

REFERENCES:

1. Manchot, W. *Z.Anorg.Chem.* 1924, *141*, 38.

COMPONENTS:	ORIGINAL MEASUREMENTS:
1. Nitrous oxide; N_2O; [10024-97-2] 2. Water; H_2O; [7732-18-5] 3. Aluminium nitrate; $Al(NO_3)_3$; [13473-90-0]	Manchot, von W.; Jahrstorfer, M.; Zepter, H. *Z. Anorg. Allg. Chem.* 1924, *141*, 45-81.
VARIABLES:	PREPARED BY:
Concentration	C. L. Young

EXPERIMENTAL VALUES:

T/K	Density, d_4^{25} of salt soln.	Conc. of salt $/10^3$ mol m^{-3} (soln.)	Conc. of salt /mol kg^{-1} (water)	S_1 $/cm^3$	S_2 $/cm^3$
298.15	1.0703	0.4795	0.4953	36.1	33.7
	1.1414	0.959	1.0235	29.3	25.7

S_1 - volume of nitrous oxide absorbed per 100 cm^3 of salt solution

S_2 - volume of nitrous oxide absorbed per 100 g of salt solution

Both S_1 and S_2 were reduced to conditions of 273.15 K and 101.3 kPa

AUXILIARY INFORMATION

METHOD/APPARATUS/PROCEDURE:	SOURCE AND PURITY OF MATERIALS:
Measurement of volume of gas absorbed by means of gas buret and pipet. Volume absorbed appears to be taken as independent of pressure. (1). Density of the aqueous solution determined by Sprengel pyknometer.	1. Prepared by heating ammonium nitrate, frozen in liquid air and then distilled. 2. Recrystallized.
	ESTIMATED ERROR:
	REFERENCES: 1. Manchot, W. *Z.Anorg.Chem.* 1924, *141*, 38.

COMPONENTS:	ORIGINAL MEASUREMENTS:
1. Nitrous oxide; N_2O; [10024-97-2] 2. Water; H_2O; [7732-18-5]; 3. Magnesium chloride; $MgCl_2$; [7786-30-3]	Sada, E.; Ando, N.; Kito, S. *J. Appl. Chem. Biotechnol.* 1972, 22, 1185-1193.
VARIABLES:	PREPARED BY:
T/K: 298.15 P/kPa: 101.325 (1 atm) Molarity of salt: mol l^{-1}	W. Gerrard/C. L. Young

EXPERIMENTAL VALUES:

T/K	Concn. of salt/mol l^{-1}	Bunsen coefficient, α
298.15	0	0.5512 (pure water)
	0.441	0.4513
	0.582	0.4254
	0.851	0.3375 *
	1.451	0.2919

* not used in calculating salt effect parameter

AUXILIARY INFORMATION

METHOD /APPARATUS/PROCEDURE:

Equilibrium established between a measured volume of gas and a measured amount of gas-free liquid in a cell fitted with a magnetic stirrer. Details in source and ref. 1.

SOURCE AND PURITY OF MATERIALS:

1. From commercial cylinder; 99.8 per cent, as attested by gas chromatography.
2. Distilled water was used.
3. Salt was of reagent grade.

ESTIMATED ERROR:

$\delta T/K = \pm 0.2$; $\delta\alpha = \pm 2\%$ (estimated by compiler).

REFERENCES:

1. Onda, K.; Sada, E.; Kobayashi, T.; Kito, S.; Ito, K. *J. Chem. Engng. Japan* 1970, 3, 18; 137.

COMPONENTS:	ORIGINAL MEASUREMENTS:
1. Nitrous oxide; N_2O; [10024-97-2] 2. Water; H_2O; [7732-18-5] 3. Magnesium sulfate; $MgSO_4$; [7487-88-9]	Manchot, von W.; Jahrstorfer, M.; Zepter, H. *Z. Anorg. Allg. Chem.* 1924, *141*, 45-81.
VARIABLES:	PREPARED BY:
Concentration	C. L. Young

EXPERIMENTAL VALUES:

T/K	Density, d_4^{25} of salt soln.	Conc. of salt $/10^3$ mol m^{-3} (soln.)	Conc. of salt /mol kg^{-1} (water)	S_1 $/cm^3$	S_2 $/cm^3$
298.15	1.0992	0.90	0.9083	29.5	26.8
	1.1925	1.78	1.8197	15.9	13.3

S_1 - volume of nitrous oxide absorbed per 100 cm^3 of salt solution

S_2 - volume of nitrous oxide absorbed per 100 g of salt solution

Both S_1 and S_2 were reduced to conditions of 273.15 K and 101.3 kPa

AUXILIARY INFORMATION

METHOD/APPARATUS/PROCEDURE:	SOURCE AND PURITY OF MATERIALS:
Measurement of volume of gas absorbed by means of gas buret and pipet. Volume absorbed appears to be taken as independent of pressure. (1) Density of the aqueous solution determined by Sprengel pyknometer.	1. Prepared by heating ammonium nitrate, frozen in liquid air and then distilled. 2. Recrystallized.
	ESTIMATED ERROR:
	REFERENCES: 1. Manchot, W. *Z.Anorg.Chem.* 1924, *141*, 38.

COMPONENTS:

1. Nitrous oxide; N_2O; [10024-97-2]
2. Water; H_2O; [7732-18-5];
3. Magnesium sulfate; $MgSO_4$; [7487-88-9]

ORIGINAL MEASUREMENTS:

Gordon, V. *Z. Phys. Chem.* 1895, *18*, 1-16.

VARIABLES:

Temperature, concentration

PREPARED BY:

W. Gerrard

EXPERIMENTAL VALUES:

T/K	Conc. of salt Weight-%	Density, ρ	Bunsen Coefficient, α	T/K	Conc. of salt Weight-%	Density, ρ	Bunsen Coefficient, α
295.46	5.9009	1.05836	0.43448	284.16	7.6585	1.07755	0.55515
291.76		1.05945	0.50647	281.26		1.0781	0.62931
288.26		1.06028	0.55857	295.46	10.7765	1.109	0.31431
284.16		1.0612	0.63936	291.76		1.1101	0.364375
281.26		1.06184	0.70626	288.26		1.1112	0.41560
295.46	7.6585	1.0753	0.38733	284.16		1.11236	0.46870
291.76		1.0762	0.44329	281.26		1.11333	0.52035
288.26		1.0768	0.48596				

Smoothing Equations

$(t = T/K - 273.15)$

For 5.9009 wt-% solution: $\alpha = 0.91034 - 0.0274t + 0.000272t^2$

For 7.6585 wt-% solution: $\alpha = 0.853488 - 0.03154t + 0.000477t^2$

For 10.7765 wt-% solution: $\alpha = 0.649261 - 0.0164252t + 0.000063t^2$

AUXILIARY INFORMATION

METHOD/APPARATUS/PROCEDURE:

Measurement of volume of gas by means of Ostwald type apparatus, gas buret and pipet.
Density of the solution determined by Sprengel pyknometer.

SOURCE AND PURITY OF MATERIALS:

1. N_2O: Self prepared and purified. Attested by combustion with hydrogen.

ESTIMATED ERROR:

REFERENCES:

COMPONENTS:	ORIGINAL MEASUREMENTS:
1. Nitrous oxide; N_2O; [10024-97-2] 2. Water; H_2O; [7732-18-5]; 3. Magnesium sulfate; $MgSO_4$; [7487-88-9]	Markham, A. E.; Kobe, K. A. *J. Amer. Chem. Soc.* 1941, *63*, 449-54.
VARIABLES:	PREPARED BY:
T/K: 273.35-313.15 Salt molality/mol kg^{-1}: 0.1-8 Partial pressure of gas/atm: 1	P. L. Long

EXPERIMENTAL VALUES:

T/K*	Salt molality/mol kg^{-1}*	Bunsen Coefficient, α*	'Solubility Coefficient', S*[a]
273.15	0	1.2970	1.2971
	0.5	0.8778	0.8777
	1	0.6019	0.6039
	2	0.2771	0.2820
298.15	0	0.5392	0.5408
	0.5	0.3840	0.3858
	1	0.2790	0.2816
	2	0.1442	0.1478
313.15	0	0.3579	0.3607
	0.5	0.2612	0.2638
	1	0.1945	0.1969
	2	0.1018	0.1049

* From the original data.

[a] The 'solubility coefficient', "refers to the volume of gas, reduced to standard conditions ... which is dissolved by the quantity of solution containing one gram of water".

AUXILIARY INFORMATION

METHOD/APPARATUS/PROCEDURE:

An Ostwald method was used. A known volume of gas was placed in contact with a known volume of gas-free liquid. After equilibrium was established by agitation, the volume of the remaining gas was measured from which the amount of gas dissolved was found. Solutions were prepared by weight, and verified by density measurements compared with the literature values. Apparatus consisted of an absorption flask with two bulbs, one twice the volume of the other, which was connected by a capillary tube to a gas buret and a manometer system.

SOURCE AND PURITY OF MATERIALS:

1. 99.7 per cent pure, no source.
2. Water was freshly boiled, distilled.
3. Analytical grade.

ESTIMATED ERROR:

$\delta T/K = \pm 0.1$ (273.35 K); ± 0.03 (above 273.35 K); $\delta\alpha/\alpha = \pm 0.02$.

REFERENCES:

COMPONENTS:

1. Nitrous oxide; N_2O; [10024-97-2]
2. Water; H_2O; [7732-18-5]
3. Magnesium nitrate; $Mg(NO_3)_2$; [10377-60-3]

ORIGINAL MEASUREMENTS:

Manchot, von W.; Jahrstorfer, M.; Zepter, H.

Z. Anorg. Allg. Chem. 1924, *141*, 45-81.

VARIABLES:

Concentration

PREPARED BY:

C. L. Young

EXPERIMENTAL VALUES:

T/K	Density, d_4^{25} of salt soln.	Conc. of salt $/10^3$ mol m^{-3} (soln.)	Conc. of salt /mol kg^{-1} (water)	S_1 $/cm^3$	S_2 $/cm^3$
298.15	1.0935	0.97	1.0215	39.2	35.8
	1.1846	1.93	2.1485	28.5	24.1

S_1 - volume of nitrous oxide absorbed per 100 cm^3 of salt solution

S_2 - volume of nitrous oxide absorbed per 100 g of salt solution

Both S_1 and S_2 were reduced to conditions of 273.15 K and 101.3 kPa

AUXILIARY INFORMATION

METHOD/APPARATUS/PROCEDURE:

Measurement of volume of gas absorbed by means of gas buret and pipet. Volume absorbed appears to be taken as independent of pressure.(1). Density of the aqueous solution determined by Sprengel pyknometer.

SOURCE AND PURITY OF MATERIALS:

1. Prepared by heating ammonium nitrate, frozen in liquid air and then distilled.

2. Recrystallized.

ESTIMATED ERROR:

REFERENCES:

1. Manchot, W. *Z.Anorg,Chem.* 1924, *141*, 38.

COMPONENTS:

1. Nitrous oxide; N_2O; [10024-97-2]
2. Water; H_2O; [7732-18-5];
3. Magnesium nitrate; $Mg(NO_3)_2$; [10377-60-3]

ORIGINAL MEASUREMENTS:

Markham, A. E.; Kobe, K. A.

J. Amer. Chem. Soc. 1941, *63*, 449-54.

VARIABLES:

T/K: 273.35-313.15
Salt molality/mol kg^{-1}: 0.1-8
Partial pressure of gas/atm: 1

PREPARED BY:

P. L. Long

EXPERIMENTAL VALUES:

T/K*	Salt molality/mol kg^{-1}*	Bunsen Coefficient, α*	'Solubility Coefficient', S*[a]
273.35	0	1.2970	1.2971
	0.5	1.0139	1.0316
	1	0.7921	0.8215
	2	0.5238	0.5664
	3	0.3618	0.4079
298.15	0	0.5392	0.5408
	0.5	0.4488	0.4593
	1	0.3850	0.4023
	2	0.2785	0.3042
	3	0.2104	0.2403
313.15	0	0.3579	0.3607
	0.5	0.3061	0.3150
	1	0.2620	0.2756
	2	0.1989	0.2188
	3	0.1556	0.1793

* From the original data.

[a] The 'solubility coefficient', "refers to the volume of gas, reduced to standard conditions ... which is dissolved by the quantity of solution containing one gram of water".

AUXILIARY INFORMATION

METHOD /APPARATUS/PROCEDURE:

An Ostwald method was used. A known volume of gas was placed in contact with a known volume of gas-free liquid. After equilibrium was established by agitation, the volume of the remaining gas was measured from which the amount of gas dissolved was found. Solutions were prepared by weight, and verified by density measurements compared with the literature values. Apparatus consisted of an absorption flask with two bulbs, one twice the volume of the other, which was connected by a capillary tube to a gas buret and a manometer system.

SOURCE AND PURITY OF MATERIALS:

1. 99.7 per cent pure, no source.
2. Water was freshly boiled, distilled.
3. Analytical grade.

ESTIMATED ERROR:

$\delta T/K = \pm 0.1$ (273.35 K); ± 0.03 (above 273.35 K); $\delta\alpha/\alpha = \pm 0.02$.

REFERENCES:

COMPONENTS:

1. Nitrous oxide; N_2O; [10024-97-2]
2. Water; H_2O; [7732-18-5];
3. Calcium chloride; $CaCl_2$; [10043-52-4]

ORIGINAL MEASUREMENTS:

Gordon, V. *Z. Phys. Chem.* 1895, *18*, 1-16.

VARIABLES:

Temperature, concentration

PREPARED BY:

W. Gerrard

EXPERIMENTAL VALUES:

T/K	Conc. of salt Weight-%	Density, ρ	Bunsen Coefficient, α	T/K	Conc. of salt Weight-%	Density, ρ	Bunsen Coefficient, α
295.46	5.7898	1.04695	0.46447	284.16	9.856	1.0864	0.56462
291.76		1.04784	0.52416	281.26		1.0873	0.61709
288.26		1.04902	0.58870	295.46	13.987	1.1209	0.30676
284.16		1.04982	0.67323	291.76		1.1221	0.33760
281.26		1.05043	0.74106	288.26		1.1234	0.37919
295.46	9.856	1.0830	0.39256	284.16		1.1247	0.42001
291.76		1.0842	0.44409	281.26		1.1256	0.46642
288.26		1.0852	0.50743				

Smoothing Equations

$(t = T/K - 273.16)$

For 5.7898 wt-% solution: $\alpha = 0.958202 - 0.029356t + 0.0003236t^2$

For 9.856 wt-% solution: $\alpha = 0.754096 - 0.01757t + 0.000081t^2$

For 13.987 wt-% solution: $\alpha = 0.58801 - 0.01638t + 0.000169t^2$

AUXILIARY INFORMATION

METHOD /APPARATUS/PROCEDURE:

Measurement of volume of gas by means of Ostwald type apparatus, gas buret and pipet (ref. 1). Density of the solution determined by Sprengel pyknometer.

SOURCE AND PURITY OF MATERIALS:

1. N_2O: Self prepared and purified. Attested by combustion with hydrogen.

ESTIMATED ERROR:

REFERENCES:

1. Timofejew, W. *Z. Phys. Chem.* 1890, *6*, 141.

COMPONENTS:	ORIGINAL MEASUREMENTS:
1. Nitrous oxide; N_2O; [10024-97-2] 2. Water; H_2O; [7732-18-5] 3. Calcium chloride; $CaCl_2$; [10043-52-4]	Manchot, von W.; Jahrstorfer, M.; Zepter, H. *Z. Anorg. Allg. Chem.* 1924, *141*, 45-81.
VARIABLES:	PREPARED BY:
Concentration	C. L. Young

EXPERIMENTAL VALUES:

T/K	Density, d_4^{25} of salt soln.	Conc. of salt /10^3 mol m^{-3} soln.)	Conc. of salt /mol kg^{-1} (water)	S_1 /cm^3	S_2 /cm^3
298.15	1.0786	0.93	0.9535	33.9	31.4
	1.1665	1.99	2.1044	20.2	17.3

S_1 - volume of nitrous oxide absorbed per 100 cm^3 of salt solution

S_2 - volume of nitrous oxide absorbed per 100 g of salt solution

Both S_1 and S_2 were reduced to conditions of 273.15 K and 101.3 kPa

AUXILIARY INFORMATION

METHOD/APPARATUS/PROCEDURE:

Measurement of volume of gas absorbed by means of gas buret and pipet. Volume absorbed appears to be taken as independent of pressure. (1). Density of the aqueous solution determined by Sprengel pyknometer.

SOURCE AND PURITY OF MATERIALS:

1. Prepared by heating ammonium nitrate, frozen in liquid air and then distilled.

2. Recrystallized.

ESTIMATED ERROR:

REFERENCES:

1. Manchot, W. *Z.Anorg.Chem.* 1924, *141*, 38.

COMPONENTS:	ORIGINAL MEASUREMENTS:
1. Nitrous oxide; N_2O; [10024-97-2] 2. Water; H_2O; [7732-18-5]; 3. Calcium chloride; $CaCl_2$; [10043-52-4]	Sada, E.; Ando, N.; Kito, S. *J. Appl. Chem. Biotechnol.* 1972, *22*, 1185-1193.
VARIABLES:	PREPARED BY:
Concentration	W. Gerrard/C. L. Young

EXPERIMENTAL VALUES:

T/K	Conc. of salt/mol l^{-1}	Bunsen coefficient, α
298.15	0	0.5512 (pure water)
	0.323	0.4705
	0.666	0.4019
	1.088	0.3275
	1.451	0.2760

AUXILIARY INFORMATION

METHOD /APPARATUS/PROCEDURE:

Equilibrium established between a measured volume of gas and a measured amount of gas-free liquid in a cell fitted with a magnetic stirrer. Details in source and ref. 1.

SOURCE AND PURITY OF MATERIALS:

1. From commercial cylinder; 99.8 per cent, as attested by gas chromatography.
2. Distilled water was used.
3. Salt was of reagent grade.

ESTIMATED ERROR:

T/K = 0.2; = 2% (estimated by compiler).

REFERENCES:

1. Onda, K.; Sada, E.; Kobayashi, T.; Kito, S.; Ito, K. *J. Chem. Engng. Japan* 1970, *3*, 18; 137.

COMPONENTS:

1. Nitrous oxide; N_2O; [10024-97-2]
2. Water; H_2O; [7732-18-5]
3. Calcium nitrate; $Ca(NO_3)_2$; [10124-37-5]

ORIGINAL MEASUREMENTS:

Manchot, von W.; Jahrstorfer, M.; Zepter, H.

Z. Anorg. Allg. Chem. 1924, *141*, 45-81.

VARIABLES:

Concentration

PREPARED BY:

C. L. Young

EXPERIMENTAL VALUES:

T/K	Density, d_4^{25} of salt soln.	Conc. of salt $/10^3$ mol m^{-3} (soln.)	Conc. of salt /mol kg^{-1} (water)	S_1 $/cm^3$	S_2 $/cm^3$
298.15	1.1503	1.365	1.474	32.2	28.0
	1.2927	2.73	3.232	19.4	15.0

S_1 - volume of nitrous oxide absorbed per 100 cm^3 of salt solution

S_2 - volume of nitrous oxide absorbed per 100 g of salt solution

Both S_1 and S_2 were reduced to conditions of 273.15 K and 101.3 kPa

AUXILIARY INFORMATION

METHOD/APPARATUS/PROCEDURE:

Measurement of volume of gas absorbed by means of gas buret and pipet. Volume absorbed appears to be taken as independent of pressure. (1). Density of the aqueous solution determined by Sprengel pyknometer.

SOURCE AND PURITY OF MATERIALS:

1. Prepared by heating ammonium nitrate, frozen in liquid air and then distilled.

2. Recrystallized.

ESTIMATED ERROR:

REFERENCES:

1. Manchot, W. *Z.Anorg.Chem.* 1924, *141*, 38.

COMPONENTS:

1. Nitrous oxide; N_2O; [10024-97-2]
2. Water; H_2O; [7732-18-5];
3. Strontium chloride; $SrCl_2$; [10476-85-4]

ORIGINAL MEASUREMENTS:

Gordon, V. *Z. Phys. Chem.* 1895, *18*, 1-16.

VARIABLES:

Temperature, concentration

PREPARED BY:

W. Gerrard

EXPERIMENTAL VALUES:

T/K	Conc. of salt Weight-%	Density, ρ	Bunsen Coefficient, α	T/K	Conc. of salt Weight-%	Density, ρ	Bunsen Coefficient, α
295.46	3.309	1.0283	0.54357	284.16	5.7315	1.0528	0.68591
291.76		1.029	0.60269	281.26		1.0537	0.75729
288.26		1.0297	0.66902	295.46	13.239	1.1226	0.36143
284.16		1.0304	0.76665	291.76		1.1238	0.41191
281.26		1.0309	0.83808	288.26		1.1251	0.44876
291.76	5.7315	1.0506	0.56234	284.16		1.1265	0.52249
288.26		1.0515	0.60944	281.26		1.1273	0.58233

Smoothing Equations

$(t = T/K - 273.15)$

For 3.309 wt-% solution: $\alpha = 1.091696 - 0.03515t + 0.000474t^2$

For 5.7315 wt-% solution: $\alpha = 1.02683 - 0.0397t + 0.000793t^2$

For 13.239 wt-% solution: $\alpha = 0.75313 - 0.023095t + 0.000248t^2$

AUXILIARY INFORMATION

METHOD/APPARATUS/PROCEDURE:

Measurement of volume of gas by means of Ostwald type apparatus, gas buret and pipet.
Density of the solution determined by Sprengel pyknometer.

SOURCE AND PURITY OF MATERIALS:

1. N_2O: Self prepared and purified. Attested by combustion with hydrogen.

ESTIMATED ERROR:

REFERENCES:

COMPONENTS:	ORIGINAL MEASUREMENTS:
1. Nitrous oxide; N_2O; [10024-97-2] 2. Water; H_2O; [7732-18-5] 3. Barium chloride; $BaCl_2$; [10361-37-2]	Manchot, von W.; Jahrstorfer, M.; Zepter, H. *Z. Anorg. Allg. Chem.* 1924, *141*, 45-81.
VARIABLES:	PREPARED BY:
Concentration	C. L. Young

EXPERIMENTAL VALUES:

T/K	Density, d_4^{25} of salt soln.	Conc. of salt $/10^3$ mol m^{-3} (soln.)	Conc. of salt /mol kg^{-1} (water)	S_1 $/cm^3$	S_2 $/cm^3$
298.15	1.1090	0.620	0.6327	37.4	33.8
	1.2290	1.313	1.3741	26.1	21.2

S_1 - volume of nitrous oxide absorbed per 100 cm^3 of salt solution

S_2 - volume of nitrous oxide absorbed per 100 g of salt solution

Both S_1 and S_2 were reduced to conditions of 273.15 K and 101.3 kPa

AUXILIARY INFORMATION

METHOD/APPARATUS/PROCEDURE:

Measurement of volume of gas absorbed by means of gas buret and pipet. Volume absorbed appears to be taken as independent of pressure. (1). Density of the aqueous solution determined by Sprengel pyknometer.

SOURCE AND PURITY OF MATERIALS:

1. Prepared by heating ammonium nitrate, frozen in liquid air and then distilled.

2. Recrystallized.

ESTIMATED ERROR:

REFERENCES:

1. Manchot, W. *Z.Anorg.Chem.* 1924, *141*, 38.

COMPONENTS:

1. Nitrous oxide; N_2O; [10024-97-2]
2. Water; H_2O; [7732-18-5];
3. Lithium chloride; LiCl; [7447-41-8]

ORIGINAL MEASUREMENTS:

Gordon, V. *Z. Phys. Chem.* 1895, *18*, 1-16.

VARIABLES:

Temperature, concentration

PREPARED BY:

W. Gerrard

EXPERIMENTAL VALUES:

T/K	Conc. of salt Weight-%	Density, ρ	Bunsen Coefficient, α	T/K	Conc. of salt Weight-%	Density, ρ	Bunsen Coefficient, α
295.46	1.346	0.0058	0.5559	295.46	3.853	1.0211	0.5005
			0.5523	291.76		1.0221	0.5643
291.76		1.0067	0.6232	288.26		1.0227	0.6268
			0.6224	284.16		1.0237	0.7090
288.26		1.0074	0.6890	281.26		1.0242	0.7917
			0.7065	295.46	11.476	1.065	0.3632
284.16		1.0081	0.7959	291.76		1.0658	0.3874
			0.7820	288.26		1.0666	0.4351
281.26		1.0085	0.8866	284.16		1.0675	0.4916
				281.26		1.0682	0.5452

Smoothing Equations

$(t = T/K - 273.16)$

For 1.345 wt-% solution: $\alpha = 1.1658234 - 0.038488t + 0.0004958t^2$

For 3.853 wt-% solution: $\alpha = 1.0343228 - 0.0333846t + 0.0004236t^2$

For 11.476 wt-% solution: $\alpha = 0.720982 - 0.0249286t + 0.0003984t^2$

AUXILIARY INFORMATION

METHOD/APPARATUS/PROCEDURE:

Measurement of volume of gas by means of Ostwald type apparatus, gas buret and pipet (ref. 1.). Density of the solution determined by Sprengel pyknometer.

SOURCE AND PURITY OF MATERIALS:

1. N_2O: Self prepared and purified. Attested by combustion with hydrogen.

ESTIMATED ERROR:

REFERENCES:

1. Timofejew, W. *Z. Phys. Chem.* 1890, *6*, 141.

COMPONENTS:

1. Nitrous oxide; N_2O; [10024-97-2]
2. Water; H_2O; [7732-18-5];
3. Lithium chloride; LiCl; [7447-41-8]

ORIGINAL MEASUREMENTS:

Geffcken, G. *Z. Phys. Chem.* 1904, *49*, 257-302.

VARIABLES:

Temperature, concentration

PREPARED BY:

W. Gerrard/C. L. Young

EXPERIMENTAL VALUES:

T/K	Conc. of salt/mol dm^{-3} (soln.)	Ostwald coefficient, L
288.16	0.558	0.6884
	0.561	0.6877
	1.057	0.6163
	1.059	0.6146
298.16	0.558	0.5276
	0.561	0.5278
	1.057	0.4760
	1.059	0.4773

AUXILIARY INFORMATION

METHOD /APPARATUS/PROCEDURE:

Measurement of volume of N_2O absorbed by the aqueous solution. Detailed description and diagram given in source.

SOURCE AND PURITY OF MATERIALS:

1. Nitrous oxide self prepared and attested.

ESTIMATED ERROR:

REFERENCES:

COMPONENTS:	ORIGINAL MEASUREMENTS:
1. Nitrous oxide; N_2O; [10024-97-2] 2. Water; H_2O; [7732-18-5] 3. Lithium chloride; LiCl; [7447-41-8]	Kreitus, I.; Gorbovitskalya, T. I. *Latv. PRS Zinat. Akad. Vestis Khim. Ser.* 1979, 664-666.
VARIABLES:	PREPARED BY:
Concentration of salt	C. L. Young

EXPERIMENTAL VALUES:

T/K	Conc. of salt /mol kg^{-1} (water)	Bunsen coefficient, α	Absorption† coefficient, S
298.15	0.0	0.55	0.55
	0.97	0.47	0.48
	2.66	0.36	0.38
	5.44	0.21	0.23
	9.10	0.12	0.15
	12.2	0.074	0.094
	15.5	0.042	0.059
	18.7	0.03	0.044
	19.8	0.026	0.039

†Volume of gas, corrected to 101.3 kPa and 273.15 K, dissolved at specified temperature and a partial pressure of 101.3 kPa, by solution containing 1 g of water.

AUXILIARY INFORMATION

METHOD/APPARATUS/PROCEDURE:

Aqueous solution saturated at 0.1013 MPa and 1 cm^3 sample analyzed by gas chromatography. Column contained molecular sieves and was operated isothermally at 250 °C using helium as carrier gas. Thermal conductivity detector used.

SOURCE AND PURITY OF MATERIALS:

1. Medical grade sample.
2. Twice distilled.
3. Twice recrystallised from laboratory reagent grade sample.

ESTIMATED ERROR:

$\delta T/K = \pm 0.1$; $\delta\alpha/\alpha = \pm 0.1$

REFERENCES:

COMPONENTS:

1. Nitrous oxide; N_2O; [10024-97-2]
2. Water; H_2O; [7732-18-5];
3. Lithium sulfate; Li_2SO_4; [10377-48-7]

ORIGINAL MEASUREMENTS:

Gordon, V. *Z. Phys. Chem.* 1895, *18*, 1-16.

VARIABLES:

Temperature, concentration

PREPARED BY:

W. Gerrard

EXPERIMENTAL VALUES:

T/K	Conc. of salt Weight-%	Density, ρ	Bunsen Coefficient, α	T/K	Conc. of salt Weight-%	Density, ρ	Bunsen Coefficient, α
295.46	2.3689	1.01873	0.52973	284.16	5.463	1.0518	0.64055
291.76		1.01954	0.59615	281.26		1.0525	0.71145
288.26		1.02024	0.66775	295.46	8.5596	1.0732	0.3912
284.16		1.02094	0.76423	291.76		1.07412	0.43526
281.26		1.02156	0.84337	288.26		1.0751	0.47611
295.46	5.463	1.04917	0.44372	284.16		1.07626	0.52889
291.76		1.050	0.49822	281.26		1.07684	0.58768
288.26		1.051	0.55552				

Smoothing Equations

$(t = T/K - 273.15)$

For 2.3689 wt-% solution: $\alpha = 1.09758 - 0.03476t + 0.0004168t^2$

For 5.463 wt-% solution: $\alpha = 0.950112 - 0.033325 + 0.000476t^2$

For 8.5596 wt-% solution: $\alpha = 0.752485 - 0.022711t + 0.00029193t^2$

AUXILIARY INFORMATION

METHOD/APPARATUS/PROCEDURE:

Measurement of volume of gas by means of Ostwald type apparatus, gas buret and pipet.
Density of the solution determined by Sprengel pyknometer.

SOURCE AND PURITY OF MATERIALS:

1. N_2O: Self prepared and purified. Attested by combustion with hydrogen.

ESTIMATED ERROR:

REFERENCES:

COMPONENTS:

1. Nitrous oxide; N_2O; [10024-97-2]
2. Water; H_2O; [7732-18-5];
3. Sodium chloride; NaCl; [7647-14-5]

ORIGINAL MEASUREMENTS:

Gordon, V. *Z. Phys. Chem.* 1895, *18*, 1-16.

VARIABLES:

Temperature, concentration

PREPARED BY:

W. Gerrard

EXPERIMENTAL VALUES:

T/K	Conc. of salt Weight-%	Density, ρ	Bunsen Coefficient, α	T/K	Conc. of salt Weight-%	Density, ρ	Bunsen Coefficient, α
295.46	8.88	1.0608	0.4091	295.46	6.196	1.04255	0.48086
			0.4013	291.76		1.0436	0.5262
291.76		1.0622	0.4577	288.26		1.0447	0.58368
			0.4558	284.16		1.0457	0.66302
288.26		1.0641	0.5046	281.26		1.04646	0.72482
			0.5112	295.46	12.782	1.0915	0.3626
284.16		1.0654	0.5790	291.76		1.0930	0.3994
			0.5790	288.26		1.0943	0.4480
281.26		1.0666	0.6412	284.16		1.0958	0.5118
			0.6437	281.26		1.0969	0.5687

Smoothing Equations

$(t = T/K - 273.16)$

For 8.88 wt-% solution: $\alpha = 0.84069 - 0.027303t + 0.0003486t^2$

For 6.196 wt-% solution: $\alpha = 0.93959 - 0.029923t + 0.00042072t^2$

For 12.782 wt-% solution: $\alpha = 0.75472 - 0.026034t + 0.00037894t^2$

AUXILIARY INFORMATION

METHOD/APPARATUS/PROCEDURE:

Measurement of volume of gas by means of Ostwald type apparatus, gas buret and pipet (ref. 1). Density of the solution determined by Sprengel pyknometer.

SOURCE AND PURITY OF MATERIALS:

1. N_2O: Self prepared and purified. Attested by combustion with hydrogen.

ESTIMATED ERROR:

REFERENCES:

1. Timofejew, W. *Z. Phys. Chem.* 1890, *6*, 141.

COMPONENTS:	ORIGINAL MEASUREMENTS:
1. Nitrous oxide; N_2O; [10024-97-2] 2. Water; H_2O; [7732-18-5]; 3. Sodium chloride; NaCl; [7647-14-5]	Roth, W. *Z. Phys. Chem.* 1897, *24*, 114-151.
VARIABLES:	**PREPARED BY:**
Temperature, concentration	W. Gerrard

EXPERIMENTAL VALUES: $t = T/K - 273.16$; density = d; α = Bunsen coefficient.

Concn. of sodium chloride, %: 0.9475
Change of d with t: $d = 1.00750 - 0.00001729t - 0.000005159t^2$
Change of α with t: $\alpha = 1.2754 - 0.045597t + 0.00063655t^2$

T/K	298.17	293.28	288.11	282.93	278.12
α	0.5332	0.6210	0.7360	0.8880	1.0649

Concn. of sodium chloride, %: 1.033
Change of d with t: $d = 1.00812 - 0.00002546t - 0.0000049403t^2$
Change of α with t: $\alpha = 1.2704 - 0.045682t + 0.0006573t^2$

T/K	298.00	293.24	288.11	283.16	278.32
α	0.5412	0.6226	0.7344	0.8746	1.0522

Concn. of sodium chloride, %: 1.754
Change of d with t: $d = 1.01362 - 0.00004550t - 0.000004672t^2$
Change of α with t: $\alpha = 1.1990 - 0.04191t + 0.00059064t^2$

T/K	298.16	293.21	288.11	283.00	278.10
α	0.5204	0.6049	0.7045	0.8380	1.0065

Concn. of sodium chloride, %: 1.862
Change of d with t: $d = 1.0145 + 0.00005068t - 0.000004778t^2$
Change of α with t: $\alpha = 1.1956 - 0.041595t + 0.00057893t^2$

T/K	298.06	293.16	288.13	283.04	278.10
α	0.5188	0.5992	0.7026	0.8297	1.0042

(continued)

AUXILIARY INFORMATION

METHOD/APPARATUS/PROCEDURE:	SOURCE AND PURITY OF MATERIALS:
Ostwald method, using gas buret and pipet. Measurement of volume of gas before and after absorption. Specific gravity of solution was determined by a Sprengel pyknometer. Vapour pressure of water, adjusted by assuming Raoult's law, was allowed for.	1. N_2O was self prepared and purified. 3. Sodium chloride was analysed by an appropriate method. **ESTIMATED ERROR:** **REFERENCES:**

COMPONENTS:	ORIGINAL MEASUREMENTS:
1. Nitrous oxide; N_2O; [10024-97-2] 2. Water; H_2O; [7732-18-5]; 3. Sodium chloride; NaCl; [7647-14-5]	Roth, W. *Z. Phys. Chem.* 1897, *24*, 114-151.

EXPERIMENTAL VALUES:

$t = T/K - 273.16$; density = d; α = Bunsen coefficient.

Concn. of sodium chloride, %: 3.718
Change of d with t: $d = 1.02864 - 0.00010466t - 0.0000041586t^2$
Change of α with t: $\alpha = 1.0678 - 0.035518t + 0.0004774t^2$

T/K	298.34	293.24	287.70	283.17	278.03
α	0.4761	0.5508	0.6523	0.7558	0.9063

Concn. of sodium chloride, %: 4.054
Change of d with t: $d = 1.03152 - 0.00009209t - 0.000004826t^2$
Change of α with t: $\alpha = 1.0907 - 0.03552t + 0.00043835t^2$

T/K	298.18	293.29	288.18	283.13	277.98
α	0.4764	0.5594	0.6561	0.7764	0.9297

Concn. of sodium chloride, %: 5.705
Change of d with t: $d = 1.04393 - 0.0001515t - 0.000003995t^2$
Change of α with t: $\alpha = 1.0025 - 0.03409t + 0.0004606t^2$

T/K	298.21	293.24	288.30	283.30	278.29
α	0.4375	0.5053	0.5919	0.7031	0.8397

Concn. of sodium chloride, %: 6.024
Change of d with t: $d = 1.04647 - 0.00017481t - 0.00003314t^2$
Change of α with t: $\alpha \quad 0.9956 - 0.032925t + 0.00043854t^2$

T/K	298.25	292.87	288.34	283.18	278.39
α	0.4456	0.5222	0.5968	0.7087	0.8354

COMPONENTS:	ORIGINAL MEASUREMENTS:
1. Nitrous oxide; N_2O; [10024-97-2] 2. Water; H_2O; [7732-18-5] 3. Sodium chloride; NaCl; [7647-14-5]	Manchot, von W.; Jahrstorfer, M.; Zepter, H. *Z. Anorg. Allg. Chem.* 1924, *141*, 45-81.
VARIABLES:	PREPARED BY:
Concentration	C. L. Young

EXPERIMENTAL VALUES:

T/K	Density, d_4^{25} of salt soln.	Conc. of salt /10^3 mol m^{-3} (soln.)	Conc. of salt /mol kg^{-1} (water)	S_1 /cm^3	S_2 /cm^3
298.15	1.0438	1.15	1.1776	39.0	37.4
	1.0874	2.31	2.4256	28.5	26.2
	1.1600	4.32	4.7606	17.2	14.8

S_1 - volume of nitrous oxide absorbed per 100 cm^3 of salt solution

S_2 - volume of nitrous oxide absorbed per 100 g of salt solution

Both S_1 and S_2 were reduced to conditions of 273.15 K and 101.3 kPa

AUXILIARY INFORMATION

METHOD/APPARATUS/PROCEDURE:

Measurement of volume of gas absorbed by means of gas buret and pipet. Volume absorbed appears to be taken as independent of pressure. (1). Density of the aqueous solution determined by Sprengel pyknometer.

SOURCE AND PURITY OF MATERIALS:

1. Prepared by heating ammonium nitrate, frozen in liquid air and then distilled.

2. Recrystallized.

ESTIMATED ERROR:

REFERENCES:

1. Manchot, W. *Z.Anorg.Chem.* 1924, *141*, 38.

COMPONENTS:	ORIGINAL MEASUREMENTS:
1. Nitrous oxide; N_2O; [10024-97-2] 2. Water; H_2O; [7732-18-5]; 3. Sodium chloride; NaCl; [7647-14-5]	Markham, A. E.; Kobe, K. A. *J. Amer. Chem. Soc.* 1941, *63*, 449-54.
VARIABLES: T/K: 273.35 - 313.15 Salt molality/mol kg^{-1}: 0.1-8 Partial pressure of gas/atm: 1	PREPARED BY: P. L. Long

EXPERIMENTAL VALUES:

T/K*	Salt molality/mol kg^{-1} *	Bunsen Coefficient, α *	'Solubility Coefficient', S *[a]
273.35	0	1.2970	1.2971
	1	0.9178	0.9327
	2	0.6675	0.6902
	3	0.5053	0.5330
298.15	0	0.5392	0.5408
	1	0.4026	0.4127
	2	0.3166	0.3299
	3	0.2502	0.2662
313.15	0	0.3579	0.3607
	1	0.2719	0.2707
	2	0.2181	0.2288
	3	0.1743	0.1866

* From the original data.

[a] The 'solubility coefficient', "refers to the volume of gas, reduced to standard conditions ... which is dissolved by the quantity of solution containing one gram of water".

AUXILIARY INFORMATION

METHOD/APPARATUS/PROCEDURE:

An Ostwald method was used. A known volume of gas was placed in contact with a known volume of gas-free liquid. After equilibrium was established by agitation, the volume of the remaining gas was measured from which the amount of gas dissolved was found. Solutions were prepared by weight, and verified by density measurements compared with the literature values. Apparatus consisted of an absorption flask with two bulbs, one twice the volume of the other, which was connected by a capillary tube to a gas buret and a manometer system.

SOURCE AND PURITY OF MATERIALS:

1. 99.7 per cent pure, no source.
2. Water was freshly boiled, distilled.
3. Analytical grade.

ESTIMATED ERROR:

$\delta T/K = \pm 0.1$ (273.35 K); ± 0.03 (above 273.35 K); $\delta\alpha/\alpha = \pm 0.02$.

REFERENCES:

COMPONENTS:	ORIGINAL MEASUREMENTS:
1. Nitrous oxide; N_2O; [10024-97-2] 2. Water; H_2O; [7732-18-5] 3. Sodium bromide; NaBr; [7647-15-6]	Manchot, von W.; Jahrstorfer, M.; Zepter, H. *Z. Anorg. Allg. Chem.* 1924, *141*, 45-81.
VARIABLES:	PREPARED BY:
Concentration	C. L. Young

EXPERIMENTAL VALUES:

T/K	Density, d_4^{25} of salt soln.	Conc. of salt $/10^3$ mol m^{-3} (soln.)	Conc. of salt /mol kg^{-1} (water)	S_1 $/cm^3$	S_2 $/cm^3$
298.15	1.0849	1.125	1.1608	40.1	37.0
	1.1645	2.17	2.3057	30.9	26.5
	1.3338	4.46	5.0983	17.8	13.3

S_1 - volume of nitrous oxide absorbed per 100 cm^3 of salt solution

S_2 - volume of nitrous oxide absorbed per 100 g of salt solution

Both S_1 and S_2 were reduced to conditions of 273.15 K and 101.3 kPa

AUXILIARY INFORMATION

METHOD/APPARATUS/PROCEDURE:

Measurement of volume of gas absorbed by means of gas buret and pipet. Volume absorbed appears to be taken as independent of pressure. (1). Density of the aqueous solution determined by Sprengel pyknometer.

SOURCE AND PURITY OF MATERIALS:

1. Prepared by heating ammonium nitrate, frozen in liquid air and then distilled.

2. Recrystallized.

ESTIMATED ERROR:

REFERENCES:

1. Manchot, W. *Z.Anorg.Chem.* 1924, *141*, 38.

COMPONENTS:	ORIGINAL MEASUREMENTS:
1. Nitrous oxide; N_2O; [10024-97-2] 2. Water; H_2O; [7732-18-5]; 3. Sodium sulfate; Na_2SO_4; [7757-82-6]	Gordon, V. *Z. Phys. Chem.* 1895, *18*, 1-16.
VARIABLES:	**PREPARED BY:**
Temperature, concentration	W. Gerrard

EXPERIMENTAL VALUES:

T/K	Conc. of salt Weight-%	Density, ρ	Bunsen Coefficient, α	T/K	Conc. of salt Weight-%	Density, ρ	Bunsen Coefficient, α
295.46	5.765	1.0514	0.4675	288.26	8.533	1.0770	0.48033
291.76		1.0525	0.5083	284.16		1.0785	0.54547
288.26		1.05374	0.57065	281.26		1.0793	0.61583
284.16		1.0548	0.67153	295.46	12.439	1.1138	0.33739
281.26		1.0555	0.72345	291.76		1.1152	0.37911
295.46	8.533	1.0747	0.3940	288.26		1.1166	0.41606
291.76		1.0758	0.43089	284.16		1.1181	0.47152

Smoothing Equations

$(t = T/K - 273.16)$

For 5.765 wt-% solution: $\alpha = 0.96489 - 0.034086t + 0.00052831t^2$

For 8.533 wt-% solution: $\alpha = 0.836072 - 0.031393t + 0.00051879t^2$

For 12.439 wt-% solution: $\alpha = 0.637428 - 0.016216t + 0.00010304t^2$

AUXILIARY INFORMATION

METHOD /APPARATUS/PROCEDURE:

Measurement of volume of gas by means of Ostwald type apparatus, gas buret and pipet (ref. 1). Density of the solution determined by Sprengel pyknometer.

SOURCE AND PURITY OF MATERIALS:

1. N_2O: Self prepared and purified. Attested by combustion with hydrogen.

ESTIMATED ERROR:

REFERENCES:

1. Timofojew, W. *Z. Phys. Chem.* 1890, *6*, 141.

COMPONENTS:	ORIGINAL MEASUREMENTS:
1. Nitrous oxide; N_2O; [10024-97-2] 2. Water; H_2O; [7732-18-5] 3. Sodium sulfate; Na_2SO_4; [7757-82-6]	Manchot, von W.; Jahrstorfer, M.; Zepter, H. *Z. Anorg. Allg. Chem.* 1924, *141*, 45-81.
VARIABLES:	**PREPARED BY:**
Concentration	C. L. Young

EXPERIMENTAL VALUES:

T/K	Density, d_4^{25} of salt soln.	Conc. of salt $/10^3$ mol m^{-3} (soln.)	Conc. of salt /mol kg^{-1} (water)	S_1 $/cm^3$	S_2 $/cm^3$
298.15	1.0550	0.4646	0.4698	36.5	34.6
	1.1141	0.977	1.0017	24.8	22.2

S_1 - volume of nitrous oxide absorbed per 100 cm^3 of salt solution

S_2 - volume of nitrous oxide absorbed per 100 g of salt solution

Both S_1 and S_2 were reduced to conditions of 273.15 K and 101.3 kPa

AUXILIARY INFORMATION

METHOD/APPARATUS/PROCEDURE:

Measurement of volume of gas absorbed by means of gas buret and pipet. Volume absorbed appears to be taken as independent of pressure. (1). Density of the aqueous solution determined by Sprengel pyknometer.

SOURCE AND PURITY OF MATERIALS:

1. Prepared by heating ammonium nitrate, frozen in liquid air and then distilled.

2. Recrystallized.

ESTIMATED ERROR:

REFERENCES:

1. Manchot, W. *Z.Anorg.Chem.* 1924, *141*, 38.

COMPONENTS:	ORIGINAL MEASUREMENTS:
1. Nitrous oxide; N_2O; [10024-97-2] 2. Water; H_2O; [7732-18-5]; 3. Sodium sulfate; Na_2SO_4; [7757-82-6]	Markham, A. E.; Kobe, K. A. *J. Amer. Chem. Soc.* 1941, *63*, 449-54.
VARIABLES:	PREPARED BY:
T/K: 273.35-313.15 Salt molality/mol kg^{-1}: 0.1-8 Partial pressure of gas/atm: 1	P. L. Long

EXPERIMENTAL VALUES:

T/K*	Salt molality/mol kg^{-1}*	Bunsen Coefficient, α*	'Solubility Coefficient', S *[a]
298.15	0	0.5392	0.5408
	0.5	0.3565	0.3612
	1	0.2476	0.2547
	1.5	0.1721	0.1797
313.15	0	0.3579	0.3607
	0.5	0.2425	0.2472
	1	0.1722	0.1791
	1.5	0.1226	0.1297

* From the original data.

[a] The 'solubility coefficient', "refers to the volume of gas, reduced to standard conditions ... which is dissolved by the quantity of solution containing one gram of water".

AUXILIARY INFORMATION

METHOD/APPARATUS/PROCEDURE:	SOURCE AND PURITY OF MATERIALS:
An Ostwald method was used. A known volume of gas was placed in contact with a known volume of gas-free liquid. After equilibrium was established by agitation, the volume of the remaining gas was measured from which the amount of gas dissolved was found. Solutions were prepared by weight, and verified by density measurements compared with the literature values. Apparatus consisted of an absorption flask with two bulbs, one twice the volume of the other, which was connected by a capillary tube to a gas buret and a manometer system.	1. 99.7 per cent pure, no source. 2. Water was freshly boiled, distilled. 3. Analytical grade. ESTIMATED ERROR: $\delta T/K = \pm 0.1$ (273.35 K); ± 0.03 (above 273.35 K); $\delta\alpha/\alpha = \pm 0.02$. REFERENCES:

COMPONENTS:	ORIGINAL MEASUREMENTS:
1. Nitrous oxide; N_2O; [10024-97-2] 2. Water; H_2O; [7732-18-5] 3. Sodium sulfate; Na_2SO_4; [7757-82-6] 4. Sulfuric acid; H_2SO_4; [7664-93-9]	Kobe, K. A.; Kenton, F. H. *Ind. Eng. Chem., Anal. Ed.* 1938, *10*, 76-77.
VARIABLES:	PREPARED BY:
T/K: 298.15 N_2O P/kPa: 101.325 (760 mmHg)	C. L. Young

EXPERIMENTAL VALUES:

Temperature t/°C	T/K	Solvent Volume cm^3	Volume Absorbed cm^3	Bunsen Coefficient α	Ostwald Coefficient L
25	298.15	49.54	7.88	0.159	0.146
		49.54	7.90		

The solvent was a mixture of 800 g H_2O
200 g Na_2SO_4 (anhydrous)
40 ml H_2SO_4 (conc., 36 N)

Thus the molality of the solution was

$$m_{Na_2SO_4}/\text{mol kg}^{-1} = 1.76$$

$$m_{H_2SO_4}/\text{mol kg}^{-1} = 0.90.$$

AUXILIARY INFORMATION

METHOD/APPARATUS/PROCEDURE:

The apparatus was described in detail in an earlier paper (1). The apparatus consists of a gas buret, a pressure compensator, and a 200 cm^3 absorption bulb and mercury leveling bulb. The absorption bulb is attached to a shaking mechanism.

The solvent and the gas are placed in the absorption bulb. The bulb was shaken until equilibrium was reached. The remaining gas was returned to the buret. The difference in final and initial volumes was taken as the gas absorbed.

SOURCE AND PURITY OF MATERIALS:

1. Nitrous oxide source not given. Purity stated to be 99+ per cent.
2. Water: distilled.
3, 4. Sodium sulfate and sulfuric acid: sources not given. Analytical grades.

ESTIMATED ERROR:

$\delta\alpha/cm^3 = \pm 0.001$

REFERENCES:

1. Kobe, K. A.; Williams, J. S. *Ind. Eng. Chem., Anal. Ed.* 1935, *7*, 37.

COMPONENTS:

1. Nitrous oxide; N_2O; [10024-97-2]
2. Water; H_2O; [7732-18-5]
3. Sodium nitrate; $NaNO_3$; [7631-99-4]

ORIGINAL MEASUREMENTS:

Knopp, W.

Z. Phys. Chem. 1904, *48*, 97-108

VARIABLES:

Concentration

PREPARED BY:

W. Gerrard

EXPERIMENTAL VALUES:

Pressure assumed to be 101.325 kPa.

T/K	Weight of salt in 100 g of solution	Conc of salt /mol dm^{-3} (soln)	Density of solution	Bunsen absorption coefficient, α
293.15	1.124	0.1336	1.00590	0.6089
	2.531	0.3052	1.01537	0.5876
	5.077	0.6286	1.03284	0.5465
	8.701	1.1200	1.05834	0.4926
			Water	0.6270

AUXILIARY INFORMATION

METHOD: /APPARATUS/PROCEDURE:

An absorption pipet and gas buret were used to measure the volume of nitrous oxide absorbed.

Densities were determined by a Sprengel pyknometer.

SOURCE AND PURITY OF MATERIALS:

1. Nitrous oxide was prepared by heating pure ammonium nitrate at 513-523 K. It was passed through aqueous ferrous sulfate, aqueous sodium hydroxide, and concentrated sulfuric acid.

2. Appeared to be of satisfactory purity.

ESTIMATED ERROR:

REFERENCES:

COMPONENTS:	ORIGINAL MEASUREMENTS:
1. Nitrous oxide; N_2O; [10024-97-2] 2. Water; H_2O; [7732-18-5] 3. Sodium nitrate; $NaNO_3$; [7631-99-4]	Manchot, von W.; Jahrstorfer, M.; Zepter, H. *Z. Anorg. Allg. Chem.* 1924, *141*, 45-81.

VARIABLES:	PREPARED BY:
Concentration	C. L. Young

EXPERIMENTAL VALUES:

T/K	Density, d_4^{25} of salt soln.	Conc. of salt $/10^3$ mol m^{-3} (soln.)	Conc. of salt /mol kg^{-1} (water)	S_1 $/cm^3$	S_2 $/cm^3$
298.15	1.0560	1.08	1.1201	42.3	40.1
	1.0677	1.31	1.3698	40.3	37.7
	1.1141	2.17	2.3343	33.5	30.1
	1.1543	3.01	3.3504	27.7	24.0
	1.2152	4.20	4.8942	21.6	17.8

S_1 - volume of nitrous oxide absorbed per 100 cm^3 of salt solution

S_2 - volume of nitrous oxide absorbed per 100 g of salt solution

Both S_1 and S_2 were reduced to conditions of 273.15 K and 101.3 kPa

AUXILIARY INFORMATION

METHOD/APPARATUS/PROCEDURE:

Measurement of volume of gas absorbed by means of gas buret and pipet. Volume absorbed appears to be taken as independent of pressure (1). Density of the aqueous solution determined by Sprengel pyknometer.

SOURCE AND PURITY OF MATERIALS:

1. Prepared by heating ammonium nitrate, frozen in liquid air and then distilled.

2. Recrystallized.

ESTIMATED ERROR:

REFERENCES:

1. Manchot, W. *Z.Anorg.Chem.* 1924, *141*, 38.

COMPONENTS:	ORIGINAL MEASUREMENTS:
1. Nitrous oxide; N_2O; [10024-97-2] 2. Water; H_2O; [7732-18-5] 3. Sodium phosphate; Na_3PO_4; [7601-54-9] or Disodium hydrogen phosphate; Na_2HPO_4; [7558-79-4]	Manchot, von W.; Jahrstorfer, M.; Zepter, H. *Z. Anorg. Allg. Chem.* 1924, *141*, 45-81.
VARIABLES:	PREPARED BY:
Concentration	C. L. Young

EXPERIMENTAL VALUES:

T/K	Density, d_4^{25} of salt soln.	Conc. of salt $/10^3$ mol m^{-3} (soln.)	Conc. of salt /mol kg^{-1} (water)	S_1 $/cm^3$	S_2 $/cm^3$
		Sodium phosphate			
298.15	1.0348	0.22	0.2203	40.7	39.3
		Disodium hydrogen phosphate			
298.15	1.0470	0.3985	0.4024	38.0	39.6

S_1 - volume of nitrous oxide in 100 cm^3 of salt solution

S_2 - volume of nitrous oxide in 100 g of salt solution

AUXILIARY INFORMATION

METHOD/APPARATUS/PROCEDURE:

Measurement of volume of gas absorbed by means of gas buret and pipet. Volume absorbed appears to be taken as independent of pressure. (1). Density of the aqueous solution determined by Sprengel pyknometer.

SOURCE AND PURITY OF MATERIALS:

1. Prepared by heating ammonium nitrate, frozen in liquid air and then distilled.

2. Recrystallized.

ESTIMATED ERROR:

REFERENCES:

1. Manchot, W. *Z.Anorg.Chem.* 1924, *141*, 38.

COMPONENTS:

1. Nitrous oxide; N_2O; [10024-97-2]
2. Water; H_2O; [7732-18-5]
3. Sodium carbonate; Na_2CO_3; [497-19-8] or Sodium bicarbonate; $NaHCO_3$; [144-55-8]

ORIGINAL MEASUREMENTS:

Hikita, H.; Asai, S.; Ishikawa, H.; Esaka, H.

J. Chem. Engng. Data, 1974, *19*, 89-92.

VARIABLES:

Concentration

PREPARED BY:

C. L. Young

EXPERIMENTAL VALUES:

T/K	Conc. of sodium carbonate /g l^{-1} (soln.)	Conc. of sodium bicarbonate /g l^{-1} (soln.)	Ionic strength /g-ion l^{-1}	Solubility, S /mol l^{-1} (soln.)
298.15	0	0	0	2.356
	0.329	0	0.987	1.753
	0.661	0	1.98	1.296
	0.992	0	2.98	1.064
	1.33	0	3.98	0.7750
	1.66	0	4.97	0.6404
	0.309	0.0630	0.990	1.729
	0.621	0.125	1.99	1.389
	0.927	0.183	2.96	1.036
	1.24	0.249	3.97	0.7923
	1.54	0.310	4.93	0.6314
	0.198	0.200	0.794	1.777
	0.395	0.398	1.58	1.429
	0.600	0.587	2.39	1.180
	0.783	0.755	3.10	0.9568
	0.0990	0.199	0.496	1.917
	0.197	0.402	0.993	1.668
	0.301	0.593	1.50	1.441
	0.395	0.801	1.99	1.186
	0.0374	0.187	0.299	2.024
	0.0740	0.373	0.595	1.807
	0.113	0.553	0.892	1.665
	0.151	0.736	1.19	1.439
	0.190	0.927	1.50	1.267

Pressure = 1 atmosphere = 1.01325 bar.

AUXILIARY INFORMATION

METHOD/APPARATUS/PROCEDURE:

Gas volumetric method similar to that used by Onda *et al.* (ref. 1). Chemical compositions of absorbing solutions determined by chemical method of Danckwerts and Kennedy (ref. 2).

SOURCE AND PURITY OF MATERIALS:

1. Commercial sample, purity better than 99.8 mole per cent.
2. Distilled and boiled.
3. Analytical grade reagent.
4. Analytical grade reagent.

ESTIMATED ERROR:

$\delta T/K = \pm 0.1$; $\delta S/mol\ l^{-1} = \pm 1.0\%$ (estimated by compiler).

REFERENCES:

1. Onda, K.; Sada, E.; Kobayashi, T T.; Kito, S.; Ito, K.; *J. Chem. Eng. Japan*, 1970, *3*, 18.
2. Dankwerts, P. V.; Kennedy, A. M.; *Chem. Eng. Sci.*, 1958, *8*, 201.

COMPONENTS:

1. Nitrous oxide; N_2O; [10024-97-2]
2. Water; H_2O; [7732-18-5];
3. Potassium hydroxide; KOH; [1310-58-3]

ORIGINAL MEASUREMENTS:

Geffcken, G. *Z. Phys. Chem.* 1904, *49*, 257-302.

VARIABLES:

Temperature, concentration

PREPARED BY:

W. Gerrard/C. L. Young

EXPERIMENTAL VALUES:

T/K	Conc. of hydroxide/mol dm^{-3} (soln.)	Ostwald coefficient, L
288.16	0.541	0.6591
	0.542	0.6595
	1.074	0.5427
	1.082	0.5392
298.15	0.541	0.5087
	0.542	0.5093
	1.074	0.4252
	1.082	0.4221

AUXILIARY INFORMATION

METHOD /APPARATUS/PROCEDURE:

Measurement of volume of N_2O absorbed by the aqueous solution. Detailed description and diagram given in source.

SOURCE AND PURITY OF MATERIALS:

1. Nitrous oxide self prepared and attested.

ESTIMATED ERROR:

REFERENCES:

COMPONENTS:	ORIGINAL MEASUREMENTS:
1. Nitrous oxide; N_2O; [10024-97-2] 2. Water; H_2O; [7732-18-5] 3. Potassium fluoride; KF; [7789-23-3]	Kreitus, I.; Abramenkov, A. *Latv. PRS Zinat. Akad. Vestis Khim. Ser.* 1980, 238.
VARIABLES:	PREPARED BY:
Concentration of salt	C. L. Young

EXPERIMENTAL VALUES:

T/K	Conc. of salt/mol kg^{-1} (solvent)	Bunsen coefficient, α
298.15	0	0.54
	1.21	0.37
	5.84	0.093
	7.78	0.079
	12.33	0.019

AUXILIARY INFORMATION

METHOD APPARATUS/PROCEDURE:

Aqueous solution saturated at 0.1013 MPa and 1 cm^3 sample analyzed by gas chromatography. Column contained molecular sieves and was operated isothermally at 250 °C using helium as carrier gas. Thermal conductivity detector used. Details in ref. (1).

SOURCE AND PURITY OF MATERIALS:

1. Medical grade sample.
2. Twice distilled.
3. Pure.

ESTIMATED ERROR:

$\delta T/K = \pm 0.1$; $\delta\alpha/\alpha = \pm 0.06$

REFERENCES:

1. Kreitus, I.; Gorbovitskalya, T. I. *Latv. PRS Zinat. Akad. Vestis Khim. Ser.* 1979, 664-666.

COMPONENTS:	ORIGINAL MEASUREMENTS:
1. Nitrous oxide; N_2O; [10024-97-2] 2. Water; H_2O; [7732-18-5]; 3. Potassium chloride; KCl; [7447-40-7]	Gordon, V. *Z. Phys. Chem.* 1895, *18*, 1-16.
VARIABLES:	**PREPARED BY:**
Temperature, concentration	W. Gerrard

EXPERIMENTAL VALUES:

T/K	Conc. of salt Weight-%	Density, ρ	Bunsen Coefficient, α	T/K	Conc. of salt Weight-%	Density, ρ	Bunsen Coefficient, α
295.46	4.899	1.029	0.5091	295.46	14.582	1.0939	0.4004
291.76		1.0302	0.5700	291.76		1.0954	0.4521
288.26		1.0309	0.6351	288.26		1.0965	0.4983
284.16		1.0318	0.7243	284.16		1.0977	0.5611
281.26		1.0324	0.7999	281.26		1.0987	0.6042
295.46	7.640	1.0461	0.4513	295.46	22.083	1.1487	0.3208
291.76		1.0475	0.5191	291.76		1.1502	0.3570
288.26		1.049	0.5893	288.26		1.1516	0.3888
284.16		1.0503	0.6661	284.16		1.1533	0.4606
281.26		1.0533	0.7329	281.26		1.1543	0.4892

Smoothing Equations

$(t = T/K - 273.16)$

For 4.899 wt-% solution: $\alpha = 1.02649 - 0.0315t + 0.000397t^2$

For 7.640 wt-% solution: $\alpha = 0.91032 - 0.0227t + 0.0000949t^2$

For 14.582 wt-% solution: $\alpha = 0.73968 - 0.017583t + 0.0001058t^2$

For 22.083 wt-% solution: $\alpha = 0.647561 - 0.0223445t + 0.0003449t^2$

AUXILIARY INFORMATION

METHOD /APPARATUS/PROCEDURE:

Measurement of volume of gas by means of Ostwald type apparatus, gas buret and pipet (ref. 1). Density of the solution determined by Sprengel pyknometer.

SOURCE AND PURITY OF MATERIALS:

1. N_2O: Self prepared and purified. Attested by combustion with hydrogen.

ESTIMATED ERROR:

REFERENCES:

1. Timofejew, W. *Z. Phys. Chem.* 1890, *6*, 141.

COMPONENTS:

1. Nitrous oxide; N_2O; [10024-97-2]
2. Water; H_2O; [7732-18-5];
3. Potassium chloride; KCl; [7447-40-7]

ORIGINAL MEASUREMENTS:

Geffcken, G. *Z. Phys. Chem.* 1904, *49*, 257-302.

VARIABLES:

Temperature, concentration

PREPARED BY:

W. Gerrard/C. L. Young

EXPERIMENTAL VALUES:

T/K	Conc. of salt/mol dm^{-3} (soln.)	Ostwald coefficient, L
288.16	0.558	0.6782
	0.559	0.6787
	1.070	0.6046
	1.102	0.6020
298.16	0.558	0.5218
	0.559	0.5217
	1.070	0.4673
	1.102	0.4639

AUXILIARY INFORMATION

METHOD/APPARATUS/PROCEDURE:

Measurement of volume of N_2O absorbed by the aqueous solution. Diagram and detailed description given in original paper.

SOURCE AND PURITY OF MATERIALS:

1. Nitrous oxide self prepared and attested.

ESTIMATED ERROR:

REFERENCES:

COMPONENTS:	ORIGINAL MEASUREMENTS:
1. Nitrous oxide; N_2O; [10024-97-2] 2. Water; H_2O; [7732-18-5] 3. Potassium chloride; KCl; [7447-40-7]	Manchot, von W.; Jahrstorfer, M.; Zepter, M. *Z. Anorg. Allg. Chem.* 1924, *141*, 45-81.
VARIABLES:	PREPARED BY:
Concentration	C. L. Young

EXPERIMENTAL VALUES:

T/K	Density, d_4^{25} of salt soln.	Conc. of salt $/10^3$ mol m^{-3} (soln.)	Conc. of salt /mol kg^{-1} (water)	S_1 $/cm^3$	S_2 $/cm^3$
298.15	1.0334	0.78	0.7998	45.3	43.8
	1.0540	1.25	1.3010	41.0	38.9
	1.0850	1.98	2.1123	35.5	32.7
	1.1385	3.21	3.5700	28.1	24.7
	1.1734	4.04	4.6321	24.0	20.4

S_1 - volume of nitrous oxide absorbed per 100 cm^3 of salt solution

S_2 - volume of nitrous oxide absorbed per 100 g of salt solution

Both S_1 and S_2 were reduced to conditions of 273.15 K and 101.3 kPa

AUXILIARY INFORMATION

METHOD/APPARATUS/PROCEDURE:

Measurement of volume of gas absorbed by means of gas buret and pipet. Volume absorbed appears to be taken as independent of pressure. (1). Density of the aqueous solution determined by Sprengel pyknometer.

SOURCE AND PURITY OF MATERIALS:

1. Prepared by heating ammonium nitrate, frozen in liquid air and then distilled.

2. Recrystallized.

ESTIMATED ERROR:

REFERENCES:

1. Manchot. W. *Z.Anorg.Chem.* 1924,*141*, 38.

COMPONENTS:

1. Nitrous oxide; N_2O; [10024-97-2]
2. Water; H_2O; [7732-18-5];
3. Potassium chloride; KCl; [7447-40-7]

ORIGINAL MEASUREMENTS:

Markham, A. E.; Kobe, K. A.

J. Amer. Chem. Soc. 1941, *63*, 449-54.

VARIABLES:

T/K: 273.35-313.15
Salt molality/mol kg^{-1}: 0.1-8
Partial pressure of gas/atm: 1

PREPARED BY:

P. L. Long

EXPERIMENTAL VALUES:

T/K*	Salt molality/mol kg^{-1}*	Bunsen Coefficient, α*	'Solubility Coefficient', S*[a]
273.35	0	1.2970	1.2971
	1	0.9880	1.0140
	2	0.7784	0.8212
	3	0.6349	0.6893
298.15	0	0.5392	0.5408
	1	0.4329	0.4466
	2	0.3580	0.3803
	3	0.3030	0.3315
313.15	0	0.3579	0.3607
	1	0.2885	0.2993
	2	0.2416	0.2613
	3	0.2077	0.2286

* From the original data.

[a] The 'solubility coefficient', "refers to the volume of gas, reduced to standard conditions ... which is dissolved by the quantity of solution containing one gram of water".

AUXILIARY INFORMATION

METHOD/APPARATUS/PROCEDURE:

An Ostwald method was used. A known volume of gas was placed in contact with a known volume of gas-free liquid. After equilibrium was established by agitation, the volume of the remaining gas was measured from which the amount of gas dissolved was found. Solutions were prepared by weight, and verified by density measurements compared with the literature values. Apparatus consisted of an absorption flask with two bulbs, one twice the volume of the other, which was connected by a capillary tube to a gas buret and a manometer system.

SOURCE AND PURITY OF MATERIALS:

1. 99.7 per cent pure, no source.
2. Water was freshly boiled, distilled.
3. Analytical grade.

ESTIMATED ERROR:

$\delta T/K = \pm 0.1$ (273.35 K); ± 0.03 (above 273.35 K); $\delta\alpha/\alpha = \pm 0.02$.

REFERENCES:

COMPONENTS:	ORIGINAL MEASUREMENTS:
1. Nitrous oxide; N_2O; [10024-97-2] 2. Water; H_2O; [7732-18-5] 3. Potassium chloride; KCl; [7447-40-7]	Kreitus, I.; Gorbovitskalya, T. I. *Latv. PRS Zinat. Akad. Vestis Khim. Ser.* 1979, 664-666.
VARIABLES:	PREPARED BY:
Concentration of salt	C. L. Young

EXPERIMENTAL VALUES:

T/K	Conc. of salt /mol kg^{-1} (solvent)	Bunsen coefficient, α	Absorption† coefficient, S
298.15	0.0	0.55	0.55
	0.51	0.49	0.50
	1.03	0.44	0.46
	2.13	0.37	0.39
	2.70	0.33	0.35
	3.61	0.29	0.32
	4.63	0.25	0.29

† Volume of gas, corrected to 101.3 kPa and 273.15 K, dissolved at specified temperature and a partial pressure of 101.3 kPa, by solution containing 1 g of water.

AUXILIARY INFORMATION

METHOD/APPARATUS/PROCEDURE:	SOURCE AND PURITY OF MATERIALS:
Aqueous solution saturated at 0.1013 MPa and 1 cm^3 sample analyzed by gas chromatography. Column contained molecular sieves and was operated isothermally at 250 °C using helium as carrier gas. Thermal conductivity detector used.	1. Medical grade sample. 2. Twice distilled. 3. Chemically pure grade.
	ESTIMATED ERROR:
	$\delta T/K = \pm 0.1$; $\delta\alpha/\alpha = \pm 0.1$
	REFERENCES:

COMPONENTS:	ORIGINAL MEASUREMENTS:
1. Nitrous oxide; N_2O; [10024-97-2] 2. Water; H_2O; [7732-18-5] 3. Potassium bromide; KBr; [7758-02-3]	Manchot, von W.; Jahrstorfer, M.; Zepter, H. *Z. Anorg. Allg. Chem.* 1924, *141*, 45-81.
VARIABLES:	**PREPARED BY:**
Concentration	C. L. Young

EXPERIMENTAL VALUES:

T/K	Density, d_4^{25} of salt soln.	Conc. of salt $/10^3$ mol m^{-3} (soln.)	Conc. of salt /mol kg^{-1} (water)	S_1 $/cm^3$	S_2 $/cm^3$
298.15	1.0891	1.11	1.1598	43.0	39.5
	1.1752	2.15	2.3387	35.1	29.9
	1.3380	4.19	4.9922	24.7	18.5

S_1 - volume of nitrous oxide absorbed per 100 cm^3 of salt solution

S_2 - volume of nitrous oxide absorbed per 100 g of salt solution

Both S_1 and S_2 were reduced to conditions of 273.15 K and 101.3 kPa

AUXILIARY INFORMATION

METHOD/APPARATUS/PROCEDURE:	SOURCE AND PURITY OF MATERIALS:
Measurement of volume of gas absorbed by means of gas buret and pipet. Volume absorbed appears to be taken as independent of pressure (1). Density of the aqueous solution determined by Sprengel pyknometer.	1. Prepared by heating ammonium nitrate, frozen in liquid air and then distilled. 2. Recrystallized.
	ESTIMATED ERROR:
	REFERENCES: 1. Manchot, W. *Z.Anorg.Chem.* 1924, *141*, 38.

COMPONENTS:

1. Nitrous oxide; N_2O; [10024-97-2]
2. Water; H_2O; [7732-18-5];
3. Potassium bromide; KBr; [7758-02-3]

ORIGINAL MEASUREMENTS:

Geffcken, G. *Z. Phys. Chem.* 1904, *49*, 257-302.

VARIABLES:

Temperature, concentration

PREPARED BY:

W. Gerrard/C. L. Young

EXPERIMENTAL VALUES:

T/K	Conc. of salt/mol dm^{-3} (soln.)	Ostwald coefficient, L
288.16	0.546	0.6877
	0.550	0.6892
	0.937	0.6352
	0.959	0.6334
298.16	0.546	0.5306
	0.550	0.5318
	0.937	0.4908
	0.959	0.4899

AUXILIARY INFORMATION

METHOD /APPARATUS/PROCEDURE:

Measurement of volume of N_2O absorbed by the aqueous solution. Detailed description and diagram given in source.

SOURCE AND PURITY OF MATERIALS:

1. Nitrous oxide self prepared and attested.

ESTIMATED ERROR:

REFERENCES:

COMPONENTS:

1. Nitrous oxide; N_2O; [10024-97-2]
2. Water; H_2O; [7732-18-5];
3. Potassium iodide; KI; [7681-11-0]

ORIGINAL MEASUREMENTS:

Geffcken, G. *Z. Phys. Chem.* 1904, *49*, 257-302.

VARIABLES:

Temperature, concentration

PREPARED BY:

W. Gerrard/C. L. Young

EXPERIMENTAL VALUES:

T/K	Conc. of salt/mol dm^{-3} (soln.)	Ostwald coefficient, L
288.16	0.550	0.6950
	0.557	0.6916
	0.886	0.6466
	0.913	0.6442
298.16	0.550	0.5367
	0.557	0.5344
	0.886	0.5025
	0.913	0.5012

AUXILIARY INFORMATION

METHOD/APPARATUS/PROCEDURE:

Measurement of volume of N_2O absorbed by the aqueous solution. Detailed description and diagram given in source.

SOURCE AND PURITY OF MATERIALS:

1. Nitrous oxide self prepared and attested.

ESTIMATED ERROR:

REFERENCES:

COMPONENTS:	ORIGINAL MEASUREMENTS:
1. Nitrous oxide; N_2O; [10024-97-2] 2. Water; H_2O; [7732-18-5]; 3. Potassium iodide; KI; [7681-11-0]	Sada, E.; Ando, N.; Kito, S. *J. Appl. Chem. Biotechnol.* 1972, *22*, 1185-1193.
VARIABLES:	PREPARED BY:
T/K: 298.15 *P*/kPa: 101.325 (atm) Molarity of salt: mol l^{-1}	W. Gerrard/C. L. Young

EXPERIMENTAL VALUES:

T/K	Concn. of salt/mol l^{-1}	Bunsen coefficient, α
298.15	0	0.5512 (pure water)
	0.644	0.4862
	1.093	0.4491
	1.661	0.4080
	2.196	0.3712
	2.859	0.3241

AUXILIARY INFORMATION

METHOD/APPARATUS/PROCEDURE:

Equilibrium established between a measured volume of gas and a measured amount of gas-free liquid in a cell fitted with a magnetic stirrer. Details in source and ref. 1.

SOURCE AND PURITY OF MATERIALS:

1. From commercial cylinder; 99.8 per cent, as attested by gas chromatography.
2. Distilled water was used.
3. Salt was of reagent grade.

ESTIMATED ERROR:

$\delta T/K = \pm 0.2$; $\delta\alpha = \pm 2\%$ (estimated by compiler).

REFERENCES:

1. Onda, K.; Sada, E.; Kobayashi, T.; Kito, S1; Ito, K. *J. Chem. Engng. Japan* 1970, *3*, 18; 137.

COMPONENTS:

1. Nitrous oxide; N_2O; [10024-97-2]
2. Water; H_2O; [7732-18-5];
3. Potassium sulfate; K_2SO_4; [7778-80-5]

ORIGINAL MEASUREMENTS:

Gordon, V. *Z. Phys. Chem.* 1895, *18*, 1-16.

VARIABLES:

Temperature, concentration

PREPARED BY:

W. Gerrard

EXPERIMENTAL VALUES:

T/K	Conc. of salt Weight-%	Density, ρ	Bunsen Coefficient, α	T/K	Conc. of salt Weight-%	Density, ρ	Bunsen Coefficient, α
295.46	2.623	1.0194	0.55560	295.46	4.784	1.0369	0.50827
291.76		1.0198	0.63128	291.76		1.0378	0.56699
288.26		1.021	0.69827	288.26		1.0387	0.63497
284.16		1.0218	0.79579	284.16		1.0395	0.73157
281.26		1.01222	0.88724	281.26		1.0403	0.81797

Smoothed Equations

(t - T/K - 273.15)

For 2.623 wt-% solution: $= 1.166991 - 0.03864t + 0.0005028t^2$

For 4.784 wt-% solution: $= 1.1033557 - 0.040109t + 0.000602t^2$

AUXILIARY INFORMATION

METHOD/APPARATUS/PROCEDURE:

Measurement of volume of gas by means of Ostwald type apparatus, gas buret and pipet.
Density of the solution determined by Sprengel pyknometer.

SOURCE AND PURITY OF MATERIALS:

1. N_2O: Self prepared and purified. Attested by combustion with hydrogen.

ESTIMATED ERROR:

REFERENCES:

COMPONENTS:	ORIGINAL MEASUREMENTS:
1. Nitrous oxide; N_2O; [10024-97-2] 2. Water; H_2O; [7732-18-5] 3. Potassium sulfate; K_2SO_4; [7778-80-5]	Manchot, von W.; Jahrstorfer, M.; Zepter, H. *Z. Anorg. Allg. Chem.* 1924, *141*, 45-81.
VARIABLES:	PREPARED BY:
Concentration	C. L. Young

EXPERIMENTAL VALUES:

T/K	Density, d_4^{25} of salt soln.	Conc. of salt /10^3 mol m^{-3} (soln.)	Conc. of salt /mol kg^{-1} (water)	S_1 /cm^3	S_2 /cm^3
298.15	1.0762	0.5991	0.6165	35.5	33.0

S_1 - volume of nitrous oxide absorbed per 100 cm^3 of salt solution

S_2 - volume of nitrous oxide absorbed per 100 g of salt solution

Both S_1 and S_2 were reduced to conditions of 273.15 K and 101.3 kPa

AUXILIARY INFORMATION

METHOD/APPARATUS/PROCEDURE:

Measurement of volume of gas absorbed by means of gas buret and pipet. Volume absorbed appears to be taken as independent of pressure. (1). Density of the aqueous solution determined by Sprengel pyknometer.

SOURCE AND PURITY OF MATERIALS:

1. Prepared by heating ammonium nitrate, frozen in liquid air and then distilled.

2. Recrystallized.

ESTIMATED ERROR:

REFERENCES:

1. Manchot, W. *Z. Anorg. Chem.* 1924, *141*, 38.

COMPONENTS:

1. Nitrous oxide; N_2O; [10024-97-2]
2. Water; H_2O; [7732-18-5]
3. Potassium nitrate; KNO_3; [7757-79-1]

ORIGINAL MEASUREMENTS:

Knopp, W. *Z. Phys. Chem.* 1904, *48*, 97-108.

VARIABLES:

Concentration

PREPARED BY:

W. Gerrard

EXPERIMENTAL VALUES:

Pressure assumed to be 101.325 kPa.

T/K	Weight of salt in 100 g of solution	Conc of salt /mol dm^3 (soln)	Density of solution	Bunsen absorption coefficient, α
293.15	1.063	0.1061	1.0049	0.6173
	2.720	0.2764	1.01534	0.6002
	5.389	0.5630	1.03231	0.5713
	10.577	1.1683	1.06644	0.5196
			Water	0.6270

AUXILIARY INFORMATION

METHOD: /APPARATUS/PROCEDURE:

An absorption pipet and gas buret were used to measure the volume of nitrous oxide absorbed.

Densities were determined by Sprengel pyknometer.

SOURCE AND PURITY OF MATERIALS:

(1) Nitrous oxide was prepared by heating pure ammonium nitrate at 523-513 K. It was passed through aqueous ferrous sulfate, aqueous sodium hydroxide, and concentrated sulfuric acid.

(2) Appeared to be of satisfactory purity.

ESTIMATED ERROR:

REFERENCES:

COMPONENTS:	ORIGINAL MEASUREMENTS:
1. Nitrous oxide; N_2O; [10024-97-2] 2. Water; H_2O; [7732-18-5] 3. Potassium nitrate; KNO_3; [7757-79-1]	Manchot, von W.; Jahrstorfer, M.; Zepter, H. *Z. Anorg. Allg. Chem.* 1924, *141*, 45-81.
VARIABLES:	PREPARED BY:
Concentration	C. L. Young

EXPERIMENTAL VALUES:

T/K	Density, d_4^{25} of salt soln.	Conc. of salt $/10^3$ mol m^{-3} (soln.)	Conc. of salt /mol kg^{-1} (water)	S_1 $/cm^3$	S_2 $/cm^3$
298.15	1.0586	1.02	1.0675	44.8	42.4
	1.1231	2.15	2.3738	38.3	34.1

S_1 - volume of nitrous oxide absorbed per 100 cm^3 of salt solution

S_2 - volume of nitrous oxide absorbed per 100 g of salt solution

Both S_1 and S_2 were reduced to conditions of 273.15 K and 101.3 kPa

AUXILIARY INFORMATION

METHOD/APPARATUS/PROCEDURE:

Measurement of volume of gas absorbed by means of gas buret and pipet. Volume absorbed appears to be taken as independent of pressure. (1). Density of the aqueous solution determined by Sprengel pyknometer.

SOURCE AND PURITY OF MATERIALS:

1. Prepared by heating ammonium nitrate, frozen in liquid air and then distilled.

2. Recrystallized.

ESTIMATED ERROR:

REFERENCES:

1. Manchot, W. *Z.Anorg.Chem.* 1924, *141*, 38.

COMPONENTS:

1. Nitrous oxide; N_2O; [10024-97-2]
2. Water; H_2O; [7732-18-5];
3. Potassium nitrate; KNO_3; [7757-79-1]

ORIGINAL MEASUREMENTS:

Markham, A. E.; Kobe, K. A.

J. Amer. Chem. Soc. 1941, *63*, 449-54.

VARIABLES:

T/K: 273.35-313.15
Salt molality/mol kg^{-1}: 0.1-8
Partial pressure of gas/atm: 1

PREPARED BY:

P. L. Long

EXPERIMENTAL VALUES:

T/K*	Salt molality/mol kg^{-1}*	Bunsen Coefficient, α*	'Solubility Coefficient', S*[a]
273.35	0	1.2970	1.2971
	0.2	1.2183	1.2267
	0.5	1.1355	1.1556
	1	1.0174	1.0545
298.15	0	0.5392	0.5408
	1	0.4552	0.4749
	2	0.3961	0.4299
	3	0.3524	0.3978
313.15	0	0.3579	0.3607
	1	0.3100	0.3254
	2	0.2761	0.3018
	3	0.2475	0.2814

* From the original data.

[a] The 'solubility coefficient', "refers to the volume of gas, reduced to standard conditions ... which is dissolved by the quantity of solution containing one gram of water".

AUXILIARY INFORMATION

METHOD/APPARATUS/PROCEDURE:

An Ostwald method was used. A known volume of gas was placed in contact with a known volume of gas-free liquid. After equilibrium was established by agitation, the volume of the remaining gas was measured from which the amount of gas dissolved was found. Solutions were prepared by weight, and verified by density measurements compared with the literature values. Apparatus consisted of an absorption flask with two bulbs, one twice the volume of the other, which was connected by a capillary tube to a gas buret and a manometer system.

SOURCE AND PURITY OF MATERIALS:

1. 99.7 per cent pure, no source.
2. Water was freshly boiled, distilled.
3. Analytical grade.

ESTIMATED ERROR:

$\delta T/K = \pm 0.1$ (273.35 K); ± 0.03 (above 273.35 K); $\delta\alpha/\alpha = \pm 0.02$.

REFERENCES:

COMPONENTS:	ORIGINAL MEASUREMENTS:
1. Nitrous oxide; N_2O; [10024-97-2] 2. Water; H_2O; [7732-18-5]; 3. Potassium nitrate; KNO_3; [7757-79-1]	Sada, E.; Ando, N.; Kito, S. *J. Appl. Chem. Biotechnol.* 1972, *22*, 1185-1193.

VARIABLES:	PREPARED BY:
T/K: 298.15 *P*/kPa: 101.325 (1 atm) Molarity of salt: mol l^{-1}	W. Gerrard/C. L. Young

EXPERIMENTAL VALUES:

T/K	Concn. of salt/mol l^{-1}	Bunsen coefficient, α
298.15	0	0.5512 (pure water)
	1.381	0.4433
	2.645	0.3626

AUXILIARY INFORMATION

METHOD/APPARATUS/PROCEDURE:

Equilibrium established between a measured volume of gas and a measured amount of gas-free liquid in a cell fitted with a magnetic stirrer. Details in source and ref. 1.

SOURCE AND PURITY OF MATERIALS:

1. From commercial cylinder; 99.8 per cent, as attested by gas chromatography.
2. Distilled water was used.
3. Salt was of reagent grade.

ESTIMATED ERROR:

$\delta T/K = \pm 0.2$; $\delta\alpha = \pm 2\%$ (estimated by compiler).

REFERENCES:

1. Onda, K.; Sada, E.; Kobayashi, T.; Kito, S.; Ito, K. *J. Chem. Engng. Japan* 1970, *3*, 18; 137.

COMPONENTS:	ORIGINAL MEASUREMENTS:
1. Nitrous oxide; N_2O; [10024-97-2] 2. Water; H_2O; [7732-18-5] 3. Potassium periodate; KIO_4; [7790-21-8]	Manchot, von W.; Jahrstorfer, M.; Zepter, H. *Z. Anorg. Allg. Chem.* 1924, *141*, 45-81.
VARIABLES:	PREPARED BY:
Concentration	C. L. Young

EXPERIMENTAL VALUES:

T/K	Density, d_4^{25} of acid soln.	Conc. of salt /10^3 mol m^{-3} (soln.)	Conc. of salt /mol kg^{-1} (water)	S_1 /cm^3	S_2 /cm^3
298.15	1.0008	0.0164	0.0164	52.1	52.1

S_1 - volume of nitrous oxide per 100 cm^3 of salt

S_2 - volume of nitrous oxide per 100 g of salt

Both S_1 and S_2 were reduced to conditions of 273.15 K and 101.3 kPa

AUXILIARY INFORMATION

METHOD/APPARATUS/PROCEDURE:

Measurement of volume of gas absorbed by means of gas buret and pipet. Volume absorbed appears to be taken as independent of pressure (1). Density of the aqueous solution determined by Sprengel pyknometer.

SOURCE AND PURITY OF MATERIALS:

1. Prepared by heating ammonium nitrate, frozen in liquid air and then distilled.

2. Recrystallized.

ESTIMATED ERROR:

REFERENCES:

1. Manchot, W. *Z.Anorg.Chem.* 1924, *141*, 38.

COMPONENTS:	ORIGINAL MEASUREMENTS:
1. Nitrous oxide; N_2O; [10024-97-2] 2. Water; H_2O; [7732-18-5] 3. Potassium carbonate; K_2CO_3; [584-08-7] 4. Potassium bicarbonate; $KHCO_3$; [298-14-6]	Joosten, G. E. H.; Danckwerts, P. V. *J. Chem. Engng. Data* 1972, *17*, 452-454.
VARIABLES:	PREPARED BY:
Concentration	C. L. Young

EXPERIMENTAL VALUES:

$T/K = 298.15$

Concentration of K_2CO_3 = Concentration of $KHCO_3$

Concentration of K_2CO_3/mol dm^{-3}	Solubility, $S \times 10^2$/mol dm^3 atm^{-1}
0.000	2.44
0.095	2.21
0.165	2.06
0.30	1.82
0.60	1.37
0.89	1.00
1.00	0.93
1.25	0.75

AUXILIARY INFORMATION

METHOD/APPARATUS/PROCEDURE:

Apparatus and procedure similar to that of Markham and Kobe (1). Few details given.

SOURCE AND PURITY OF MATERIALS:

No details given.

ESTIMATED ERROR:

$\delta T/K = \pm 0.02$; $\delta S/S = \pm 0.015$.

REFERENCES:

1. Markham, A. E.; Kobe, K. A. *J. Amer. Chem. Soc.* 1941, *63*, 449.

COMPONENTS:	ORIGINAL MEASUREMENTS:
1. Nitrous oxide; N_2O; [10024-97-2] 2. Water; H_2O; [7732-18-5]; 3. Rubidium chloride; RbCl; [7791-11-9]	Geffcken, G. *Z. Phys. Chem.* 1904, *49*, 257-302.
VARIABLES:	PREPARED BY:
Temperature, concentration	W. Gerrard/C. L. Young

EXPERIMENTAL VALUES:

T/K	Conc. of salt/mol dm^{-3} (soln.)	Ostwald coefficient, L
288.16	0.439	0.7050
	0.444	0.7053
	0.977	0.6306
	0.993	0.6276
298.16	0.439	0.5399
	0.444	0.5386
	0.977	0.4873
	0.993	0.4846

AUXILIARY INFORMATION

METHOD/APPARATUS/PROCEDURE:

Measurement of volume of N_2O absorbed by the aqueous solution. Detailed description and diagram given in source.

SOURCE AND PURITY OF MATERIALS:

1. Nitrous ocide self prepared and attested.

ESTIMATED ERROR:

REFERENCES:

COMPONENTS:

1. Nitrous oxide; N_2O; [10024-97-2]
2. Water; H_2O; [7732-18-5];
3. Caesium chloride; CsCl; [7647-17-8]

ORIGINAL MEASUREMENTS:

Geffcken, G. *Z. Phys. Chem.* 1904, *49*, 257-302.

VARIABLES:

Temperature, concentration

PREPARED BY:

W. Gerrard/C. L. Young

EXPERIMENTAL VALUES:

T/K	Conc. of salt/mol dm^{-3} (soln.)	Ostwald coefficient, L
288.16	0.514	0.7074
	0.545	0.7036
298.16	0.514	0.5428
	0.545	0.5406

AUXILIARY INFORMATION

METHOD/APPARATUS/PROCEDURE:

Measurement of volume of N_2O absorbed by the aqueous solution. Detailed description and diagram given in source.

SOURCE AND PURITY OF MATERIALS:

1. Nitrous oxide self prepared and attested.

ESTIMATED ERROR:

REFERENCES:

COMPONENTS:	ORIGINAL MEASUREMENTS:
1. Nitrous oxide; N_2O; [10024-47-2] 2. Water; H_2O; [7732-18-5] 3. Cesium chloride; CsCl; [7647-17-8]	Kreitus, I.; Abramenkov, A. *Latv. PRS Zinat. Akad. Vestis Khim. Ser.* 1980, 238.
VARIABLES:	PREPARED BY:
Concentration of salt	C. L. Young

EXPERIMENTAL VALUES:

T/K	Conc. of salt/mol kg^{-1} (water)	Bunsen coefficient, α
298.15	0	0.54
	0.50	0.49
	0.93	0.43
	2.45	0.32
	4.85	0.23
	7.25	0.14

AUXILIARY INFORMATION

METHOD APPARATUS/PROCEDURE:

Aqueous solution saturated at 0.1013 MPa and 1 cm^3 sample analyzed by gas chromatography. Column contained molecular sieves and was operated isothermally at 250 °C using helium as carrier gas. Thermal conductivity detector used.
Details in ref. (1).

SOURCE AND PURITY OF MATERIALS:

1. Medical grade sample.
2. Twice distilled.
3. Pure

ESTIMATED ERROR:

$\delta T/K = \pm 0.1$; $\delta\alpha/\alpha = \pm 0.06$

REFERENCES:

1. Kreitus, I.; Gorbovitskalya, T. I.
Latv. PRS Zinat. Akad. Vestis Khim. Ser.
1979, 664-666.

COMPONENTS:	EVALUATOR:
1. Nitrous oxide; N_2O; [10024-97-2] 2. Water; H_2O; [7732-18-5] 3. Weak electrolytes and Nonelectrolytes	Colin L. Young, School of Chemistry, University of Melbourne, Parkville, Victoria 3052, Australia. February 1981

CRITICAL EVALUATION:

The most extensive studies of these systems has been undertaken by Sada and coworker (1), (2), (3), (4) and (5). Comparison with data of other workers for different systems indicates that Sada's data are fairly reliable, hence all data given in references (1) to (5) are classified as tentative. Roth's data (6) on oxalic acid, glycerol and urea are of fairly low precision and are classified as doubtful. Roth's data on electrolyte solution also studied by other workers indicate that the data are not as reliable as most other more recent data (see Electrolyte solution evaluation).

The data of Knopp (7) appear to be of fairly good accuracy. The data for solubility in propanoic acid are consistent with Sada *et al.* data (3). Therefore Knopp's data are classified as tentative.

In general it is not possible to fit the solubility data for non-electrolyte or weak electrolyte solutions with equations of the Sechenow type. Markham and Kobe (8) suggested an alternative equation:

$$\frac{\alpha}{\alpha_o} = ac + \frac{1}{1 + bc}$$

where c is the molarity and a and b are constants for electrolyte solutions.

Values of a and b calculated from the measurements of Sada *et al.* (3) are given below.

	$a/dm^3\ mol^{-1}$	$b/dm^3\ mol^{-1}$
Formic acid	0.0716	0.0776
Acetic acid	0.1231	0.1298
Propanoic acid	0.1529	0.1523
Oxalic acid	0.0335	0.0683

References

1. Sada, E.; Kito, S.; Ito, Y. *Ind. Eng. Chem. Fundam.* 1975, *14*, 232.
2. Sada, E.; Kumazawa, H.; Butt, M. A. *J. Chem. Engng. Data* 1977, *22*, 277.
3. Sada, E.; Kito, S.; Ito, Y.; *J. Chem. Eng. Japan* 1974, *7*, 57.
4. Sada, E.; Kito, S. *Kagaku Kogaku* 1972, *36*, 218.
5. Sada, E.; Kumazawa, H.; Butt, M. A. *J. Chem. Engng. Data* 1978, *23*, 161.
6. Roth, W. *Z. Phys. Chem.* 1897, *24*, 114.
7. Knopp, W. *Z. Phys. Chem.* 1904, *48*, 97.
8. Markham, A. E.; Kobe, K. A. *J. Am. Chem. Soc.* 1941, *63*, 449.

COMPONENTS:	ORIGINAL MEASUREMENTS:
1. Nitrogen oxide (Nitrous oxide); N_2O; [10024-97-2]; 2. Water; H_2O; [7732-18-5]; 3. Methanol; CH_3OH; [67-56-1]	Sada, E.; Kito, S.; Ito, Y. *Ind. Eng. Chem. Fundam.* 1975, *14*, 232-237.
VARIABLES:	PREPARED BY:
Mole fraction of the alcohol	W. Gerrard/C. L. Young

EXPERIMENTAL VALUES:

T/K	Mole fraction of alcohol	Henry's law constant atm *	x_{N_2O} †	Mole fraction of alcohol	Henry's law constant arm *	x_{N_2O} †
298.16	0.0	2320.1	0.000431	0.398	921.6	0.001085
	0.022	2208.7	0.000453	0.408	892.8	0.00112
	0.048	2106.7	0.000475	0.435	827.1	0.00121
	0.054	2090.7	0.000478	0.595	515.7	0.00194
	0.060	2062.1	0.000485	0.670	420.8	0.00238
	0.077	2025.7	0.000494	0.690	401.1	0.00249
	0.095	1980.2	0.000505	0.770	327.7	0.00305
	0.140	1814.0	0.000551	0.826	295.6	0.00338
	0.155	1670.8	0.000600	0.870	255.2	0.00392
	0.175	1724.3	0.000580	0.928	222.1	0.00450
	0.197	1631.6	0.000613	0.932	219.6	0.00455
	0.204	1609.0	0.000621	1.0	190.6	0.00525
	0.208	(1951.1)#	0.000314			

* This Henry's law constant appears to have been derived by dividing the observed, but unspecified, pressure of N_2O in atm, by the mole fraction x_{N_2O} for that pressure.

† Calculated by the compiler by: 1/(Henry's law constant). The value refers to 1 atm (101.325 kPa).

\# Appears to be in error.

AUXILIARY INFORMATION

METHOD/APPARATUS/PROCEDURE:	SOURCE AND PURITY OF MATERIALS:
Equilibrium established between a measured volume of gas and a measured amount of gas-free liquid in a cell fitted with a magnetic stirrer. The densities of the mixed liquids were determined by an Ostwald-type pyknometer.	1. Nitrous oxide was used from a commercial cylinder (Japan), and stated to be of a purity better than 99.8%, as attested by gas-chromatography. 2. The water was carefully distilled. 3. The purity of the alcohol was stated to be satisfactory (ref. 2).
	ESTIMATED ERROR:
	REFERENCES:
	1. Onda, K.; Sada, E.; Kobayashi, T.; Kito, S.; Ito, K. *J. Chem. Eng. Japan* 1970, *3*, 18; 137. 2. Sada, E.; Kito, S.; Ito, Y. *J. Chem. Eng. Japan* 1974, *7*, 57.

COMPONENTS:

1. Nitrogen oxide (Nitrous oxide); N_2O; [10024-97-2];
2. Water; H_2O; [7732-18-5];
3. Ethanol; C_2H_6O; [64-17-5]

ORIGINAL MEASUREMENTS:

Sada, E.; Kito, S.; Ito, Y. *Ind. Eng. Chem. Fundam.* 1975, *14*, 232-237.

VARIABLES:

Mole fraction of the alcohol

PREPARED BY:

W. Gerrard/C. L. Y oung

EXPERIMENTAL VALUES:

T/K	Mole fraction of alcohol	Henry's law constant atm *	x_{N_2O} †	Mole fraction of alcohol	Henry's law constant atm *	x_{N_2O} †
298.16	0.0	2320.1	0.000431	0.442	539.1	0.00185
	0.023	2201.8	0.000454	0.537	437.8	0.00228
	0.036	2155.6	0.000469	0.651	303.6	0.00329
	0.058	2102.3	0.000476	0.731	256.6	0.00390
	0.077	2083.1	0.000480	0.783	216.5	0.00462
	0.130	1889.7	0.000529	0.797	218.5	0.00458
	0.192	1500.8	0.000666	0.853	193.3	0.00517
	0.197	1458.6	0.000686	0.868	189.6	0.00527
	0.251	1127.7	0.000887	0.978	151.1	0.00662
	0.349	761.1	0.001314	1.0	145.8	0.00686

* This Henry's law constant appears to have been derived by dividing the observed, but unspecified, pressure of N_2O in atm, by the mole fraction, x_{N_2O} for that pressure.

† Calculated by the compiler by: 1/(Henry's law constant). The value refers to 1 atm (101.325 kPa).

AUXILIARY INFORMATION

METHOD /APPARATUS/PROCEDURE:

A gas volumetric method (ref. 1) was used. The densities of the mixture of liquids were determined by an Ostwald-type pyknometer. Equilibrium established between a measured volume of gas and a measured amount of gas-free liquid in a cell fitted with a magnetic stirrer.

SOURCE AND PURITY OF MATERIALS:

1. Nitrous oxide was used from a commercial cylinder (Japan), and stated to be of a purity better than 99.8%, as attested by gas-chromatography.
2. The water was carefully distilled.
3. The purity of the alcohol was stated to be satisfactory (ref. 2).

ESTIMATED ERROR:

REFERENCES:

1. Onda, K.; Sada, E.; Kobayashi, T.; Kito, S.; Ito, K. *J. Chem. Eng. Japan* 1970, *3*, 18; 137.
2. Sada, E.; Kito, S.; Ito, Y. *J. Chem. Eng. Japan* 1974, *7*, 57.

COMPONENTS:

1. Nitrogen oxide (Nitrous oxide); N_2O; [10024-97-2];
2. Water; H_2O; [7732-18-5];
3. 1-Propanol; C_3H_8O; [71-23-8]

ORIGINAL MEASUREMENTS:

Sada, E.; Kito, S.; Ito, Y.; *Ind. Eng. Chem. Fundam.* 1975, *14*, 232-237.

VARIABLES:

Mole fraction of the alcohol

PREPARED BY:

W. Gerrard/C. L. Young

EXPERIMENTAL VALUES:

T/K	Mole fraction of alcohol	Henry's law constant atm *	x_{N_2O} †	Mole fraction of alcohol	Henry's law constant atm *	x_{N_2O} †
298.16	0.0	2320.1	0.000431	0.517	302.1	0.00331
	0.040	2131.6	0.000469	0.655	224.6	0.00445
	0.085	1809.6	0.000553	0.707	204.5	0.00489
	0.167	1053.5	0.000949	0.852	160.4	0.00623
	0.267	636.5	0.00157	1.0	125.8	0.00795

* This Henry's law constant appears to have been derived by dividing the observed, but unspecified, pressure of N_2O in atm, by the mole fraction, x_{N_2O} for that pressure.

† Calculated by the compiler by: 1/(Henry's law constant). The value refers to 1 atm (101.325 kPa).

AUXILIARY INFORMATION

METHOD/APPARATUS/PROCEDURE:

A gas volumetric method (ref. 1) was used. The densities of the mixture of liquids were determined by an Ostwald-type pyknometer. Equilibrium established between a measured volume of gas and a measured amount of gas-free liquid in a cell fitted with a magnetic stirrer.

SOURCE AND PURITY OF MATERIALS:

1. Nitrous oxide was used from a commercial cylinder (Japan), and stated to be of a purity better than 99.8%, as attested by gas-chromatography.
2. The water was carefully distilled.
3. The purity of the alcohol was stated to be satisfactory (ref. 2).

ESTIMATED ERROR:

REFERENCES:

1. Onda, K.; Sada, E.; Kobayashi, T.; Kito, S.; Ito, K. *J. Chem. Eng. Japan* 1970, *3*, 18; 137.
2. Sada, E.; Kito, S.; Ito, Y. *J. Chem. Eng. Japan* 1974, *7*, 57.

COMPONENTS:	ORIGINAL MEASUREMENTS:
1. Nitrous oxide; N_2O; [10024-97-2] 2. Water; H_2O; [7732-18-5] 3. 1-Propanol; C_3H_8O; [71-23-8]	Laddha, S. S.; Diaz, J. M.; Danckwerts, P. V. *Chem. Eng. Sci.* 1981, *36*, 228-229.
VARIABLES:	PREPARED BY:
Composition of liquid	C. L. Young

EXPERIMENTAL VALUES:

T/K	Mole fraction of component 3	10^{10} Solubility, S /mol cm^{-3} Pa^{-1}
298.2	0.02	2.00
	0.04	1.68
	0.06	1.53

AUXILIARY INFORMATION

METHOD APPARATUS/PROCEDURE:	SOURCE AND PURITY OF MATERIALS:
Apparatus consisted of two vessels of about 600 cm^3 capacity. Each vessel filled with gas at a pressure "somewhat less than atmospheric". A measured volume of water admitted to one vessel and an equal amount of mixture added to other vessel. Liquids stirred and pressure difference between flasks measured. From this measurement and a knowledge of the absolute pressures and exact volume of each vessel, it was possible to establish the solubility. Corrections were made for partial pressure of liquid.	No details given. ESTIMATED ERROR: $\delta T/K = \pm 0.1$; $\delta S_{N_2O} = \pm 3\%$ (estimated by compiler). REFERENCES:

COMPONENTS:

1. Nitrogen oxide (Nitrous oxide); N_2O; [10024-97-2];
2. Water; H_2O; [7732-18-5];
3. 2-Propanol; C_3H_8O; [67-63-0]

ORIGINAL MEASUREMENTS:

Sada, E.; Kito, S.; Ito, Y.
Ind. Eng. Chem. Fundam. 1975, *14*, 232-237.

VARIABLES:

Mole fraction of the alcohol

PREPARED BY:

W. Gerrard/ C. L. Young

EXPERIMENTAL VALUES:

T/K	Mole fraction of alcohol	Henry's law constant, atm *	x_{N_2O} †	Mole fraction of alcohol	Henry's law constant, atm *	x_{N_2O} †
298.16	0.0	2320.1	0.000431	0.338	524.4	0.00191
	0.008	2273.3	0.000440	0.433	382.0	0.00262
	0.028	2230.0	0.000448	0.572	266.3	0.00376
	0.051	2237.7	0.000447	0.708	201.0	0.00498
	0.070	2164.7	0.000462	0.760	182.9	0.00547
	0.076	2096.6	0.000477	0.865	148.5	0.00673
	0.140	1471.3	0.000680	1.0	125.8	0.00795
	0.267	718.1	0.00139			

* This Henry's law constant appears to have been derived by dividing the observed, but unspecified, pressure of N_2O in atm, by the mole fraction, x_{N_2O} for that pressure.

† Calculated by the compiler by: 1/(Henry's law constant). The value refers to 1 atm (101.325 kPa).

AUXILIARY INFORMATION

METHOD/APPARATUS/PROCEDURE:

A gas volumetric method (ref. 1) was used. The densities of the mixture of liquids were determined by an Ostwald-type pyknometer. Equilibrium established between a measured volume of gas and a measured amount of gas-free liquid in a cell fitted with a magnetic stirrer.

SOURCE AND PURITY OF MATERIALS:

1. Nitrous oxide was used from a commercial cylinder (Japan), and stated to be of a purity better than 99.8%, as attested by gas-chromatography.
2. The water was carefully distilled.
3. The purity of the alcohol was stated to be satisfactory (ref. 2).

ESTIMATED ERROR:

REFERENCES:

1. Onda, K.; Sada, E.; Kobayashi, T. Kito, S.; Ito, K. *J. Chem. Eng. Japan* 1970, *3*, 18; 137.
2. Sada, E.; Kito, S.; Ito, Y. *J. Chem. Eng. Japan* 1974, *7*, 57.

COMPONENTS:	ORIGINAL MEASUREMENTS:
1. Nitrous oxide; N_2O; [10024-97-2] 2. Water; H_2O; [7732-18-5] 3. 1,2-Ethanediol; $C_2H_6O_2$; [107-21-1]	Laddha, S. S.; Diaz, J. M.; Danckwerts, P. V. *Chem. Eng. Sci.* 1981, *36*, 228-229.
VARIABLES:	PREPARED BY:
Composition of liquid	C. L. Young

EXPERIMENTAL VALUES:

T/K	Mole fraction of component 3	10^{10} Solubility, S /mol cm^{-3} Pa^{-1}
298.2	0.02	2.29
	0.04	2.22
	0.06	2.13

AUXILIARY INFORMATION

METHOD APPARATUS/PROCEDURE:

Apparatus consisted of two vessels of about 600 cm^3 capacity. Each vessel filled with gas at a pressure "somewhat less than atmospheric". A measured volume of water admitted to one vessel and an equal amount of mixture added to other vessel. Liquids stirred and pressure difference between flasks measured. From this measurement and a knowledge of the absolute pressures and exact volume of each vessel, it was possible to establish the solubility. Corrections were made for partial pressure of liquid.

SOURCE AND PURITY OF MATERIALS:

No details given.

ESTIMATED ERROR:

$\delta T/K = \pm 0.1$; $\delta S_{N_2O} = \pm 3\%$ (estimated by compiler).

REFERENCES:

COMPONENTS:	ORIGINAL MEASUREMENTS:
1. Nitrous oxide; N_2O; [10024-97-2] 2. Water; H_2O; [7732-18-5] 3. 1,2,3-Propanetriol; $C_3H_8O_3$; [56-81-5]	Laddha, S. S.; Diaz, J. M. Danckwerts, P. V. *Chem. Eng. Sci.* 1981, *36*, 228-229.
VARIABLES:	PREPARED BY:
Composition of liquid	C. L. Young

EXPERIMENTAL VALUES:

T/K	Mole fraction of component 3	10^{10} Solubility, S /mol cm^{-3} Pa^{-1}
298.2	0.02	2.18
	0.04	2.05
	0.06	1.88

AUXILIARY INFORMATION

METHOD APPARATUS/PROCEDURE:

Apparatus consisted of two vessels of about 600 cm^3 capacity. Each vessel filled with gas at a pressure "somewhat less than atmospheric". A measured volume of water admitted to one vessel and an equal amount of mixture added to other vessel. Liquids stirred and pressure difference between flasks measured. From this measurement and a knowledge of the absolute pressures and exact volume of each vessel, it was possible to establish the solubility. Corrections were made for partial pressure of liquid.

SOURCE AND PURITY OF MATERIALS:

No details given.

ESTIMATED ERROR:

$\delta T/K = \pm 0.1$; $\delta S_{N_2O} = \pm 3\%$ (estimated by compiler).

REFERENCES:

COMPONENTS:	ORIGINAL MEASUREMENTS:
1. Nitrogen oxide (Nitrous oxide); N_2O; [10024-97-2] 2. Water; H_2O; [7732-18-5]; 3. 1,2,3-Propanetriol (Glycerol); $(CH_2OH)_2CHOH$; [56-81-5]	Roth, W. *Z. Phys. Chem.* 1897, *24*, 114-151.
VARIABLES: Temperature, concentration	PREPARED BY: W. Gerrard

EXPERIMENTAL VALUES: t = T/K - 273.16; density = d; α = Bunsen coefficient.

Concn. of glycerol, %: 3.376
Change of d with t: $d = 1.00858 - 0.000007042t - 0.000005447t^2$
Change of α with t: $\alpha = 1.3205 - 0.045803t + 0.0006068t^2$

T/K	298.29	293.23	288.09	283.47	278.51
α	0.5527	0.6451	0.7719	0.9041	1.0928

Concn. of glycerol, %: 3.544
Change of d with t: $d = 1.00895 - 0.000004004t - 0.000005582t^2$
Change of α with t: $\alpha = 1.2906 - 0.043676t + 0.0005731t^2$

T/K	298.26	293.37	288.31	283.27	278.19
α	0.5553	0.6429	0.7604	0.8984	1.0854

Concn. of glycerol, %: 6.338
Change of d with t: $d = 1.01611 - 0.00003711t - 0.0000052202t^2$
Change of α with t: $\alpha = 1.2709 - 0.04242t + 0.00053176t^2$

T/K	298.40	293.24	288.31	283.23	278.19
α	0.5390	0.6347	0.7503	0.8932	1.0710

Concn. of glycerol, %: 7.114
Change of d with t: $d = 1.01810 - 0.000046138t - 0.0000050224t^2$
Change of α with t: $\alpha = 1.2285 - 0.040685t + 0.00052632t^2$

T/K	298.08	293.17	288.21	283.10	278.12
α	0.5415	0.6242	0.7354	0.8661	1.0396

(continued)

AUXILIARY INFORMATION

METHOD/APPARATUS/PROCEDURE:	SOURCE AND PURITY OF MATERIALS:
Ostwald method, using gas buret and pipet. Measurement of volume of gas before and after absorption. Specific gravity of solution was determined by a Sprengel pyknometer. Vapour pressure of water, adjusted by assuming Raoult's law, was allowed for.	1. N_2O was self prepared and purified. 2. Glycerol was analysed by an appropriate method.
	ESTIMATED ERROR:
	REFERENCES:

COMPONENTS:	ORIGINAL MEASUREMENTS:
1. Nitrogen oxide (Nitrous oxide); N_2O; [10024-97-2]; 2. Water; H_2O; [7732-18-5]; 3. 1,2,3-Propanetriol (Glycerol); $(CH_2OH)_2CHOH$; [56-81-5]	Roth, W. *Z. Phys. Chem.* 1897, *24*, 114-151.

EXPERIMENTAL VALUES:

t = T/K - 273.16; density = *d*; α = Bunsen coefficient.

Concn. of glycerol, %: 11.483
Change of *d* with t: $d = 1.02926 - 0.000084093t - 0.0000047447t^2$
Change of α with t: $\alpha = 1.1837 - 0.04000t + 0.00054926t^2$

T/K	298.04	293.14	288.23	283.18	277.90
α	0.5285	0.6029	0.7056	0.8336	1.0064

Concn. of glycerol, %: 12.756
Change of *d* with t: $d = 1.03259 - 0.000094282t - 0.0000047244t^2$
Change of α with t: $\alpha = 1.1833 - 0.03911t + 0.00051373t^2$

T/K	298.24	293.24	288.17	282.83	278.06
α	0.5255	0.6056	0.7120	0.8493	1.0040

Concn. of glycerol, %: 16.175
Change of *d* with t: $d = 1.04145 - 0.00012475t - 0.000004543t^2$
Change of α with t: $\alpha = 1.1375 - 0.036345t + 0.00044917t^2$

T/K	298.14	293.24	288.13	283.00	277.98
α	0.5099	0.5916	0.6941	0.8199	0.9728

Concn. of glycerol, %: 16.313
Change of *d* with t: $d = 1.04192 - 0.0001389t - 0.000004059t^2$
Change of α with t: $\alpha = 1.1243 - 0.037362t + 0.0005068t^2$

T/K	298.14	293.16	288.31	283.30	277.81
α	0.5073	0.5768	0.6746	0.7953	0.9615

COMPONENTS:	ORIGINAL MEASUREMENTS:
1. Nitrous oxide; N_2O; [10024-97-2] 2. Water; H_2O; [7732-18-5] 3. 2,2'-Oxybisethanol (Diethylene glycol); $C_4H_{10}O_3$; [111-46-6]	Laddha, S. S.; Diaz, J. M.; Danckwerts, P. V. *Chem. Eng. Sci.* 1981, *36*, 228-229.
VARIABLES:	PREPARED BY:
Composition of liquid	C. L. Young

EXPERIMENTAL VALUES:

T/K	Mole fraction of component 3	10^{10} Solubility, S /mol cm^{-3} Pa^{-1}
298.2	0.02	2.27
	0.04	2.18
	0.06	2.09

AUXILIARY INFORMATION

METHOD APPARATUS/PROCEDURE:

Apparatus consisted of two vessels of about 600 cm^3 capacity. Each vessel filled with gas at a pressure "somewhat less than atmospheric". A measured volume of water admitted to one vessel and an equal amount of mixture added to other vessel. Liquids stirred and pressure difference between flasks measured. From this measurement and a knowledge of the absolute pressures and exact volume of each vessel, it was possible to establish the solubility. Corrections were made for partial pressure of liquid.

SOURCE AND PURITY OF MATERIALS:

No details given.

ESTIMATED ERROR:

$\delta T/K = \pm 0.1$; $\delta S_{N_2O} = \pm 3\%$ (estimated by compiler).

REFERENCES:

COMPONENTS:

1. Nitrous oxide; N_2O; [10024-97-2]
2. Water; H_2O; [7732-18-5]
3. 1,5-Pentanediol; $C_5H_{12}O_2$; [111-29-5]

ORIGINAL MEASUREMENTS:

Laddha, S. S.; Diaz, J. M.; Danckwerts, P. C.
Chem. Eng. Sci.
1981, *36*, 228-229.

VARIABLES:

Composition of liquid

PREPARED BY:

C. L. Young

EXPERIMENTAL VALUES:

T/K	Mole fraction of component 3	10^{10} Solubility, S /mol cm^{-3} Pa^{-1}
298.2	0.02	2.32
	0.04	2.26
	0.06	2.21

AUXILIARY INFORMATION

METHOD APPARATUS/PROCEDURE:

Apparatus consisted of two vessels of about 600 cm^3 capacity. Each vessel filled with gas at a pressure "somewhat less than atmospheric". A measured volume of water admitted to one vessel and an equal amount of mixture added to other vessel. Liquids stirred and pressure difference between flasks measured. From this measurement and a knowledge of the absolute pressures and exact volume of each vessel, it was possible to establish the solubility. Corrections were made for partial pressure of liquid.

SOURCE AND PURITY OF MATERIALS:

No details given.

ESTIMATED ERROR:

$\delta T/K = \pm 0.1$; $\delta S_{N_2O} = \pm 3\%$ (estimated by compiler).

REFERENCES:

COMPONENTS:

1. Nitrogen oxide (Nitrous oxide); N_2O; [10024-97-2];
2. Water; H_2O; [7732-18-5];
3. Formic acid; CH_2O_2; [64-18-6]

ORIGINAL MEASUREMENTS:

Sada, E.; Kito, S.; Ito, Y.
J. Chem. Eng. Japan 1974, *7*, 57-59.

VARIABLES:

T/K: 298.15 *P*/kPa: 101.325 (1 atm)
Molarity of acid: mol l^{-1}

PREPARED BY:

W. Gerrard/C. L. Young

EXPERIMENTAL VALUES:

Pressure = 101.325 kPa

T/K	Molarity of acid mol l^{-1}	Bunsen Coefficient α	T/K	Molarity of acid mol l^{-1}	Bunsen Coefficient α
298.15	0	0.5512	298.15	0	0.5512
	0.6234	0.5481		2.4451	0.5579
	1.3125	0.5529		2.8999	0.5653
	1.4033	0.5542			

AUXILIARY INFORMATION

METHOD/APPARATUS/PROCEDURE:

Equilibrium established between a measured volume of gas and a measured amount of gas-free liquid in a cell fitted with a magnetic stirrer. Concentration of the organic acid was determined by a volumetric titration with sodium hydroxide to a phenolphthalein end-point. Details of apparatus and procedure in ref. 1.

SOURCE AND PURITY OF MATERIALS:

1. High purity nitrous oxide was used; supplied by Showa Denko Co. Ltd.; attested to be 99.8 per cent by gas chromatography.
2. Distilled and degassed; *d* and n_D given.
3. Reagent grade was used; *d* and n_D given.

ESTIMATED ERROR:

$\delta T/K = \pm 0.2$; $\delta\alpha = \pm 2\%$ (estimated by compiler).

REFERENCES:

1. Onda, K.; Sada, E.; Kobayashi, T.; Kito, S.; Ito, K. *J. Chem. Eng. Japan* 1970, *3*, 18; 137.

COMPONENTS:	ORIGINAL MEASUREMENTS:
1. Nitrogen oxide (Nitrous oxide); N_2O; [10024-97-2] 2. Water; H_2O; [7732-18-5] 3. Ethanedioic acid (Oxalic acid); $(COOH)_2$; [144-62-7]	Roth, W. *Z. Phys. Chem.* 1897, *24*, 114-151.
VARIABLES:	PREPARED BY:
Temperature, concentration	W. Gerrard

EXPERIMENTAL VALUES: $t = T/K - 273.16$; density = d; α = Bunsen coefficient.

Concn. of oxalic acid, %: 0.7746
Change of d with t: $d = 1.00440 - 0.000001414t - 0.000005449t^2$
Change of α with t: $\alpha = 1.3667 - 0.048565t + 0.0006894t^2$

T/K	298.33	293.26	288.29	283.28	278.29
α	0.5810	0.6714	0.7897	0.9465	1.1357

Concn. of oxalic acid, %: 0.8497
Change of d with t: $d = 1.00453 + 0.00003186t - 0.0000064535t^2$
Change of α with t: $\alpha = 1.3759 - 0.048714t + 0.00066483t^2$

T/K	298.04	293.33	288.21	283.10	278.10
α	0.5754	0.6719	0.7933	0.9509	1.1514

Concn. of oxalic acid, %: 3.326
Change of d with t: $d = 1.01754 - 0.000047524t - 0.000005253t^2$
Change of α with t: $\alpha = 1.3178 - 0.046596t + 0.0006572t^2$

T/K	298.27	293.21	288.25	283.28	278.73
α	0.5621	0.6482	0.7642	0.9130	1.0787

Concn. of oxalic acid, %: 3.640
Change of d with t: $d = 1.01911 - 0.000056536t - 0.0000050445t^2$
Change of α with t: $\alpha = 1.3338 - 0.046913t + 0.0006380t^2$

T/K	298.25	293.37	288.41	283.29	278.17
α	0.5584	0.6493	0.7668	0.9309	1.1148

(continued)

AUXILIARY INFORMATION

METHOD /APPARATUS/PROCEDURE:

Ostwald method, using gas buret and pipet. Measurement of volume of gas before and after absorption. Specific gravity of solution was determined by a Sprengel pyknometer. Vapour pressure of water, adjusted by assuming Raoult's law, was allowed for.

SOURCE AND PURITY OF MATERIALS:

1. N_2O was self prepared and purified.
2. Oxalic acid was analysed by an appropriate method.

ESTIMATED ERROR:

REFERENCES:

COMPONENTS:	ORIGINAL MEASUREMENTS:
1. Nitrogen oxide (Nitrous oxide); N_2O; [10024-97-2] 2. Water; H_2O; [7732-18-5] 3. Ethanedioic acid (Oxalic acid); $(COOH)_2$; [144-62-7]	Roth, W. *Z. Phys. Chem.* 1897, *24*, 114-151.

EXPERIMENTAL VALUES:

$t = T/K - 273.16$; density = d; α = Bunsen coefficient.

Concn. of oxalic acid, %: 4.130
Change of d with t: $d = 1.02164 - 0.000058537t - 0.0000052552t^2$
Change of α with t: $\alpha = 1.3189 - 0.044307t + 0.00057285t^2$

T/K	298.21	293.30	288.25	283.24	278.15
α	0.5685	0.6615	0.7808	0.9247	1.1121

COMPONENTS:

1. Nitrogen oxide (Nitrous oxide); N_2O; [10024-97-2];
2. Water; H_2O; [7732-18-5];
3. Ethanedioic acid, (Oxalic acid); $C_2H_2O_4$; [144-62-7]

ORIGINAL MEASUREMENTS:

Sada, E.; Kito, S.; Ito, Y.

J. Chem. Eng. Japan 1974, *7*, 57-59.

VARIABLES:

T/K: 298.15 *P*/kPa: 101.325 (1 atm)

Molarity of acid: mol l^{-1}

PREPARED BY:

W. Gerrard/C. L. Young

EXPERIMENTAL VALUES:

Pressure = 101.325 kPa

T/K	Molarity of acid mol l^{-1}	Bunsen Coefficient α
298.15	0	0.5512
	0.2091	0.5471
	0.2433	0.5468
	0.5227	0.5418
	0.8434	0.5380
	0.8567	0.5353

AUXILIARY INFORMATION

METHOD/APPARATUS/PROCEDURE:

Equilibrium established between a measured volume of gas and a measured amount of gas-free liquid in a cell fitted with a magnetic stirrer. Concentration of the organic acid was determined by a volumetric titration with sodium hydroxide to a phenolphthalein end-point. Details of apparatus and procedure in ref. 1.

SOURCE AND PURITY OF MATERIALS:

1. High purity nitrous oxide was used; supplied by Showa Denko Co. Ltd.; attested to be 99.8 per cent by gas chromatography.
2. Distilled and degassed; *d* and n_D given.
3. Reagent grade was used; *d* and n_D given.

ESTIMATED ERROR:

$\delta T/K = \pm 0.2$; $\delta\alpha = \pm 2\%$ (estimated by compiler).

REFERENCES:

1. Onda, K.; Sada, E.; Kobayashi, T.; Kito, S.; Ito, K. *J. Chem. Eng. Japan* 1970, *3*, 18; 137.

COMPONENTS:	ORIGINAL MEASUREMENTS:
1. Nitrogen oxide (Nitrous oxide); N_2O; [10024-97-2] 2. Water; H_2O; [7732-18-5] 3. Acetic acid; $C_2H_4O_2$; [64-19-7]	Sada, E.; Kito, S.; Ito, Y. *J. Chem. Eng. Japan* 1974, *7*, 57-59.

VARIABLES:	PREPARED BY:
T/K: 298.15 *P*/kPa: 101.325 (1 atm) Molarity of acid: mol l^{-1}	W. Gerrard/C. L. Young

EXPERIMENTAL VALUES:

Pressure=101.325 kPa

T/K	Molarity of acid mol l^{-1}	Bunsen Coefficient α	T/K	Molarity of acid mol l^{-1}	Bunsen Coefficient α
298.15	0	0.5512	298.15	0	0.5512
	0.3096	0.5476		2.2750	0.5815
	0.5012	0.5472		2.6174	0.5897
	0.5466	0.5458		2.6975	0.5899
	0.6188	0.5472		3.2019	0.6071
	1.1204	0.5610		3.5392	0.6193
	1.8731	0.5742		3.7865	0.6227
	2.2123	0.5814			

AUXILIARY INFORMATION

METHOD /APPARATUS/PROCEDURE:

Equilibrium established between a measured volume of gas and a measured amount of gas-free liquid in a cell fitted with a magnetic stirrer. Concentration of the organic acid was determined by a volumetric titration with sodium hydroxide to a phenolphthalein end-point. Details of apparatus and procedure in ref. 1.

SOURCE AND PURITY OF MATERIALS:

1. High purity nitrous oxide was used; supplied by Showa Denko Co. Ltd.; attested to be 99.8 mole per cent by gas chromatography.
2. Distilled and degassed; *d* and n_D given.
3. Reagent grade was used; *d* and n_D given.

ESTIMATED ERROR:

$\delta T/K = \pm 0.2$; $\delta\alpha = \pm 2\%$ (estimated by compiler).

REFERENCES:

1. Onda, K.; Sada, E.; Kobayashi, T.; Kito, S.; Ito, K. *J. Chem. Eng. Japan* 1970, *3*, 18; 137.

COMPONENTS:

1. Nitrogen oxide; (Nitrous oxide); N_2O; [10024-97-2]

2. Water; H_2O; [7732-18-5]

3. Propanoic acid (Propionic acid); $C_3H_6O_2$; [79-09-4]

ORIGINAL MEASUREMENTS:

Knopp, W.

Z. Phys. Chem. 1904, *48*, 97-108

VARIABLES:

PREPARED BY:

C.L. Young

EXPERIMENTAL VALUES:

Pressure assumed to be 101.325 kPa.

T/K	Weight of acid in 100 g of solution.	Conc of salt /mol dm^3 (soln)	Density of solution	Bunsen absorption coefficient, α
293.16	1.492	0.2045	0.99964	0.6323
	5.702	0.816	1.00349	0.6369
	13.680	2.140	1.01061	0.6504
	15.011	2.385	1.01190	0.6534
	25.589	4.645	1.01933	0.7219
			Water	0.6270

Calculated by compiler: Mole fraction, x_1, of N_2O in water: 0.000506. Mole fraction of N_2O in the solution containing 4.645 moles of propionic acid in 1 dm^3 of solution : 0.000768.

AUXILIARY INFORMATION

METHOD /APPARATUS/PROCEDURE:

An absorption pipet and gas buret were used to measure the volume of gas absorbed.

Densities were determined by a Sprengel pyknometer.

SOURCE AND PURITY OF MATERIALS:

1. Nitrous oxide was prepared by heating pure ammonium nitrate at 513-523K. It was passed through aqueous ferrous sulfate, aqueous sodium hydroxide, and concentrated sulfuric acid.

3. Analytically attested.

ESTIMATED ERROR:

REFERENCES:

COMPONENTS:	ORIGINAL MEASUREMENTS:
1. Nitrogen oxide (Nitrous oxide); N_2O; [10024-97-2] 2. Water; H_2O; [7732-18-5] 3. Propanoic acid, (Propionic acid); $C_3H_6O_2$; [79-09-4]	Sada, E.; Kito, S.; Ito, Y. *J. Chem. Eng. Japan* 1974, *7*, 57-59.
VARIABLES:	PREPARED BY:
T/K: 298.15 *P*/kPa: 101.325 (1 atm) Molarity of acid: mol l^{-1}	W. Gerrard/C. L. Young

EXPERIMENTAL VALUES:

Pressure = 101.325 kPa

T/K	Molarity of acid mol l^{-1}	Bunsen Coefficient α
298.15	0	0.5512
	0.4391	0.5583
	0.8473	0.5612
	1.3523	0.5711
	2.3654	0.5985
	3.2994	0.6481

AUXILIARY INFORMATION

METHOD/APPARATUS/PROCEDURE:

Equilibrium established between a measured volume of gas and a measured amount of gas-free liquid in a cell fitted with a magnetic stirrer. Concentration of the organic acid was determined by a volumetric titration with sodium hydroxide to a phenolphthalein end-point. Details of apparatus and procedure in ref. 1.

SOURCE AND PURITY OF MATERIALS:

1. High purity nitrous oxide was used; supplied by Showa Denko Co. Ltd.; attested to be 99.8 mole per cent by gas chromatography.
2. Distilled and degassed; *d* and n_D given.
3. Reagent grade was used; *d* and n_D given.

ESTIMATED ERROR:

$\delta T/K = \pm 0.2$; $\delta\alpha = \pm 2\%$ (estimated by compiler).

REFERENCES:

1. Onda, K.; Sada, E.; Kobayashi, T.; Kito, S.; Ito, K. *J. Chem. Eng. Japan* 1970, *3*, 18; 137.

COMPONENTS:

1. Nitrogen oxide (Nitrous oxide); N_2O; [10024-97-2];
2. Water; H_2O; [7732-18-5];
3. Urea; $CO(NH_2)_2$; [57-13-6]

ORIGINAL MEASUREMENTS:

Roth, W.

Z. Phys. Chem. 1897, *24*, 114-151.

VARIABLES:

Temperature, concentration

PREPARED BY:

W. Gerrard

EXPERIMENTAL VALUES: $t = T/K - 273.16$; density = d; α = Bunsen coefficient.

Concn. of urea, %: 3.288
Change of d with t: $d = 1.01013 - 0.000058765t - 0.0000045803t^2$
Change of α with t: $\alpha = 1.3252 - 0.046462t + 0.0006493t^2$

T/K	297.56	293.17	287.75	282.81	280.78	278.10
α	0.5781	0.6591	0.7855	0.9369	1.0092	1.1114

Concn. of urea, %: 3.336
Change of d with t: $d = 1.01059 - 0.00007987t - 0.000003545t^2$
Change of α with t: $\alpha = 1.3141 - 0.04637t + 0.00066066t^2$

T/K	297.52	293.20	287.70	282.70	278.40
α	0.5765	0.6501	0.7806	0.9276	1.0892

Concn. of urea, % = 4.670
Change of d with t: $d = 1.01458 - 0.00011876t - 0.000003346t^2$
Change of α with t: $\alpha = 1.2920 - 0.04236t + 0.0005399t^2$

T/K	298.04	287.84	277.81
α	0.5721	0.7864	1.1066

Concn. of urea, % = 4.963
Change of d with t: $d = 1.01513 - 0.00007098t - 0.000004811t^2$
Change of α with t: $\alpha = 1.2927 - 0.04387t + 0.00057684t^2$

T/K	297.64	293.20	287.82	282.83	278.84
α	0.5643	0.6516	0.7734	0.9177	1.0621

(continued)

AUXILIARY INFORMATION

METHOD /APPARATUS/PROCEDURE:

Ostwald method, using gas buret and pipet. Measurement of volume of gas before and after absorption. Specific gravity of solution was determined by a Sprengel pyknometer. Vapour pressure of water, adjusted by assuming Raoult's law, was allowed for.

SOURCE AND PURITY OF MATERIALS:

1. N_2O was self prepared and purified.

3. Urea was analysed by an appropriate method.

ESTIMATED ERROR:

REFERENCES:

COMPONENTS:	ORIGINAL MEASUREMENTS:
1. Nitrogen oxide (Nitrous oxide); N_2O; [10024-97-2]; 2. Water; H_2O; [7732-18-5]; 3. Urea; $CO(NH_2)_2$; [57-13-6]	Roth, W. *Z. Phys. Chem.* 1897, *24*, 114-151.

EXPERIMENTAL VALUES:

t = T/K - 273.16; density = *d*; α = Bunsen coefficient.

Concn. of urea, %: 5.288
Change of *d* with t: $d = 1.01640 - 0.00007353t - 0.000004942t^2$
Change of α with t: $\alpha = 1.3217 - 0.046054t + 0.0006497t^2$

T/K	297.75	293.14	287.80	282.69	278.69	276.70
α	0.5846	0.6699	0.7867	0.9350	1.0788	1.1668

Concn. of urea, %: 6.249
Change of *d* with t: $d = 1.01931 - 0.0001160t - 0.000003861t^2$
Change of α with t: $\alpha = 1.3063 - 0.045862t + 0.00063275t^2$

T/K	297.21	292.19	287.53	283.61
α	0.5693	0.6627	0.7811	0.8962

Concn. of urea, %: 6.483
Change of *d* with t: $d = 1.01964 - 0.00010387t - 0.0000042285t^2$
Change of α with t: $\alpha = 1.2917 - 0.04482t + 0.0006258t^2$

T/K	297.04	292.14	287.50	283.70	279.01	274.70
α	0.5734	0.6712	0.7777	0.8868	1.0451	1.2242

Concn. of urea, %: 7.262
Change of *d* with t: $d = 1.02214 - 0.0001099t - 0.000004543t^2$
Change of α with t: $\alpha = 1.3393 - 0.04750t + 0.0006866t^2$

T/K	297.63	292.46	287.67	283.63	280.79
α	0.5881	0.6830	0.7946	0.9191	1.0169

Concn. of urea, %: 7.330
Change of *d* with t: $d = 1.02200 - 0.00007845t - 0.000005349t^2$
Change of α with t: $\alpha = 1.2872 - 0.043684t + 0.00058266t^2$

T/K	297.54	292.34	287.63	283.72	280.86
α	0.5685	0.6691	0.7771	0.8924	0.9854

Concn. of urea, %: 9.931
Change of *d* with t: $d = 1.03007 - 0.00015102t - 0.000004114t^2$
Change of α with t: $\alpha = 1.2528 - 0.040516t + 0.0005196t^2$

T/K	297.66	292.49	287.78	283.50	278.31
α	0.5721	0.6635	0.7715	0.8915	1.0578

Concn. of urea, %: 10.000
Change of *d* with t: $d = 1.03029 - 0.00014818t - 0.000004305t^2$
Change of α with t: $\alpha = 1.2772 - 0.04381t + 0.00062565t^2$

T/K	297.28	292.36	287.69	283.77	279.83	276.90	275.91
α	0.5844	0.6758	0.7727	0.8819	1.0054	1.1165	1.1615

COMPONENTS:

1. Nitrous oxide; N_2O; [10024-97-2]
2. Water; H_2O; [7732-18-5]
3. Urea; CH_4NO; [57-13-6]

ORIGINAL MEASUREMENTS:

Manchot, von W.; Jahrstorfer, M.; Zepter, H.
Z. Anorg. Allg. Chem.
1924, *141*, 45.

VARIABLES:

Concentration

PREPARED BY:

C. L. Young

EXPERIMENTAL VALUES:

T/K	Density, d_4^{25} of soln.	Conc. of soln. /mol dm^{-3}	Bunsen coefficient, α
298.15	1.0134	0.97	51.0
	1.0287	1.95	49.2
	1.0619	4.05	46.3
	1.0905	5.89	44.5

AUXILIARY INFORMATION

METHOD/APPARATUS/PROCEDURE:

Measurement of volume of gas absorbed by means of gas buret and pipet. Volume absorbed appears to be taken as independent of pressure (1). Density of the aqueous solution determined by Sprengel pykometer.

SOURCE AND PURITY OF MATERIALS:

1. Prepared by heating ammonium nitrate, frozen in liquid air and then distilled.

2. No details given.

ESTIMATED ERROR:

$\delta\alpha = \pm 0.1$.

REFERENCES:

1. Manchot, W.
Z. Anorg. Chem.
1924, *141*, 38.

COMPONENTS:	ORIGINAL MEASUREMENTS:
1. Nitrous oxide; N_2O; [10024-97-2] 2. Water; H_2O; [7732-18-5] 3. 2-Aminoethanol, (monoethanolamine); C_2H_7NO; [141-43-5]	Sada, E.; Kito, S. *Kagaku Kogaku*, 1972, *36*, 218-20.
VARIABLES:	PREPARED BY:
Temperature, concentration	W.Gerrard / C.L. Young

EXPERIMENTAL VALUES:

T/K	Conc. of monoethanolamine /mol l^{-1} (soln)	Bunsen coefficient, L
288.15	0	0.7500
	1.0783	0.7435
	1.6839	0.7344
	2.4272	0.7219
	2.4420	0.7176
	3.1821	0.6991
	4.4639	0.6541
	5.2185	0.6315
298.15	0	0.5512
	1.0853	0.5490
	2.0571	0.5406
	2.8975	0.5340
	4.1214	0.5111
	4.8359	0.4977
	5.8611	0.4762

AUXILIARY INFORMATION

METHOD APPARATUS/PROCEDURE:	SOURCE AND PURITY OF MATERIALS:
Equilibrium established between measured volume of gas and a measured amount of gas-free liquid in a cell fitted with a magnetic stirrer. Amount of gas absorbed estimated from change in volume of gas. Concentration of amine determined by titration. Details in source and ref. (1).	1. Commercial sample, purity 99.8 mole per cent. 2/3 Of satisfactory purity.
	ESTIMATED ERROR:
	$\delta T/K = \pm 0.2$; $\delta \alpha = \pm 2\%$. (estimated by compiler).
	REFERENCES:
	1. Onda, K.; Sada, E.; Kobayashi, T. *J. Chem. Eng. Japan.* 1970, *3*, 18 and 137.

COMPONENTS:

1. Nitrous oxide; N_2O; [10024-97-2]
2. Water; H_2O; [7732-18-5];
3. 1,2-Ethanediamine (Ethylenediamine); $C_2H_8N_2$; [107-15-3]

ORIGINAL MEASUREMENTS:

Sada, E.; Kumazawa, H.; Butt, M.A.
J. Chem. Engng. Data 1977, *22*, 277-278.

VARIABLES:

Composition

PREPARED BY:

C. L. Young

EXPERIMENTAL VALUES:

T/K	Conc. of amine/mol l^{-1}	Bunsen coefficient, α
298.15	0.0	0.5512
	0.805	0.5276
	1.473	0.5106
	1.871	0.4936
	2.267	0.4728
	2.371	0.4687
	2.738	0.4415
	3.133	0.4211

Pressure = 1 atmosphere = 1.01325×10^5 Pa.

AUXILIARY INFORMATION

METHOD/APPARATUS/PROCEDURE:

Equilibrium established between measured volume of gas and a measured amount of gas-free liquid in a cell fitted with a magnetic stirrer. Amount of gas absorbed estimated from change in volume of gas. Concentration of amine determined by titration. Details in source and ref. 1.

SOURCE AND PURITY OF MATERIALS:

1. Commercial sample, minimum purity 99.8 mole per cent.
2. Distilled and degassed.
3. Reagent grade of guaranteed quality.

ESTIMATED ERROR:

$\delta T/K = \pm 0.2$; $\delta\alpha = \pm 2\%$ (estimated by compiler).

REFERENCES:

1. Onda, K.; Sada, E.; Kobayashi, T.; Kito, S.; Ito, K. *J. Chem. Engng. Japan* 1970, *3*, 18.

COMPONENTS:	ORIGINAL MEASUREMENTS:
1. Nitrous oxide; N_2O; [10024-97-2] 2. Water; H_2O; [7732-18-5] 3. 1-Amino-2-propanol, (*iso*-propanolamine); C_3H_9NO; [78-96-6]	Sada, E.; Kumazawa, H.; Butt, M. A. *J. Chem. Engng. Data* 1978, *23*, 161-163.
VARIABLES:	PREPARED BY:
Temperature, composition	C. L. Young

EXPERIMENTAL VALUES:

T/K	Conc/mol l^{-1}	Bunsen Coefficient, α
298.15	0.0	0.5512
	0.255	0.5470
	0.501	0.5353
	0.761	0.5340
	1.194	0.5296
	1.695	0.5219
	2.282	0.4960
	2.928	0.4754
	3.400	0.4576
	3.736	0.4387

Pressure = 1 atmosphere = 1.01325×10^5 Pa.

AUXILIARY INFORMATION

METHOD/APPARATUS/PROCEDURE:

Equilibrium established between measured volume of gas and measured amount of gas-free liquid in a cell fitted with a magnetic stirrer. Amount of gas absorbed estimated from change in volume of gas. Concentration of amine determined by titration. Details in source and refs. 1 and 2.

SOURCE AND PURITY OF MATERIALS:

1. Commercial sample, minimum purity 99.8 mole per cent.
2. Distilled and degassed.
3. Reagent grade of guaranteed quality.

ESTIMATED ERROR:

$\delta T/K = \pm 0.2$; $\delta\alpha = \pm 2\%$ (estimated by compiler).

REFERENCES:

1. Sada, E.; Kumazawa, H; Butt, M. A. *J. Chem. Engng. Data* 1977, *22*, 277
2. Onda, S.; Sada, E.; Kobayashi, T.; Kito, S.; Ito, K. *J. Chem. Engng. Japan* 1970, *3*, 18.

COMPONENTS:

1. Nitrous oxide; N_2O; [10024-97-2]
2. Water; H_2O; [7732-18-5]
3. 2,2′-Iminobisethanol,(diethanolamine) $C_4H_{11}NO_2$; [111-42-2]

ORIGINAL MEASUREMENTS:

Sada, E.; Kumazawa, H.; Butt, M. A.; *J. Chem. Engng. Data* 1977, *22*, 277-278.

VARIABLES:

Composition

PREPARED BY:

C. L. Young

EXPERIMENTAL VALUES:

T/K	Conc. of amine/mol l^{-1}	Bunsen coefficient, α
298.15	0.0	0.5512
	0.449	0.5480
	0.996	0.5406
	2.026	0.5205
	2.313	0.5114
	3.081	0.4918

Pressure = 1 atmosphere = 1.01325×10^5 Pa.

AUXILIARY INFORMATION

METHOD/APPARATUS/PROCEDURE:

Equilibrium established between a measured volume of gas and a measured amount of gas-free liquid in a cell fitted with a magnetic stirrer. Amount of gas absorbed estimated from change in volume of gas. Concentration of amine determined by titration. Details in source and ref. 1.

SOURCE AND PURITY OF MATERIALS:

1. Commercial sample, minimum purity 99.8 mole per cent.
2. Distilled and degassed.
3. Reagent grade of guaranteed quality.

ESTIMATED ERROR:

$\delta T/K = \pm 0.2$; $\delta\alpha = \pm 2\%$ (estimated by compiler).

REFERENCES:

1. Onda, K.; Sada, E.; Kobayashi, T.; Kito, S.; Ito, K. *J. Chem. Engng. Japan* 1970, *3*, 18.

COMPONENTS:

1. Nitrous oxide; N_2O; [10024-97-2]
2. Water; H_2O; [7732-18-5]
3. 1,1´-Iminobis-2-propanol, (Diisopropanolamine); $C_6H_{15}NO_2$; [110-97-4]

ORIGINAL MEASUREMENTS:

Sada, E.; Kumazawa, H.; Butt, M. A. *J. Chem. Engng. Data* 1978, *23*, 161-163.

VARIABLES:

Temperature, composition

PREPARED BY:

C. L. Young

EXPERIMENTAL VALUES:

T/K	Amine Conc/mol l^{-1}	Bunsen Coefficient, α
298.15	0.0	0.5512
	0.255	0.5386
	0.486	0.5330
	0.915	0.5145
	1.356	0.4968
	1.584	0.4775
	1.950	0.4512
	2.379	0.4144
	2.528	0.4091
	2.619	0.3883
	2.918	0.2947

Pressure = 1 atmosphere = 1.01325×10^5 Pa.

AUXILIARY INFORMATION

METHOD /APPARATUS/PROCEDURE:

Equilibrium established between measured volume of gas and measured amount of gas-free liquid in a cell fitted with a magnetic stirrer. Amount of gas absorbed estimated from change in volume of gas. Concentration of amine determined by titration. Details in source and refs. 1 and 2.

SOURCE AND PURITY OF MATERIALS:

1. Commercial sample, minimum purity 99.8 mole per cent.
2. Distilled and degassed.
3. Reagent grade of guaranteed quality.

ESTIMATED ERROR:

$\delta T/K = \pm 0.2$; $\delta\alpha = \pm 2\%$ (estimated by compiler).

REFERENCES:

1. Sada, E.; Kumazawa, H.; Butt, M. A. *J. Chem. Engng. Data* 1977, *22*, 277.
2. Onda, S.; Sada, E.; Kobayashi, T.; Kito, S.; Ito, K. *J. Chem. Engng. Japan* 1970, *3*, 18.

COMPONENTS:

1. Nitrous oxide; N_2O; [10024-97-2]
2. Water; H_2O; [7732-18-5]
3. 2,2´,2´´-Nitrilotrisethanol, (Triethanolamine); $C_6H_{15}NO_3$; [102-71-6]

ORIGINAL MEASUREMENTS:

Sada, E.; Kumazawa, H.; Butt, M. A.; *J. Chem. Engng. Data* 1977, *22*, 277-278.

VARIABLES:

Composition

PREPARED BY:

C. L. Young

EXPERIMENTAL VALUES:

T/K	Conc. of amine/mol l^{-1}	Bunsen coefficient, α
298.15	0.0	0.5512
	0.628	0.4997
	0.874	0.4774
	1.293	0.4327
	2.160	0.3825
	2.912	0.3170

Pressure = 1 atmosphere = 1.01325×10^5 Pa.

AUXILIARY INFORMATION

METHOD/APPARATUS/PROCEDURE:

Equilibrium established between a measured volume of gas and a measured amount of gas-free liquid in a cell fitted with a magnetic stirrer. Amount of gas absorbed estimated from change in volume of gas. Concentration of amine determined by titration. Details in source and ref. 1.

SOURCE AND PURITY OF MATERIALS:

1. Commercial sample, minimum purity 99.8 mole per cent.
2. Distilled and degassed.
3. Reagent grade of guaranteed quality.

ESTIMATED ERROR:

$\delta T/K = \pm 0.2$; $\delta\alpha = \pm 2\%$ (estimated by compiler).

REFERENCES:

1. Onda, K.; Sada, E.; Kobayashi, T.; Kito, S.; Ito, K. *J. Chem. Engng. Japan* 1970, *3*, 18.

COMPONENTS:	ORIGINAL MEASUREMENTS:
1. Nitrous oxide; N_2O; [10024-97-2] 2. Water; H_2O; [7732-18-5] 3. 2,2,2-Trichloro-1,1-ethanediol, (Chloral hydrate); $C_2H_3Cl_3O_2$; [302-17-0]	Knopp, W. *Z. Phys. Chem.* 1904, *48*, 97-108
VARIABLES:	PREPARED BY:
Concentration	W. Gerrard

EXPERIMENTAL VALUES:

T/K	Weight of chloral hydrate in 100 g of solution	Conc. of chloral hydrate/mol dm^{-3} (soln).	Density of solution / kg dm^{-3}	Bunsen absorption coefficient, α
293.15	2.947	0.184	1.01124	0.6182
	6.848	0.445	1.02907	0.6128
	13.48	0.942	1.06110	0.5960
	16.15	1.165	1.07407	0.5891
	19.60	1.474	1.09224	0.5793
	24.02	1.911	1.11602	0.5675
			Water	0.6270

Calculated by compiler: Mole fraction, x_1, of N_2O in water = 0.000506

Mole fraction, x_1, of N_2O in the solution containing 1.911 moles chloral hydrate in 1 dm^3 of solution = 0.000547. Thus, although α decreases as the molarity of chloral hydrate increases, the x_1 value increases.

AUXILIARY INFORMATION

METHOD APPARATUS/PROCEDURE:

An absorption pipet and a gas buret were used to measure the volume of nitrous oxide absorbed.

The densities of the solutions were determined by a Sprengel pyknometer.

SOURCE AND PURITY OF MATERIALS:

1. Nitrous oxide was prepared by heating pure ammonium nitrate at 513-523 K. It was passed in order through aqueous ferrous sulfate, sodium hydroxide, and concentrated sulfuric acid.

3. Attested analytically.

ESTIMATED ERROR:

REFERENCES:

COMPONENTS:

1. Nitrous oxide; N_2O; [10024-97-2]
2. Water; H_2O; [7732-18-5]
3. Colloids;

ORIGINAL MEASUREMENTS:

Geffcken, G.; *Z. Phys. Chem.* 1904, *49*, 257-302.

VARIABLES:

PREPARED BY:

W. Gerrard

EXPERIMENTAL VALUES:

T = 298.15 K Pressure = 101.325 kPa.

Colloid	Conc.of colloid /g dm^{-3}	Ostwald coefficient
Arsenious sulfide; As_2S_3; [1303-33-9]	39.6	0.5819
	42.4	0.5833
Ferric hydroxide; $Fe(OH)_3$; [1309-33-7]	47.7	0.5799
	47.9	0.5787
	Water alone	0.5942

AUXILIARY INFORMATION

METHOD APPARATUS/PROCEDURE:

Volume of nitrous oxide absorbed was measured by means of a gas buret and absorption pipet based on the technique of Ostwald. Detailed description and diagram were given.

SOURCE AND PURITY OF MATERIALS:

1. Nitrous oxide was self prepared and attested.

2,3. Of satisfactory purity.

ESTIMATED ERROR:

REFERENCES:

COMPONENTS:	ORIGINAL MEASUREMENTS:
1. Nitrous oxide; N_2O; [10024-97-2] 2. Water; H_2O; [7732-18-5] 3. Ferric hydroxide ; $Fe(OH)_3$; [1309-33-7]	Findlay, A.; Creighton, H.J.M. *J. Chem. Soc.* 1910, *97*, 536-561.
VARIABLES:	PREPARED BY:
Pressure, concentration	W. Gerrard

EXPERIMENTAL VALUES:

Solubility, *S*, expressed as $\frac{\text{concentration of the gas in the liquid phase}}{\text{concentration of the gas in the gaseous phase}}$

T/K = 298.16

Weight of colloid in 100 cm^3 of solution,g.	Density of solution	Pressure of gas /kPa	Solubility, *S*
0.625	1.001	101.057	0.590
		112.789	0.586
		124.521	0.584
		134.653	0.588
		149.452	0.588
		184.382	0.588
1.49	1.008	97.857	0.586
		110.389	0.579
		124.654	0.577
		143.719	0.581
		161.984	0.585
		190.914	0.586
4.061	1.029	100.523	0.578
		111.322	0.573
		117.722	0.571
		153.719	0.574
		161.051	0.579
		181.049	0.580

AUXILIARY INFORMATION

METHOD APPARATUS/PROCEDURE:

Gas buret and absorption pipet similar to that of Geffcken (1), except that the manometer tube was longer to give the higher pressures.

Concentration of the colloid was determined by precipitation with ammonium sulfate.

SOURCE AND PURITY OF MATERIALS:

1. Self prepared and purified; not attested.

2. & 3. The colloid was self prepared from ammonium carbonate and ferric chloride in water. The product was dialysed.

ESTIMATED ERROR:

Stated to be ∓ 0.25%

REFERENCES:

1. Geffcken, G. *Z. Phys. Chem.* 1904, *49*, 257.

COMPONENTS:

1. Nitrous oxide; N_2O; [10024-97-2]

2. Water; H_2O; [7732-18-5]

3. Ferric hydroxide, (colloidal); $Fe(OH)_3$; [1309-33-7]

ORIGINAL MEASUREMENTS:

Findlay, A.; Howell, O.R.;

J. Chem. Soc. 1914, *105*, 291-8

VARIABLES:

Pressure, concentration

PREPARED BY:

W. Gerrard

EXPERIMENTAL VALUES: Temperature not stated: presumably 298.16 T/K

Solubility, S, given as $\dfrac{\text{concentration of the gas in the liquid phase}}{\text{concentration of the gas in the gaseous phase}}$

Conc. of colloid $/10^{-2}$g cm^{-3}	Density of solution	p_{N_2O} + /kPa	S	p_{N_2O} + /kPa	S	p_{N_2O} + /kPa	S
0.43	1.001	38.823	0.594	54.541	0.594	76.619	0.591
0.43	1.001	86.458	0.589	102.323	0.583	137.293	0.580
0.43	1.001	37.250	0.594	53.608	0.592	74.846	0.591
0.43	1.001	89.178	0.588	104.776	0.583	139.146	0.580
0.92	1.003	38.263	0.589	55.328	0.587	76.206	0.584
0.92	1.003	90.844	0.582	105.578	0.578	140.599	0.576
0.92	1.003	37.730	0.590	54.461	0.586	75.272	0.584
0.92	1.003	86.058	0.582	103.456	0.579	136.880	0.574
3.82	1.027	34.143	0.583	49.715	0.581	72.486	0.580
3.82	1.027	84.472	0.577	101.963	0.572	135.226	0.568
3.82	1.027	33.050	0.583	48.435	0.582	69.913	0.579
3.82	1.027	86.178	0.576	99.710	0.573	131.627	0.568

+ p_{N_2O} is the pressure of N_2O over the solution.

AUXILIARY INFORMATION

METHOD /APPARATUS/PROCEDURE :

Measurement of volume of N_2O by gas buret and pipet (1).

SOURCE AND PURITY OF MATERIALS:

1. Self prepared and purified (2).

2. Ferric hydroxide solution was prepared by method of Noyes (3), and dialysed.

ESTIMATED ERROR:

REFERENCES:

1. Findlay, A.; Williams, T. *J. Chem. Soc.* 1913, *103*, 636.
2. Findlay, A.; Creighton, H.J.M. *J. Chem. Soc.* 1910, *97*, 536.
3. Noyes, A.A.; *J. Am. Chem. Soc.* 1905, *37*, 94.

COMPONENTS:	ORIGINAL MEASUREMENTS:
1. Nitrous oxide; N_2O; [10024-97-2] 2. Water; H_2O; [7732-18-5] 3. Arsenious sulfide; As_2S_3, (colloidal); [1303-33-9]	Findlay, A.; Creighton, H.J.M. *J. Chem. Soc.* 1910, *97*, 536-561.
VARIABLES:	PREPARED BY:
Pressure, concentration	W. Gerrard.

EXPERIMENTAL VALUES:

Solubility, *S*, expressed as $\frac{\text{concentration of the gas in the liquid phase}}{\text{concentration of the gas in the gaseous phase}}$

T/K = 298.16

Weight of colloid in 100 cm^3 of solution, g.	Density of solution	Pressure of gas /kPa	Solubility, *S*.
1.85	1.004	99.457	0.591
		109.322	0.590
		123.188	0.590
		140.653	0.592
		159.451	0.593
		179.449	0.593
2.29	1.007	99.457	0.590
		113.322	0.586
		134.120	0.588
		147.985	0.589
		161.184	0.589
		173.316	0.590

AUXILIARY INFORMATION

METHOD/APPARATUS/PROCEDURE:

Gas buret and absorption pipet similar to that of Geffcken (1), except that the manometer tube was longer to give the higher pressures.

Concentration of the colloid was determined by precipitation with hydrochloric acid, drying and weighing.

SOURCE AND PURITY OF MATERIALS:

1. Self prepared and purified; not attested.
2. & 3. Self made by the action of hydrogen sulfide on the purest arsenious oxide in water. The excess of hydrogen sulfide was removed by a stream of hydrogen, and the product filtered.

ESTIMATED ERROR:

Stated to be ∓ 0.25%

REFERENCES:

1. Geffcken, G. *Z. Phys. Chem.* 1904, *49*, 257.

COMPONENTS:

1. Nitrous oxide; N_2O; [10024-97-2]
2. Water; H_2O; [7732-18-5]
3. Silica, SiO_2; (in suspension); [7631-86-9]

ORIGINAL MEASUREMENTS:

Findlay, A.; Creighton, H.J.M.

J. Chem. Soc. 1910, *97*, 536-561.

VARIABLES:

Pressure

PREPARED BY:

W. Gerrard.

EXPERIMENTAL VALUES:

Solubility, *S*, expressed as $\frac{\text{concentration of the gas in the liquid phase}}{\text{concentration of the gas in the gaseous phase}}$

T/K = 298.16

100 cm^3 of liquid contained 0.30 g of silica, SiO_2: density, 1.000

Pressure of N_2O/kPa	Solubility, *S*
97.324	0.592
112.789	0.593
127.987	0.59
144.119	0.597
163.184	0.597
181.982	0.600
197.447	0.602

AUXILIARY INFORMATION

METHOD/APPARATUS/PROCEDURE:

Gas buret and absorption pipet similar to that of Geffcken (1), except that the manometer tube was longer to give the higher pressure.

SOURCE AND PURITY OF MATERIALS:

1. Self prepared and purified; not attested.

3. Used Kahlbaum's pure silica.

ESTIMATED ERROR:

Stated to be ∓ 0.25%

REFERENCES:

1. Geffcken, G.; *Z. Phys. Chem.* 1904, *49*, 257.

COMPONENTS:

1. Nitrous oxide; N_2O; [10024-97-2]
2. Water; H_2O; [7732-18-5];
3. Silicic acid (as SiO_2); [1342-98-2]

ORIGINAL MEASUREMENTS:

Findlay, A.; Creighton, H. J. M.
J. Chem. Soc. 1910, *97*, 536-561.

VARIABLES:

Pressure, concentration

PREPARED BY:

W. Gerrard

EXPERIMENTAL VALUES:

T/K 298.16

Solubility, *s*, expressed as $\frac{\text{concentration of the gas in the liquid phase}}{\text{concentrarion of the gas in the gaseous phase}}$

Conc. of colloid 10^2g (SiO_2) cm^{-3} (soln.)	Density of solution	Pressure of gas kPa	Solubility, *s*
1.87	1.001	99.723	0.596
		109.989	0.598
		122.788	0.598
		139.453	0.600
		162.250	0.602
		179.849	0.604
3.63	1.005	98.790	0.601
		113.055	0.602
		132.520	0.605
		149.585	0.607
		162.250	0.608
		185.848	0.609

AUXILIARY INFORMATION

METHOD/APPARATUS/PROCEDURE:

Gas buret and absorption pipet similar to that of Geffcken (ref. 1), except that the manometer tube was longer to give the higher pressures.

SOURCE AND PURITY OF MATERIALS:

1. Nitrous oxide self prepared and purified, not attested.
3. Kahlbaum's pure silica was dissolved in aqueous potassium hydroxide; the clear solution was poured into hydrochloric acid and the product was dialysed.

ESTIMATED ERROR:

Stated to be ±0.25%.

REFERENCES:

1. Geffcken, G. *Z. Phys. Chem.* 1904, *49*, 257.

COMPONENTS:	ORIGINAL MEASUREMENTS:
1. Nitrous oxide; N_2O; [10024-97-2] 2. Water; H_2O; [7732-18-5]; 3. Silicic acid (colloidal); [1343-98-2]	Findlay, A.; Howell, O. R. *J. Chem. Soc.* 1914, *105*, 291-8.
VARIABLES:	PREPARED BY:
Pressure, concentration	W. Gerrard

EXPERIMENTAL VALUES:

Temperature not stated: presumably 298.16 T/K

Solubility, s, given as $\frac{\text{concentration of the gas in the liquid phase}}{\text{concentration of the gas in the gaseous phase}}$

Conc. of colloid 10^2g cm^{-3} (soln.)	Density of solution	P_{N_2O}† kPa	s	P_{N_2O}† kPa	s	P_{N_2O}† kPa	s
1.62	1.000	33.893	0.590	49.208	0.587	71.566	0.589
		88.524	0.589	101.963	0.588	134.267	0.592
		34.010	0.594	49.382	0.591	71.673	0.589
		90.124	0.588	102.043	0.588	136.733	0.591
3.5	1.004	33.357	0.600	50.022	0.595	72.859	0.594
		91.658	0.593	101.150	0.594	137.813	0.598
		33.343	0.596	50.208	0.595	74.286	0.593
		87.138	0.594	101.003	0.595	133.853	0.598

† P_{N_2O} is the pressure of N_2O over the solution.

AUXILIARY INFORMATION

METHOD /APPARATUS/PROCEDURE:

Measurement of volume of N_2O by gas buret and pipet (ref. 1).

SOURCE AND PURITY OF MATERIALS:

1. Nitrous oxide self prepared and purified, see ref. 2.
3. Pure silica was dissolved in aqueous potassium hydroxide; the clear solution was poured into hydrochloric, and the product was dialysed.

ESTIMATED ERROR:

REFERENCES:

1. Findlay, A.; Williams, T. *J. Chem. Soc.* 1913, *103*, 636.
2. Findlay, A.; Creighton, H. J. M. *J. Chem. Soc.* 1910, *97*, 536.

COMPONENTS:

1. Nitrous oxide; N_2O; [10024-97-2]
2. Water; H_2O; [7732-18-5];
3. Gelatin (colloidal)

ORIGINAL MEASUREMENTS:

Findlay, S.; Creighton, H. J. M.
J. Chem. Soc. 1910, *97*, 536-561.

VARIABLES:

Pressure, concentration

PREPARED BY:

W. Gerrard

EXPERIMENTAL VALUES:

T/K 298.16

Solubility, *s*, expressed as $\frac{\text{concentration of the gas in the liquid phase}}{\text{concentration of the gas in the gaseous phase}}$

Conc. of colloid 10^2g cm^{-3} (soln.)	Density of solution	Pressure of gas kPa	Solubility, *s*
1.31	0.999	97.457	0.589
		113.189	0.590
		124.921	0.590
		142.519	0.592
		156.784	0.592
		177.049	0.592
3.09	1.003	97.324	0.581
		114.389	0.582
		126.654	0.584
		145.185	0.586
		163.984	0.588
		183.048	0.588
6.06	1.008	97.324	0.560
		113.322	0.563
		128.121	0.566
		146.252	0.568
		166.250	0.570
		183.848	0.571

AUXILIARY INFORMATION

METHOD/APPARATUS/PROCEDURE:

Gas buret and absorption pipet similar to that of Geffcken (ref. 1), except that the manometer tube was longer to give the higher pressures.

SOURCE AND PURITY OF MATERIALS:

1. Nitrous oxide self prepared and purified, not attested.

3. French sheet gelatin, free from salts, was used.

ESTIMATED ERROR:

Stated to be ±0.25%.

REFERENCES:

1. Geffcken, G. *Z. Phys. Chem.* 1904, *49*, 257.

COMPONENTS:	ORIGINAL MEASUREMENTS:
1. Nitrous oxide; N_2O; [10024-97-2] 2. Water; H_2O; [7732-18-5]; 3. Gelatin (colloidal)	Findlay, A.; Howell, O. R.; *J. Chem. Soc.* 1914, *105*, 291-8.
VARIABLES:	PREPARED BY:
Pressure, concentration	W. Gerrard

EXPERIMENTAL VALUES:

Temperature not stated: presumably 298.16 T/K

Solubility, s, given as $\frac{\text{concentration of the gas in the liquid phase}}{\text{concentration of the gas in the gaseous phase}}$

Conc. of colloid 10^2g cm^{-3} (soln.)	Density of solution	P_{N_2O}† kPa	s	P_{N_2O}† kPa	s	P_{N_2O}† kPa	s
1.45	1.000	34.197	0.582	49.622	0.581	70.726	0.577
		83.178	0.575	100.657	0.579	134.600	0.581
3.12	1.004	33.930	0.577	48.968	0.576	70.700	0.568
		84.312	0.569	100.083	0.572	133.320	0.576
1.45	1.000	34.783	0.581	50.582	0.582	72.326	0.575
		86.191	0.577	101.776	0.579	137.600	0.579
3.16	1.004	38.369	0.577	55.261	0.574	75.886	0.570
		87.791	0.570	106.163	0.572	140.519	0.576
6.10	1.008	34.357	0.556	50.795	0.556	72.846	0.548
		84.978	0.546	101.710	0.550	137.306	0.557
6.14		34.530	0.556	50.742	0.555	72.859	0.548
		85.298	0.546	101.270	0.550	-	-

† P_{N_2O} is the pressure of N_2O over the solution.

AUXILIARY INFORMATION

METHOD /APPARATUS/PROCEDURE:	SOURCE AND PURITY OF MATERIALS:
Measurement of volume of N_2O by gas buret and pipet (ref. 1).	1. Nitrous oxide self prepared and purified, see ref. 2. 3. French gelatin, free from salts, was used.
	ESTIMATED ERROR:
	REFERENCES:

1. Findlay, A.; Williams, T. *J. Chem. Soc.* 1913, *103*, 636.

2. Findlay, A.; Creighton, H. J. M. *J. Chem. Soc.* 1910, *97*, 536.

COMPONENTS:

1. Nitrous oxide; N_2O; [10024-97-2]
2. Water; H_2O; [7732-18-5]
3. Gelatin (colloidal)

ORIGINAL MEASUREMENTS:

Shkol'nikova, R. I.
Uchenye Zapiski Leningrad. Gosudart
1959, *No. 18, Part 272*, 64-86.

VARIABLES:

Temperature, concentration of colloid

PREPARED BY:

W. Gerrard

EXPERIMENTAL VALUES:

T/K	Conc. of colloid/%	α	L
283.15	1	0.5340	0.5535
288.15		0.4890	0.5159
293.15		0.4370	0.4690
298.15		0.4150	0.4530
303.15		0.3950	0.4384
308.15		0.3200	0.3610
313.15		0.2610	0.2993
283.15	5	0.4852	0.5030
288.15		0.4321	0.4482
293.15		0.3809	0.4088
298.15		0.3614	0.3945
303.15		0.3430	0.3807
308.15		0.2921	0.3290
313.15		0.2410	0.2763

α = the Bunsen absorption coefficient.

$L = \alpha \times T/K/273$, but defined as the ratio of the concentration of the gas in the liquid phase to that in the gas phase.

Partial pressure of gas was taken to be 760 mmHg.

760 mmHg = 1 atm = 101.325 kPa.

(cont.)

AUXILIARY INFORMATION

METHOD/APPARATUS/PROCEDURE:

The volume of N_2O absorbed was measured by the use of a gas buret and pipet.

SOURCE AND PURITY OF MATERIALS:

1. Purity of nitrous oxide stated to be 99.6-99.7%; contained N_2, 0.4-0.3%.
2. Water may be taken as of satisfactory purity.
3. The gelatin solution was dialized.

ESTIMATED ERROR:

Reproducibility stated to be within ±0.2%.

REFERENCES:

COMPONENTS:	ORIGINAL MEASUREMENTS:
1. Nitrous oxide; N_2O; [10024-97-2]	Shkol'nikova, R. I.
2. Water; H_2O; [7732-18-5]	*Uchenye Zapiski Leningrad. Gosudart*
3. Gelatin (colloidal)	1959, *No. 18, Part 272*, 64-86.

EXPERIMENTAL VALUES:

T/K	Conc. of colloid/%	α	L
283.15	10	0.4300	0.4458
288.15		0.3850	0.4060
293.15		0.3366	0.3613
298.15		0.3194	0.3486
303.15		0.3030	0.3363
308.15		0.2710	0.3057
313.15		0.2280	0.2614

Conc. of colliod/%	A/cal mol^{-1}	B/cal mol^{-1}
0 (water)	5760	5161
1	4700	4100
5	4090	3520
10	3730	3140

Where A and B are defined by the equations:

$$\frac{d \ln \alpha}{dT} = - \frac{A}{RT^2}$$

$$\frac{d \ln L}{dT} = - \frac{B}{RT^2}$$

COMPONENTS:	ORIGINAL MEASUREMENTS:
1. Nitrous oxide; N_2O; [10024-97-2] 2. Water; H_2O; [7732-18-5] 3. Starch, (colloidal)	Findlay, A.; Creighton, H.J.M. *J. Chem. Soc.* 1910, *97*, 536-561.
VARIABLES:	PREPARED BY:
Pressure, concentration	W. Gerrard

EXPERIMENTAL VALUES:

$$\text{Solubility expressed as } \frac{\text{concentration of the gas in the liquid phase}}{\text{concentration of the gas in the gaseous phase}} = S$$

p_{N_2O} is the pressure of N_2O over the solution. Temperature = 298.16 K.

Conc. of soln. $/10^{-2}$ g cm^{-3}	Density of Solution							
2.50	1.009	p_{N_2O}/kPa	98.923	116.122	135.986	155.451	171.183	192.114
		S	0.580	0.576	0.575	0.578	0.581	0.582
6.89	1.021	p_{N_2O}/kPa	98.923	113.055	123.854	139.453	168.117	184.115
		S	0.561	0.554	0.553	0.554	0.562	0.567
10.00	1.030	p_{N_2O}/kPa	98.923	114.655	126.387	142.786	165.183	179.982
		S	0.550	0.544	0.545	0.545	0.553	0.555
13.73	1.040	p_{N_2O}/kPa	98.523	111.456	130.920	151.452	166.917	184.915
		S	0.537	0.532	0.530	0.535	0.536	0.538

AUXILIARY INFORMATION

METHOD: /APPARATUS/PROCEDURE:

Gas buret and pipet similar to that of Geffcken (1), except that the manometer tube was longer to give the higher pressures.

SOURCE AND PURITY OF MATERIALS:

1. Self prepared and purified: not attested.

2. Used Kahlbaum's pure soluble starch.

ESTIMATED ERROR:

$\delta S/S$ Stated to be $\mp$ 0.25%.

REFERENCES:

1. Geffcken, G. *Z. Phys. Chem.* 1904, *49*, 257.

COMPONENTS:

1. Nitrous oxide; N_2O; [10024-97-2]
2. Water; H_2O; [7732-18-5];
3. Starch (colloidal)

ORIGINAL MEASUREMENTS:

Findlay, A.; Howell, O. R.
J. Chem. Soc. 1914, *105*, 291-8.

VARIABLES:

Pressure, concentration

PREPARED BY:

W. Gerrard

EXPERIMENTAL VALUES:

Temperature not stated: presumably 298.16 T/K

Solubility, s, given as $\frac{\text{concentration of the gas in the liquid phase}}{\text{concentration of the gas in the gaseous phase}}$

Conc. of colloid 10^2g cm^{-3} (soln.)	Density of solution	$N_2O^†$ kPa	s	$N_2O^†$ kPa	s	$N_2O^†$ kPa	s
6.76	1.023	38.023	0.565	55.328	0.563	75.552	0.560
		87.658	0.560	102.696	0.553	140.519	0.550
6.70		35.143	0.566	49.395	0.563	69.953	0.561
		86.205	0.558	100.070	0.554	132.987	0.550
9.58	1.030	35.676	0.554	49.822	0.551	67.230	0.549
		83.685	0.546	99.603	0.541	136.520	0.539
9.40	1.029	38.729	0.551	55.528	0.550	76.899	0.548
		87.858	0.543	103.430	0.540	133.800	0.537
13.62	1.039	37.903	0.541	55.781	0.537	81.965	0.535
		93.751	0.532	112.389	0.528	-	-
13.60		35.170	0.541	50.475	0.539	66.247	0.536
		83.192	0.534	100.670	0.530	129.787	0.525

† p_{N_2O} is the pressure of N_2O over the solution.

AUXILIARY INFORMATION

METHOD /APPARATUS/PROCEDURE:

Measurement of volume of N_2O by gas buret and pipet (ref. 1).

SOURCE AND PURITY OF MATERIALS:

1. Nitrous oxide self prepared and purified, see ref. 2.
3. Starch: Kahlbaum's pure soluble.

ESTIMATED ERROR:

REFERENCES:

1. Findlay, A.; Williams, T. *J. Chem. Soc.* 1913, *103*, 636.

2. Findlay, A.; Creighton, H. J. M. *J. Chem. Soc.* 1910, *97*, 536.

COMPONENTS:	ORIGINAL MEASUREMENTS:
1. Nitrous oxide; N_2O; [10024-97-2] 2. Water; H_2O; [7732-18-5] 3. Charcoal in suspension.	Findlay, A.; Creighton, H.J.M. *J. Chem. Soc.* 1910, *97*, 536-561
VARIABLES:	PREPARED BY:
Pressure	W. Gerrard

EXPERIMENTAL VALUES:

Solubility, *S*, expressed as $\frac{\text{concentration of the gas in the liquid phase}}{\text{concentration of the gas in the gaseous phase}}$

T/K = 298.16

100 cm^3 of liquid contained 0.227 g of charcoal; density = 1.000.

Pressure of N_2O/kPa	97.190	109.856	124.788	137.853	167.183	180.782
Solubility, *S*	0.596	0.600	0.618	0.635	0.648	0.674

AUXILIARY INFORMATION

METHOD/APPARATUS/PROCEDURE:

Gas buret and absorption pipet similar to that of Geffcken (1), except that the manometer tube was longer to give the higher pressures.

SOURCE AND PURITY OF MATERIALS:

1. Self prepared and purified; not attested.

3. Used Kahlbaum's well powdered bone charcoal.

ESTIMATED ERROR:

Stated to be ∓ 0.25%

REFERENCES:

1. Geffcken, G. *Z. Phys. Chem.* 1904, *49*, 257.

COMPONENTS:	ORIGINAL MEASUREMENTS:
1. Nitrous oxide; N_2O; [10024-97-2] 2. Water; H_2O; [7732-18-5] 3. Charcoal suspension,	Findlay, A.; Howell, O.R. *J. Chem. Soc.* 1914, *105*, 291-8
VARIABLES:	PREPARED BY:
Pressure	W. Gerrard

EXPERIMENTAL VALUES:

Solubility, S, given as $\frac{\text{concentration of the gas in the liquid phase}}{\text{concentration of the gas in the gaseous phase}}$

Temperature not stated: presumably 298.16 T/K

3.0 g of charcoal in 100 cm^3 water	p_{N_2O}+ /kPa	S	p_{N_2O}+ /kPa	S	p_{N_2O}+ /kPa	S
	33.677	0.580	48.902	0.586	70.353	0.587
	83.538	0.588	99.963	0.588	131.920	0.609
	34.0366	0.583	49.902	0.581	72.699	0.586
	85.191	0.591	101.363	0.588	133.493	0.610

+ p_{N_2O} is the pressure of N O over the liquid.

NOTE: No data were given for a suspension of silica; but it was stated that the values did not differ appreciably from those in pure water.

AUXILIARY INFORMATION

METHOD /APPARATUS/PROCEDURE:	SOURCE AND PURITY OF MATERIALS:
Measurement of volume of N_2O by gas buret and pipet (1).	1. Self prepared and purified (2). 2. Finely powdered animal charcoal was boiled with water, dried at 373.16 K, and then heated in a vacuum almost to redness.
	ESTIMATED ERROR:
	REFERENCES:
	1. Findlay, A.; Williams. *J. Chem. Soc.* 1913, *103*, 636. 2. Findlay, A.; Creighton, H.J.M. *J. Chem. Soc.* 1910, *97*, 536.

COMPONENTS:	EVALUATOR:
1. Nitrous oxide; N_2O; [10024-97-2] 2. Organic liquids	Colin Young School of Chemistry University of Melbourne, Victoria, Australia: 3052. March 1980.

Most solubility data on this gas are at temperatures in the range 273K to 323K and near atmospheric pressure. Nitrous oxide has a boiling point at 1 atmosphere pressure of 184.67K and a critical temperature of 309.6K. At 273K the vapor pressure of nitrous oxide is approximately 31.5 atm or 3.2 MPa.

The main sources of data for non-aqueous solvents are those of Kunerth (1), Horiuti (2), Yen and McKetta (3), Hsu and Campbell (4). Makranczy *et al* (5), and Sada *et al* (6). In general terms the data of ref. (2), (3), (5) and (6) appear to be reliable. The data of ref. (4) are usually in agreement with other data on the same system but not always. The extensive data of Kunerth (1) appears to have an unusual temperature dependence the solubility appears to decrease slightly too rapidly with increasing temperature.

ALKANES:

Pentane: There is reasonable agreement between the data of Hsu and Campbell (4) and Makranczy *et al* (5) for this system and both are classified as tentative.

Heptane: The solubility in this alkane has been studied by Yen and McKetta (3) and Makranczy *et al* (5) There is good agreement between the two sets of data at 298K but at 313K the Makranczy *et al* (5) data indicates the solubility is higher than reported by Yen and McKetta (3). It appears likely that the data of Makranczy *et al* (5) might be slighly in error at the higher temperature. The other data in the Makranczy *et al* (5) paper are also slightly suspect when comparison is made for systems on which there is more extensive data, see Helium and alkane evaluation (7). The data of Yen and McKetta (3) are classified as tentative.

2,2,4-Trimethylpentane: Yen and McKetta (3) data are classified as tentative since there are no other data with which to compare their data and their data on other systems appears reliable.

Other alkanes: The data of Makranczy *et al* (5) are classified as tentative however the comments under heptane regarding these workers data are brought to the users attention. Probably the data at 298K is more reliable than that at 313K.

Benzene: The data of Horiuti (2) and Yen and McKetta (3) are in good agreement and are classified as tentative.

ALCOHOLS

Methanol: The data of Kunerth (1) Hsu and Campbell (4) and Sada *et al.* (6) are in fair agreement. Only Kunerth (1) values are over a range of temperature and the temperature dependence of his results is unusual. All three sets of data are classified as tentative. The mole fraction solubility of Makranczy *et al.* (8) is considerably greater than that of the other three studies and is classified as doubtful.

Ethanol: This system has been studied by five groups. The data of Kunerth (1) has an unusual temperature dependence and the mole fraction data, interpolated to 298.15 K, are considerably lower than that of any other worker. They are therefore classified as doubtful. The mole fraction solubility at 298.15 K calculated from the results of Makranczy *et al.* (8) Hsu and Campbell (4) and Carius (9) agree within experimental error and are classified as tentative. The results of Carius (9) are over 120 years old and the agreement with recent data is surprising. (Weiss (10) also remarked on the surprising agreement between Carius' results and recent data for the system nitrous oxide and water). The mole fraction solubility calculated from the results of Sada *et al.* are about 8-10% lower than the values from references (4) (8) and (9). Such a large discrepancy is greater than can be accounted for by the estimated experimental error.

2-Propanone (Acetone)

The solubility of nitrous oxide in 2-propanone has been studied by Horiuti (1), Hsu and Campbell (4) and Kunerth (1). The Hsu and Campbell isolated value is in reasonable agreement with the other two sets of data. There is fair agreement between Kunerth's and Horiuti's datum at 298K but Kunerth's data have an unusual temperature dependence and there is a fairly wide discrepancy between the two sets of data at lower temperatures. The data of Horiuti (1) and Hsu and Campbell (4) are classified as tentative whereas that of Kunerth (1) is less reliable.

Acetic acid and Pentyl ester

There is good agreement between the data of Kunerth (1) and the isolated value at 294.3 K of Hsu and Campbell (4) for acetic acid. Therefore both workers values are classified as tentative. Similarly the results of both these workers are in good agreement for the pentyl-ester.

Halogenated hydrocarbons.

Tetrachloromethane (Carbon tetrachloride). The data of Horiuti (2) and Yen and McKetta (3) are in good agreement whereas the isolated value of Hsu and Campbell (4) is widely different. The values of Horiuti (2) and Yen and McKetta (3) are therefore classified as tentative whereas that of Hsu and Campbell is classified as doubtful.

Trichloromethane (Chloroform). The temperature dependence of values of Kunerth (1) is unusual but the isolated value of Hsu and Campbell (4) is in agreement with Kunerth's data and therefore both sets of data are classified as tentative.

1,2-Dibromoethane. There is good agreement between the values of Kunerth (1) and that of Hsu and Campbell (4) on this system and therefore both sets of data are classified as tentative.

REFERENCES:

1. Kunerth, W. *Phys. Rev.* 1922, *19*, 512.
2. Horiuti, J. *Sci. Pap. Inst. Phys. Chem. Res. (Japan)*, 1931-2, *17*, 125.
3. Yen, L.C.; McKetta, J.J.Jr. *J. Chem. Eng. Data.* 1962, *7*, 288.
4. Hsu, H.; Campbell, D. *Aerosol Age.* 1964, *December*, 34.
5. Makranczy, J.; Megyery-Balog, K.; Rusz, L.; Patyi, L. *Hung. J. Ind. Chem.* 1976, *4*, *(2)*, 269.
6. Sada, E.; Kito, S.; Ito, Y. *Eng. Chem. Fundam.* 1975, *14*, 232.
7. *Helium and Neon Gas Solubilities, Solubility Data Series*, Vol. 1. ed. H.L. Clever, Pergamon, Oxford. 1979
8. Makranczy, J.; Rusz, L.; Balog-Megyery, K. *Hung. J. Ind. Chem.* 1979, *7*, 41.
9. Carius, L. *Annalen*, 1855, *94*, 129.
10. Weiss, R.F.; Price, B.A. *Marine Chem.* 1980, *8*, 347.

* * *

COMPONENTS:	ORIGINAL MEASUREMENTS:
1. Nitrous oxide; N_2O; [10024-97-2] 2. Pentane; C_5H_{12}; [109-66-0]	Hsu, H.; Campbell, D. *Aerosol Age*, 1964, *December*, 34.
VARIABLES:	PREPARED BY: W. Gerrard / C.L. Young

EXPERIMENTAL VALUES:

T/K	Ostwald coefficient, L	Mole fraction* of nitrous oxide in liquid, x_{N_2O}
294.3	4.13	0.0194

* calculated by compiler as for a partial pressure of gas of 101.325 kPa. Molar volume of gas at 294.24K taken as 23967cm^3 based on the density (1) 1.9775 g dm^{-3} at 101.325 kPa and 273.15K.

AUXILIARY INFORMATION

METHOD/APPARATUS/PROCEDURE:

Copper equilibrium cell fitted with Bourdon gauge and thermometer. Total amount of gas needed to attain given, but not stated, pressure measure. Ostwald coefficient calculated from knowledge of volume of liquid and container.

SOURCE AND PURITY OF MATERIALS:

No details given.

ESTIMATED ERROR:

$\delta T/K = \pm 0.5$; $\delta x_{N_2O} = \pm 3\%$.

(estimated by compiler).

REFERENCES:

1. Kaye, G.W.C.; Laby, T.H. *Tables of Physical and Chemical Constants*, Longmans, London, *1966*.

COMPONENTS:	ORIGINAL MEASUREMENTS:
1. Nitrous oxide; N_2O; [10024-97-2] 2. Pentane; C_5H_{12}; [109-66-0]	Makranczy, J.; Megyery-Balog, K.; Rusz, L.; Patyi, L. *Hung. J. Ind. Chem.* 1976, *4(2)*, 269-280.
VARIABLES: T/K: 298.15 - 313.15 P/kPa: 101.325 (1 atm)	PREPARED BY: S. A. Johnson

EXPERIMENTAL VALUES:

T/K	Mole fraction x_{N_2O}	Bunsen coefficient α	Ostwald Coefficient L
298.15	0.01906	3.750	4.093
313.15	0.01682	3.217	3.688

Mole fractions and Bunsen coefficients were calculated by the compiler.

AUXILIARY INFORMATION

METHOD/APPARATUS/PROCEDURE:

Volumetric method, described in detail in reference (1).

SOURCE AND PURITY OF MATERIALS:

Analytical grade reagents of Hungarian and foreign origin were used (both liquids and gases). No further information.

ESTIMATED ERROR:

$\delta x_{N_2O}/x_{N_2O} = \pm 0.03$.

REFERENCES:

1. Bodor, E.; Bor, Gy.; Mohai, B.; Sipos, G. *Veszprémi Vegyipari Egyetem Köslemények* 1957, *1*, 55; *CA 55*, 3175h.

COMPONENTS:

1. Nitrous oxide; N_2O; [10024-97-2]
2. Hexane; C_6H_{14}; [110-54-3]

ORIGINAL MEASUREMENTS:

Makranczy, J.; Megyery-Balog, K.; Rusz, L; Patyi, L.
Hung. J. Ind. Chem. 1976, *4(2)*, 269-280.

VARIABLES:

T/K: 298.15 - 313.15
P/kPa: 101.325 (1 atm)

PREPARED BY:

S. A. Johnson

EXPERIMENTAL VALUES:

T/K	Mole fraction x_{N_2O}	Bunsen coefficient α	Ostwald coefficient L
298.15	0.01833	3.180	3.471
313.15	0.01619	2.744	3.146

Mole fractions and Bunsen coefficients were calculated by the compiler.

AUXILIARY INFORMATION

METHOD /APPARATUS/PROCEDURE:

Volumetric method, described in detail in reference (1).

SOURCE AND PURITY OF MATERIALS:

Analytical grade reagents of Hungarian and foreign origin were used (both liquids and gases).
No further information.

ESTIMATED ERROR:

$\delta x_{N_2O}/x_{N_2O} = \pm 0.03.$

REFERENCES:

1. Bodor, E.; Bor, Gy.; Mohai, B.; Sipos, G. *Veszprêmi Vegyipari Egyetem Köslemênyei* 1957, *1*, 55; *CA 55*, 3175h.

COMPONENTS:	ORIGINAL MEASUREMENTS:
1. Nitrous oxide; N_2O; [10024-97-2] 2. Hexane; C_6H_{14}; [110-54-3]	Patyi, L.; Furmer, I. E.; Makranczy, J.; Sadilenko, A. S.; Stepanova, Z. G.; Berengarten, M. G. *Zh. Prikl. Khim.* 1978, *51*, 1296-1300.
VARIABLES:	PREPARED BY: C. L. Young

EXPERIMENTAL VALUES:

T/K	α^*	Mole fraction of nitrous oxide at a partial pressure of 101.325 kPa
298.15	3.19	0.01851

* volume of gas (measured at 101.325 kPa and 273.15 K) dissolved by one volume of hexane.

AUXILIARY INFORMATION

METHOD/APPARATUS/PROCEDURE:

Volumetric method. Pressure measured when known amounts of gas were added, in increments, to a known amount of liquid in a vessel of known dimensions. Corrections were made for the partial pressure of solvent. Details in ref. (1).

SOURCE AND PURITY OF MATERIALS:

Purity better than 99 mole per cent as determined by gas chromatography.

ESTIMATED ERROR:

$\delta T/K = \pm 0.1$; $\delta\alpha = \pm 4\%$ or less.

REFERENCES:

1. Bodor, E.; Bor, G. J.; Mohai, B.; Sipos, G. *Veszpremi. Vegyip. Egy. Kozl.* 1957, *1*, 55.

COMPONENTS:	ORIGINAL MEASUREMENTS:
1. Nitrous oxide; N_2O; [10024-97-2] 2. Heptane; C_7H_{16}; [142-82-5]	Yen, L. C.; McKetta, J. J., Jr. *J. Chem. Eng. Data* 1962, *7*, 288-289.
VARIABLES: Temperature	PREPARED BY: W. Gerrard/C. L. Young

EXPERIMENTAL VALUES:

T/K	Bunsen coefficient	Mole fraction of nitrous oxide, x_{N_2O}
273.16	4.136	0.02589
283.16	3.538	0.02250
293.16	3.042	0.01963
303.16	2.564	0.01681
313.16	2.244	0.01492

The partial pressure of the gas was 101.325 kPa.

Smoothed Data

$$\Delta G° = -RT \ln x_{N_2O} = (-9928 + 66.655 \times T/K) \text{ J mol}^{-1}$$

(Std. deviation = 22 J mol^{-1})

T/K	Mole fraction of nitrous oxide, x_{N_2O}
273.16	0.02612
283.16	0.02238
293.16	0.01938
303.16	0.01694
313.16	0.01494

AUXILIARY INFORMATION

METHOD/APPARATUS/PROCEDURE:

Measurement of volume of gas absorbed by a known volume of degassed liquid at a partial pressure of 101.325 kPa. The vapor pressure of the liquid was allowed for. Gas buret and absorption pipet. Modified form of apparatus and technique used by Markham and Kobe (1).

SOURCE AND PURITY OF MATERIALS:

1. Gas of 98 per cent purity was purified at 193.16 K. Mass spectrograph then showed purity of 99.5 per cent.
2. Phillips Petroleum pure grade sample freshly fractionated.

ESTIMATED ERROR:

$\delta T/K = \pm 0.1$; $\delta x_{N_2O} = \pm 2\%$ (estimated by compiler).

REFERENCES:

(1) Markham, A. E.; Kobe, K. A.; *J. Am. Chem. Soc.* 1941, *63*, 449.

COMPONENTS:	ORIGINAL MEASUREMENTS:
1. Nitrous oxide; N_2O; [10024-97-2] 2. Heptane; C_7H_{16}; [142-82-5]	Makranczy, J.; Megyery-Balog, K.; Rusz, L.; Patyi, L. *Hung. J. Ind. Chem.* 1976, *4(2)*, 269-280.
VARIABLES:	**PREPARED BY:**
T/K: 298.15 - 313.15 *P*/kPa: 101.325 (1 atm)	S. A. Johnson

EXPERIMENTAL VALUES:

T/K	Mole fraction x_{N_2O}	Bunsen coefficient α	Ostwald coefficient L
298.15	0.01796	2.780	3.034
313.15	0.01561	2.366	2.712

Mole fractions and Bunsen coefficients were calculated by the compiler.

AUXILIARY INFORMATION

METHOD /APPARATUS/PROCEDURE:

Volumetric method, described in detail in reference (1).

SOURCE AND PURITY OF MATERIALS:

Analytical grade reagents of Hungarian and foreign origin were used (both liquids and gases). No further information.

ESTIMATED ERROR:

$\delta x_{N_2O}/x_{N_2O} = \pm 0.03.$

REFERENCES:

1. Bodor, E.; Bor, Gy.; Mohai, B.; Sipos, G. *Veszprémi Vegyipari Egyetem Közleményei* 1957, *1*, 55; *CA 55*, 3175h.

COMPONENTS:	ORIGINAL MEASUREMENTS:
1. Nitrous oxide; N_2O; [10024-97-2] 2. Octane; C_8H_{18}; [111-65-9]	Makranczy, J.; Megyery-Balog, K.; Rusz, L.; Patyi, L. *Hung. J. Ind. Chem.* 1976, *4(2)*, 269-280.
VARIABLES: T/K: 298.15 - 313.15 *P*/kPa: 101.325 (1 atm)	PREPARED BY: S. A. Johnson

EXPERIMENTAL VALUES:

T/K	Mole fraction x_{N_2O}	Bunsen coefficient α	Ostwald coefficient L
298.15	0.01791	2.500	2.729
313.15	0.01540	2.107	2.415

Mole fractions and Bunsen coefficients were calculated by the compiler.

AUXILIARY INFORMATION

METHOD /APPARATUS/PROCEDURE:

Volumetric method, described in detail in reference (1).

SOURCE AND PURITY OF MATERIALS:

Analytical grade reagents of Hungarian and foreign origin were used (both liquids and gases). No further information.

ESTIMATED ERROR:

$\delta x_{N_2O}/x_{N_2O}$ = ±0.03.

REFERENCES:

1. Bodor, E.; Bor, Gy.; Mohai, B.; Sipos, G. *Veszprémi Vegyipari Egyetem Közleményei* 1957, *1*, 55; *CA 55*, 3175h.

COMPONENTS:

1. Nitrous oxide; N_2O; [10024-97-2]
2. 2,2,4-Trimethylpentane; C_8H_{18}; [540-84-1]

ORIGINAL MEASUREMENTS:

Yen, L. C.; McKetta, J. J., Jr.
J. Chem. Eng. Data 1962, *7*, 288-289.

VARIABLES:

Temperature

PREPARED BY:

W. Gerrard/C. L. Young

EXPERIMENTAL VALUES:

T/K	Bunsen coefficient	Mole fraction of nitrous oxide, x_{N_2O}
273.16	4.587	0.03217
283.16	3.786	0.02701
293.16	3.155	0.02287
303.16	2.679	0.01972
313.16	2.273	0.01698

The partial pressure of the gas was 101.325 kPa.

Smoothed Data

$$\Delta G° = -RT \ln x_{N_2O} = (-11346 + 70.101 \times T/K) \text{ J mol}^{-1}$$

(Std. deviation = 6.3 J mol^{-1})

T/K	Mole fraction of nitrous oxide, x_{N_2O}
273.16	0.03220
283.16	0.02700
293.16	0.02290
303.16	0.01964
313.16	0.01701

AUXILIARY INFORMATION

METHOD/APPARATUS/PROCEDURE:

Measurement of volume of gas absorbed by a known volume of degassed liquid at a partial pressure of 101.325 kPa. The vapor pressure of the liquid was allowed for. Gas buret and absorption pipet. Modified form of apparatus and technique used by Markham and Kobe (1).

SOURCE AND PURITY OF MATERIALS:

1. Gas of 98 per cent purity was purified at 193.16 K. Mass spectrograph then showed purity of 99.5 per cent.
2. Phillips Petroleum pure grade sample, freshly fractionated.

ESTIMATED ERROR:

$\delta T/K = \pm 0.1$; $\delta x_{N_2O} = \pm 2\%$ (estimated by compiler).

REFERENCES:

(1) Markham, A. E.; Kobe, K. A.; *J. Am. Chem. Soc.* 1941, *63*, 449.

COMPONENTS:	ORIGINAL MEASUREMENTS:
1. Nitrous oxide; N_2O; [10024-97-2] 2. Nonane; C_9H_{20}; [111-84-2]	Makranczy, J.; Megyery-Balog, K.; Rusz, L.; Patyi, L. *Hung. J. Ind. Chem.* 1976, *4(2)*, 269-280.
VARIABLES: T/K: 298.15 - 313.15 P/kPa: 101.325 (1 atm)	PREPARED BY: S. A. Johnson

EXPERIMENTAL VALUES:

T/K	Mole fraction x_{N_2O}	Bunsen coefficient α	Ostwald coefficient L
298.15	0.01788	2.270	2.478
313.15	0.01507	1.878	2.153

Mole fractions and Bunsen coefficients were calculated by the compiler.

AUXILIARY INFORMATION

METHOD /APPARATUS/PROCEDURE:

Volumetric method, described in detail in reference (1).

SOURCE AND PURITY OF MATERIALS:

Analytical grade reagents of Hungarian and foreign origin were used (both liquids and gases). No further information.

ESTIMATED ERROR:

$\delta x_{N_2O}/x_{N_2O} = \pm 0.03$.

REFERENCES:

1. Bodor, E.; Bor, Gy.; Mohai, B.; Sipos, G. *Veszprémi Vegyipari Egyetem Kösleményei* 1957, *1*, 55; *CA 55*, 3175h.

COMPONENTS:	ORIGINAL MEASUREMENTS:
1. Nitrous oxide; N_2O; [10024-97-2] 2. Decane; $C_{10}H_{22}$: [124-18-5]	Makranczy, J.; Megyery-Balog, K.; Rusz, L.; Patyi, L. *Hung. J. Ind. Chem.* 1976, *4(2)*, 269-280.
VARIABLES: T/K: 298.15 - 313.15 P/kPa: 101.325 (1 atm)	PREPARED BY: S. A. Johnson

EXPERIMENTAL VALUES:

T/K	Mole fraction x_{N_2O}	Bunsen coefficient α	Ostwald coefficient L
298.15	0.01786	2.080	2.270
313.15	0.01490	1.703	1.952

Mole fractions and Bunsen coefficients were calculated by the compiler.

AUXILIARY INFORMATION

METHOD /APPARATUS/PROCEDURE:

Volumetric method, described in detail in reference (1).

SOURCE AND PURITY OF MATERIALS:

Analytical grade reagents of Hungarian and foreign origin were used (both liquids and gases). No further information.

ESTIMATED ERROR:

$\delta x_{N_2O}/x_{N_2O} = \pm 0.03.$

REFERENCES:

1. Bodor, E.; Bor, Gy.; Mohai, B.; Sipos, G. *Veszprémi Vegyipari Egyetem Köslemériyei* 1957, *1*, 55; *CA 55*, 3175h.

COMPONENTS:	ORIGINAL MEASUREMENTS:
1. Nitrous oxide; N_2O; [10024-97-2] 2. Undecane; $C_{11}H_{24}$; [1120-21-4]	Makranczy, J.; Megyery-Balog, K.; Rusz, L.; Patyi, L. *Hung. J. Ind. Chem.* 1976, *4(2)*, 269-280.
VARIABLES: T/K: 298.15 - 313.15 *P*/kPa: 101.325 (1 atm)	PREPARED BY: S. A. Johnson

EXPERIMENTAL VALUES:

T/K	Mole fraction x_{N_2O}	Bunsen coefficient α	Ostwald coefficient L
298.15	0.01758	1.890	2.063
313.15	0.01489	1.573	1.803

Mole fractions and Bunsen coefficients were calculated by the compiler.

AUXILIARY INFORMATION

METHOD /APPARATUS/PROCEDURE:

Volumetric method, described in detail in reference (1).

SOURCE AND PURITY OF MATERIALS:

Analytical grade reagents of Hungarian and foreign origin were used (both liquids and gases). No further information.

ESTIMATED ERROR:

$\delta x_{N_2O}/x_{N_2O} = \pm 0.03$.

REFERENCES:

1. Bodor, E.; Bor, Gy.; Mohai, B.; Sipos, G. *Veszprémi Vegyipari Egyetem Kösleményei* 1957, *1*, 55; *CA 55*, 3175h.

COMPONENTS:	ORIGINAL MEASUREMENTS:
1. Nitrous oxide; N_2O; [10024-97-2] 2. Dodecane; $C_{12}H_{26}$; [112-40-3]	Makranczy, J.; Megyery-Balog, K.; Rusz, L.; Patyi, L. *Hung. J. Ind. Chem.* 1976, *4(2)*, 269-280.
VARIABLES: T/K: 298.15 - 313.15 *P*/kPa: 101.325 (1 atm)	PREPARED BY: S. A. Johnson

EXPERIMENTAL VALUES:

T/K	Mole fraction x_{N_2O}	Bunsen coefficient α	Ostwald coefficient L
298.15	0.01723	1.720	1.877
313.15	0.01472	1.444	1.655

Mole fractions and Bunsen coefficients were calculated by the compiler.

AUXILIARY INFORMATION

METHOD/APPARATUS/PROCEDURE:

Volumetric method, described in detail in reference (1).

SOURCE AND PURITY OF MATERIALS:

Analytical grade reagents of Hungarian and foreign origin were used (both liquids and gases). No further information.

ESTIMATED ERROR:

$\delta x_{N_2O}/x_{N_2O} = \pm 0.03$.

REFERENCES:

1. Bodor, E.; Bor, Gy.; Mohai, B.; Sipos, G. *Veszprémi Vegyipari Egyetem Közleményei* 1957, *1*, 55; *CA 55*, 3175h.

COMPONENTS:	ORIGINAL MEASUREMENTS:
1. Nitrous oxide; N_2O; [10024-97-2] 2. Tridecane, $C_{13}H_{28}$; [629-50-5]	Makranczy, J.; Megyery-Balog, K.; Rusz, L.; Patyi, L. *Hung. J. Ind. Chem.* 1976, *4(2)*, 269-280.
VARIABLES:	PREPARED BY:
T/K: 298.15 - 313.15 P/kPa: 101.325 (1 atm)	S. A. Johnson

EXPERIMENTAL VALUES:

T/K	Mole fraction x_{N_2O}	Bunsen coefficient α	Ostwald coefficient L
298.15	0.01750	1.630	1.779
313.15	0.01456	1.333	1.528

Mole fractions and Bunsen coefficients were calculated by the compiler.

AUXILIARY INFORMATION

METHOD/APPARATUS/PROCEDURE:

Volumetric method, described in detail in reference (1).

SOURCE AND PURITY OF MATERIALS:

Analytical grade reagents of Hungarian and foreign origin were used (both liquids and gases). No further information.

ESTIMATED ERROR:

$\delta x_{N_2O}/x_{N_2O}$ = ±0.03.

REFERENCES:

1. Bodor, E.; Bor, Gy.; Mohai, B.; Sipos, G. *Veszprémi Vegyipari Egyetem Közleményei* 1957, *1*, 55; *CA 55*, 3175h.

COMPONENTS:	ORIGINAL MEASUREMENTS:
1. Nitrous oxide; N_2O; [10024-97-2] 2. Tetradecane, $C_{14}H_{30}$; [629-59-4]	Makranczy, J.; Megyery-Balog, K.; Rusz, L.; Patyi, L. *Hung. J. Ind. Chem.* 1976, *4(2)*, 269-280.
VARIABLES: T/K: 298.15 - 313.15 *P*/kPa: 101.325 (1 atm)	PREPARED BY: S. A. Johnson

EXPERIMENTAL VALUES:

T/K	Mole fraction x_{N_2O}	Bunsen coefficient α	Ostwald coefficient L
298.15	0.01718	1.500	1.637
313.15	0.01445	1.239	1.421

Mole fractions and Bunsen coefficients were calculated by the compiler.

AUXILIARY INFORMATION

METHOD/APPARATUS/PROCEDURE:

Volumetric method, described in detail in reference (1).

SOURCE AND PURITY OF MATERIALS:

Analytical grade reagents of Hungarian and foreign origin were used (both liquids and gases). No further information.

ESTIMATED ERROR:

$\delta x_{N_2O}/x_{N_2O} = \pm 0.03$

REFERENCES:

1. Bodor, E.; Bor, Gy.; Mohai, B.; Sipos, G. *Veszprémi Vegyipari Egyetem Köslerményei* 1957, *1*, 55; *CA 55*, 3175h.

COMPONENTS:	ORIGINAL MEASUREMENTS:
1. Nitrous oxide; N_2O; [10024-97-2] 2. Pentadecane; $C_{15}H_{32}$; [629-62-9]	Makranczy, J.; Megyery-Balog, K.; Rusz, L.; Patyi, L. *Hung. J. Ind. Chem.* 1976, *4(2)*, 269-280.
VARIABLES: T/K: 298.15 - 313.15 *P*/kPa: 101.325 (1 atm)	PREPARED BY: S. A. Johnson

EXPERIMENTAL VALUES:

T/K	Mole fraction x_{N_2O}	Bunsen coefficient α	Ostwald coefficient L
298.15	0.01729	1.420	1.550
313.15	0.01429	1.154	1.323

Mole fractions and Bunsen coefficients were calculated by the compiler.

AUXILIARY INFORMATION

METHOD/APPARATUS/PROCEDURE:

Volumetric method, described in detail in reference (1).

SOURCE AND PURITY OF MATERIALS:

Analytical grade reagents of Hungarian and foreign origin were used (both liquid and gases). No further information.

ESTIMATED ERROR:

$\delta x_{N_2O}/x_{N_2O} = \pm 0.03$.

REFERENCES:

1. Bodor, E.; Bor, Gy.; Mohai, B.; Sipos, G. *Veszprémi Vegyipari Egyetem Köslemenyei* 1957, *1*, 55; *CA 55*, 3175h.

COMPONENTS:	ORIGINAL MEASUREMENTS:
1. Nitrous oxide; N_2O; [10024-97-2] 2. Hexadecane; $C_{16}H_{34}$; [544-76-3]	Makranczy, J.; Megyery-Balog, K.; Rusz, L.; Patyi, L. *Hung. J. Ind. Chem.* 1976, *4(2)*, 269-280.
VARIABLES: T/K: 298.15 - 313.15 P/kPa: 101.325 (1 atm)	PREPARED BY: S. A. Johnson

EXPERIMENTAL VALUES:

T/K	Mole fraction x_{N_2O}	Bunsen coefficient α	Ostwald coefficient L
298.15	0.01753	1.360	1.484
313.15	0.01435	1.095	1.255

Mole fractions and Bunsen coefficients were calculated by the compiler.

AUXILIARY INFORMATION

METHOD/APPARATUS/PROCEDURE:

Volumetric method, described in detail in reference (1).

SOURCE AND PURITY OF MATERIALS:

Analytical grade reagents of Hungarian and foreign origin were used (both liquids and gases). No further information.

ESTIMATED ERROR:

$\delta x_{N_2O}/x_{N_2O} = \pm 0.03$.

REFERENCES:

1. Bodor, E.; Bor, Gy.; Mohai, B.; Sipos, G. *Veszprémi Vegyipari Egyetem Közlemënyei* 1957, *1*, 55; *CA 55*, 3175h.

COMPONENTS:

1. Nitrous oxide; N_2O; [10024-97-2]

2. Cyclohexane; C_6H_{12}; [110-82-7]

ORIGINAL MEASUREMENTS:

Patyi, L.; Furmer; I. E.; Makranczy, J.; Sadilenko, A. S.; Stepanova, Z. G.; Berengarten, M. G.
Zh. Prikl. Khim. 1978, *51*, 1296-1300.

VARIABLES:

PREPARED BY:

C. L. Young

EXPERIMENTAL VALUES:

T/K	α^*	Mole fraction of nitrous oxide at a partial pressure of 101.325 kPa
298.15	1.78	0.00853

* volume of gas (measured at 101.325 kPa and 273.15 K) dissolved by one volume of cyclohexane.

AUXILIARY INFORMATION

METHOD/APPARATUS/PROCEDURE:

Volumetric method. Pressure measured when known amounts of gas were added, in increments, to a known amount of liquid in a vessel of known dimensions. Corrections were made for the partial pressure of solvent. Details in ref. (1).

SOURCE AND PURITY OF MATERIALS:

Purity better than 99 mole per cent as determined by gas chromatography.

ESTIMATED ERROR:

$\delta T/K = \pm 0.1$; $\delta\alpha = \pm 4\%$ or less.

REFERENCES:

1. Bodor, E.; Bor, G. J.; Mohai, B.; Sipos, G. *Veszpremi. Vegyip. Egy. Kozl.* 1957, *1*, 55.

COMPONENTS:	ORIGINAL MEASUREMENTS:
1. Nitrous oxide; N_2O; [10024-97-2] 2. Benzene; C_6H_6; [71-43-2]	Horiuti, J. *Sci. Pap. Inst. Phys. Chem. Res. (Japan)*[†] 1931/32, *17*, 125-256.
VARIABLES: Temperature	PREPARED BY: M. E. Derrick

EXPERIMENTAL VALUES: Total pressure is 1 atm for original measurements.

T/K	$10^4 x_{N_2O}$	Ostwald coefficient, L *	Bunsen coefficient, α
313.15	109.4	3.123	2.724
298.15	132.9	3.686	3.377
283.15	165.5	4.453	4.296

* original data

Smoothed Data

$\Delta G° = -RT \ln x_{N_2O} = (-10178.1 + 70.0506T/K)$ J mol^{-1}

Std. dev. $\Delta G°$ = 2.86 J mol^{-1}. Coef. Corr. = 1.00

$\Delta H°$/J mol^{-1} = -10178.1; $\Delta S°$/J mol^{-1} K^{-1} = -70.0506

T/K	$\Delta G°$/J mol^{-1}	x_{N_2O}
283.15	9656.69	0.01654
288.15	10006.9	0.01535
293.15	10357.2	0.01427
298.15	10707.4	0.01331
303.15	11057.7	0.01244
308.15	11408.0	0.01165
313.15	11758.2	0.01093

AUXILIARY INFORMATION

METHOD/APPARATUS/PROCEDURE:

Composed of a gas buret, a solvent reservoir, and an absorption pipet. Volume of pipet is determined at various heights of the meniscus using a weighed quantity of water, measuring the height of the meniscus using a cathetometer. Dry gas is introduced into degassed solvent. System is mixed using a magnetic stirrer until saturation occurs. Care is taken to prevent solvent vapor from mixing with gas in the gas buret. Volume of gas determined from gas buret reading; volume of liquid determined from height of meniscus in the absorption pipet.

SOURCE AND PURITY OF MATERIALS:

1. Nitrous oxide prepared from hydroxylamine sulfate and sodium nitrite both from Kahlbaum (extra pure grade). Gas washed to remove nitric oxide and dried. Fractionated.
2. C_6H_6 Extra pure, free from sulfur sample from Merck refluxed with sodium amalgam. B.P. at normal *P*, 80.18 °C.

ESTIMATED ERROR:

$\delta T/K = \pm 0.1$.

REFERENCES:

† Also reported in: Horiuti, J. *Bull. Inst. Phys. Chem. Rev., Tokyo* 1928, *7(2)*, 119-172.

COMPONENTS:	ORIGINAL MEASUREMENTS:
1. Nitrous oxide; N_2O; [10024-97-2] 2. Benzene, C_6H_6; [71-43-2] (Incorrectly given as C_6H_{16} in original paper.)	Yen, L. C.; McKetta, J. J., Jr. *J. Chem. Eng. Data* 1962, *7*, 288-289.
VARIABLES:	PREPARED BY:
Temperature	W. Gerrard/C. L. Young

EXPERIMENTAL VALUES:

T/K	Bunsen coefficient	Mole fraction of nitrous oxide, x_{N_2O}
283.16	4.256	0.01652
293.16	3.675	0.01446
303.16	3.173	0.01266
313.16	2.786	0.01127

The partial pressure of the gas was 101.325 kPa.

Smoothed Data

$$\Delta G^\circ = -RT \ln x_{N_2O} = (-9941 + 67.449 \times T/K)\ \mathrm{J\ mol^{-1}}$$
(Std. deviation = 5.5 J mol^{-1})

T/K	Mole fraction of nitrous oxide, x_{N_2O}
283.16	0.01654
293.16	0.01442
303.16	0.01269
313.16	0.01126

AUXILIARY INFORMATION

METHOD /APPARATUS/PROCEDURE:

Measurement of volume of gas absorbed by a known volume of degassed liquid at a partial pressure of 101.325 kPa. The vapor pressure of the liquid was allowed for. Gas buret and absorption pipet. Modified form of apparatus and technique used by Markham and Kobe (1).

SOURCE AND PURITY OF MATERIALS:

1. Gas of 98 per cent purity was purified at 193.16 K. Mass spectrograph then showed purity of 99.5 per cent.
2. J. T. Baker Chemical Co. sample; Baker analyzed grade freshly fractionated.

ESTIMATED ERROR:

$\delta T/K = \pm 0.1$; $\delta x_{N_2O} = \pm 2\%$ (estimated by compiler).

REFERENCES:

(1) Markham, A. E.; Kobe, K. A. *J. Am. Chem. Soc.* 1941, *63*, 449.

COMPONENTS:	ORIGINAL MEASUREMENTS:
1. Nitrous oxide; N_2O; [10024-97-2] 2. Benzene; C_6H_6; [71-43-2]	Patyi, L.; Furmer, I. E.; Makranczy, J.; Sadilenko, A. S.; Stepanova, Z. G.; Berengarten, M. G. *Zh. Prikl. Khim.* 1978, *51*, 1296-1300.
VARIABLES:	PREPARED BY: C. L. Young

EXPERIMENTAL VALUES:

T/K	α^*	Mole fraction of nitrous oxide at a partial pressure of 101.325 kPa
298.15	3.30	0.01309

* volume of gas (measured at 101.325 kPa and 273.15 K) dissolved by one volume of benzene.

AUXILIARY INFORMATION

METHOD/APPARATUS/PROCEDURE:

Volumetric method. Pressure measured when known amounts of gas were added, in increments, to a known amount of liquid in a vessel of known dimensions. Corrections were made for the partial pressure of solvent. Details in ref. (1).

SOURCE AND PURITY OF MATERIALS:

Purity better than 99 mole per cent as determined by gas chromatography.

ESTIMATED ERROR:

$\delta T/K = \pm 0.1$; $\delta\alpha = \pm 4\%$ or less.

REFERENCES:

1. Bodor, E.; Bor, G. J.; Mohai, B.; Sipos, G. *Veszpremi. Vegyip. Egy. Kozl.* 1957, *1*, 55.

COMPONENTS:	ORIGINAL MEASUREMENTS:
1. Nitrous oxide; N_2O; [10024-97-2] 2. Methanol; CH_3OH; [67-56-1]	Kunerth, W. *Phys. Rev.* 1922, *19*, 512-524.
VARIABLES:	PREPARED BY:
Temperature	W. Gerrard/C. L. Young

EXPERIMENTAL VALUES:

T/K	Ostwald coefficient, L	Temperature coefficient, dL/Ldt	Mole fraction, x_{N_2O}*
291.16	3.38	0.0040	0.00572
293.16	3.32	0.0058	0.00560
295.16	3.27	0.0077	0.00549
297.16	3.23	0.0097	0.00540
299.16	3.19	0.0118	0.00531
301.16	3.07	0.0138	0.00508
303.16	2.98	0.0157	0.00491
305.16	2.88	0.0177	0.00473
307.16	-	0.0197	-
309.16	-	0.0217	-

* The x_{N_2O} values were calculated by the compiler, the molar volume of N_2O being taken as 22257 × (T/273.16) cm^3, based on the standard density of 1.9775 g dm^{-3} at 273.16 K and 101.325 kPa (ref. 1). The L values were for a total pressure equal to the prevailing barometric pressure. Kunerth's value for x_{N_2O} at 293.16 K was 0.0053.

(cont.)

AUXILIARY INFORMATION

METHOD/APPARATUS/PROCEDURE:

The volume of gas absorbed by a measured volume of degassed liquid was measured at a total pressure equal to the prevailing barometric pressure. Based on the technique of McDaniel (ref. 2). Apparatus consisted of a gas buret attached to leveling tube containing mercury, and to pipet. A layer of liquid was held over the mercury layer in the buret to saturate the gas with vapor of the liquid. Buret and pipet were kept at a common temperature by electric coils.

SOURCE AND PURITY OF MATERIALS:

1. Gas (S.S.White Dental Co.) of 99.7 per cent purity was frozen to remove volatile gases before being passed over P_2O_5. Density was found to be 1.968 g dm^{-3} at 273.16 K and 101.325 kPa.

2. Attested by b.p. and density.

ESTIMATED ERROR:

$\delta T/K = \pm 0.1$; $\delta x_{N_2O} = \pm 2\%$

(estimated by compiler).

REFERENCES:

1. Kaye, G. W. C. and Laby, T. H.; *Tables of Physical and Chemical Constants*, Longmans, London, 1966.

2. McDaniel, A. S.; *J. Phys. Chem.* 1911, *15*, 587.

COMPONENTS:	ORIGINAL MEASUREMENTS:
1. Nitrous oxide; N_2O; [10024-97-2] 2. Methanol; CH_3OH; [67-56-1]	Kunerth, W. *Phys. Rev.* 1922, *19*, 512-524.

EXPERIMENTAL VALUES:

Smoothed Data

$$\Delta G° = -RT \ln x_{N_2O} = -(120034 - 794.83(T/K) + 1.4615(T/K)^2) \text{ J mol}^{-1}$$

(Std. deviation = 12.50 J mol^{-1})

T/K	Mole fraction of nitrous oxide, x_{N_2O}
291.16	0.00569
293.16	0.00562
295.16	0.00552
297.16	0.00540
299.16	0.00526
301.16	0.00510
303.16	0.00492
305.16	0.00473

COMPONENTS:	ORIGINAL MEASUREMENTS:
1. Nitrous oxide; N_2O; [10024-97-2] 2. Alcohols	Hsu, H.; Campbell, D. *Aerosol Age.* 1964, *December*, 34.
VARIABLES:	PREPARED BY: W. Gerrard / C.L. Young

EXPERIMENTAL VALUES:

Solvent	T/K	Ostwald coefficient, L	Mole fraction* of nitrous oxide in liquid, x_{N_2O}
Methanol; CH_4O; [67-56-1]	294.3	3.20	0.0054
Ethanol; C_2H_6O; [64-17-5]	294.3	2.96	0.0072
3-Methyl-1-butanol, , (*iso*-amyl alcohol); $C_5H_{12}O$; [123-51-3]	294.3	2.3	0.0103
Cyclohexanol; $C_6H_{12}O$; [108-93-0]	294.3	0.23	0.0010
1,2,3-Propanetriol, (Glycerol); $C_3H_8O_3$; [56-81-5]	294.3	1.2	0.00365

* calculated by compiler as for a partial pressure of gas of 101.325 kPa. Molar volume of gas at 294.26K taken as 23967 cm^3 based on the density (1) 1.9775 g dm^{-3} at 101.325 kPa and 273.15K.

AUXILIARY INFORMATION

METHOD/APPARATUS/PROCEDURE:

Copper equilibrium cell fitted with Bourdon gauge and thermometer. Total amount of gas needed to attain given but not stated, pressure measure. Ostwald coefficient calculated from knowledge of volume of liquid and container.

SOURCE AND PURITY OF MATERIALS:

No details given.

ESTIMATED ERROR:

$\delta T/K = \pm 0.5$; $\delta x_{N_2O} = \pm 3\%$.

(estimated by compiler).

REFERENCES:

1. Kaye, G.W.C.; Laby, T.H. *Tables of Physical and Chemical Constants*, Longmans, London, 1966

COMPONENTS:

1. Nitrous oxide; N_2O; [10024-97-2]

2. Alcohols

ORIGINAL MEASUREMENTS:

Sada, E.; Kito, S.; Ito, Y.

Eng. Chem. Fundam, 1975, *14*, 232-237

VARIABLES:

PREPARED BY:

W. Gerrard/C.L. Young

EXPERIMENTAL VALUES:

Solvent	T/K	Henry's law constant /atm	Mole fraction* of nitrous oxide in liquid, x_{N_2O}
Methanol; CH_4O; [67-56-1]	298.15	190.6	0.00525
Ethanol; C_2H_6O; [64-17-5]	298.15	145.8	0.00686
1-Propanol; C_3H_8O; [71-23-8]	298.15	125.8	0.00795
2-Propanol; C_3H_8O; [67-63-0]	298.15	125.8	0.00795

* Calculated by compiler for a partial pressure of nitrous oxide = 101.325 kPa.

NOTE: The Henry's law constant appears to have been derived by dividing the observed, but unspecified, pressure in atm by the calculated mole fraction, x_1, for that pressure.

The mole fraction, x_1, was calculated by 1/ (Henry's law constant).

AUXILIARY INFORMATION

METHOD:/APPARATUS/PROCEDURE:

A gas volumetric method (1) was used.

Equilibrium established between a measured volume of gas and a measured amount of gas-free liquid in a cell fitted with a magnetic stirrer.

SOURCE AND PURITY OF MATERIALS:

Nitrous oxide was used from a commercial cylinder (Japan), and stated to be of a purity better than 99.8%, as attested by gas-chromatography.

The purity of the alcohol was stated to be satisfactory (2).

ESTIMATED ERROR:

REFERENCES:

(1) Onda, K.; Sada, E.; Kobayashi, T.; Kito, S.; Ito, K. *J. Chem. Eng. Jpn.* 1970, *3*,18; 137.

(2) Sada, E., Kito, S.; Ito, Y. *J. Chem. Eng. Jpn.* 1974, *7*, 57.

COMPONENTS:	ORIGINAL MEASUREMENTS:
1. Nitrous oxide; N_2O; [10024-97-2] 2. Methanol; CH_4O; [67-56-1]	Makranczy, J.; Rusz, L.; Balog-Megyery, K. *Hung. J. Ind. Chem.* 1979, *7*, 41-6.
VARIABLES:	PREPARED BY: C.L. Young

EXPERIMENTAL VALUES:

T/K	P^+/kPa	Ostwald coefficient	Mole fraction of nitrous oxide *, x_{N_2O}
298.15	101.3	3.755	0.00626

* calculated by compiler using a molar volume of nitrous oxide of 0.02430 m^3 at 298.15K

\+ partial pressure of nitrous oxide.

AUXILIARY INFORMATION

METHOD/APPARATUS/PROCEDURE:

Volumetric method. The apparatus of Bodor, Bor, Mohai and Sipos (1) was used.

SOURCE AND PURITY OF MATERIALS:

No details given.

ESTIMATED ERROR:

$\delta x_{N_2O} = \pm 3\%$

REFERENCES:

1. Bodor, E.; Bor, Gy.; Mohai, B.; Sipos. G. *Veszpremi Vegyip. Egy. Kozl.* 1957, *1*, 55. *Chem. Abstr.* 1961, *55*, 3175

COMPONENTS:	ORIGINAL MEASUREMENTS:
1. Nitrous oxide; N_2O; [10024-97-2] 2. Ethanol, C_2H_6OH; [64-17-5]	Carius, L., *Annalen*, 1855, *94*, 129-166.
VARIABLES:	**PREPARED BY:**
Temperature	W. Gerrard

EXPERIMENTAL VALUES:

T/K	Bunsen coefficient, α	Mole fraction, x_1 (Calculated by compiler)*
275.45	4.0262	0.010345
280.15	3.7069	0.009575
284.75	3.4219	0.008882
291.35	3.1105	0.008128
296.15	2.8861	0.007581

Bunsen coefficient, $\alpha = 4.17805 - 0.0698160\ t + 0.0006090\ t^2$

(From 273.15 to 298.15 K) $t = T/K - 273.15$.

Henrich (1) used Carius's data to give the modified smoothing equation :

$$\alpha = 4.1902 - 0.074389\ t + 0.00078226\ t^2$$

Note: Henrich did not give any experimental data.

* The gas molecular volume of N_2O at 273.15 K and 101.325 kPa was taken to be 22250 cm^3, based on the standard density of 1.9775 g/dm^3.

AUXILIARY INFORMATION

METHOD/APPARATUS/PROCEDURE:	SOURCE AND PURITY OF MATERIALS:
Measurement of volume by the Bunsen gas buret. Gas placed in an absorption tube and its pressure and volume determined. Liquid was then added, the system shaken until volume constant. Pressure and volume redetermined.	1. Self prepared from pure ammonium nitrate. 2. Distilled "absolute" $d^{20} = 0.792$.
	ESTIMATED ERROR:
	REFERENCES: 1. Henrich, F.; *Z. Phys. Chem.* 1892, *9*, 435.

COMPONENTS:

1. Nitrous oxide; N_2O; [10024-97-2]

2. Ethanol; C_2H_6O; [64-17-5]

ORIGINAL MEASUREMENTS:

Kunerth, W.

Phys. Rev. 1922, *19*, 512-524.

VARIABLES:

Temperature

PREPARED BY:

W. Gerrard/C. L. Young

EXPERIMENTAL VALUES:

T/K	Ostwald coefficient, L	Temperature coefficient, dL/Ldt	Mole fraction, x_{N_2O}*
291.16	3.07	0.0116	0.00748
293.16	2.99	0.0123	0.00725
295.16	2.91	0.0132	0.00702
297.16	2.85	0.0140	0.00685
299.16	2.77	0.0148	0.00663
301.16	2.68	0.0157	0.00640
303.16	2.61	0.0167	0.00620
305.16	2.52	0.0174	0.00596
307.16	2.43	0.0184	0.00572
309.16	2.33	0.0190	0.00546

* The x_{N_2O} values were calculated by the compiler, the molar volume of N_2O being taken as 22257 × (T/273.16) cm^3, based on the standard density of 1.9775 g dm^{-3} at 273.16 K and 101.325 kPa (ref. 1). The L values were for a total pressure equal to the prevailing barometric pressure. Kunerth's value for x_{N_2O} at 293.16 K was 0.0072.

(cont.)

AUXILIARY INFORMATION

METHOD/APPARATUS/PROCEDURE:

The volume of gas absorbed by a measured volume of degassed liquid was measured at a total pressure equal to the prevailing barometric pressure. Based on the technique of McDaniel (ref. 2). Apparatus consisted of a gas buret attached to leveling tube containing mercury, and to pipet. A layer of liquid was held over the mercury layer in the buret to saturate the gas with vapor of the liquid. Buret and pipet were kept at a common temperature by electric coils.

SOURCE AND PURITY OF MATERIALS:

1. Gas (S.S.White Dental Co.) of 99.7 per cent purity was frozen to remove volatile gases before being passed over P_2O_5. Density was found to be 1.968 g dm^{-3} at 273.16 K and 101.325 kPa.

2. Attested by b.p. and density

DATA CLASS:

ESTIMATED ERROR:

$\delta T/K = \pm 0.1$; $\delta x_{N_2O} = \pm 2\%$

(estimated by compiler).

REFERENCES:

1. Kaye, G. W. C. and Laby, T. H. *Tables of Physical and Chemical Constants*, Longmans, London, 1966.

2. McDaniel, A. S.; *J. Phys. Chem.* 1911, *15*, 587.

COMPONENTS:	ORIGINAL MEASUREMENTS:
1. Nitrous oxide; N_2O; [10024-97-2] 2. Ethanol; C_2H_6O; [64-17-5]	Kunerth, W. *Phys. Rev.* 1922, *19*, 512-524.

EXPERIMENTAL VALUES:

Smoothed Data

$$\Delta G° = -RT \ln x_{N_2O} = -(56414.4 - 376.86(T/K) + 0.7688(T/K)^2) \text{ J mol}^{-1}$$

(Std. deviation = 7.3 J mol^{-1})

T/K	Mole fraction of nitrous oxide, x_{N_2O}
291.16	0.00745
293.16	0.00726
295.16	0.00706
297.16	0.00685
299.16	0.00663
301.16	0.00641
303.16	0.00618
305.16	0.00595
307.16	0.00571
309.16	0.00548

COMPONENTS:

1. Nitrous oxide; N_2O; [10024-97-2]
2. Ethanol; C_2H_6O; [64-17-5]

ORIGINAL MEASUREMENTS:

Makranczy, J.; Rusz, L.; Balog-Megyery, K.

Hung. J. Ind. Chem. 1979, *7*, 41-6

VARIABLES:

PREPARED BY:

C.L. Young

EXPERIMENTAL VALUES:

T/K	P^+/kPa	Ostwald coefficient	Mole fraction of nitrous oxide*, x_{N_2O}
298.15	101.3	3.078	0.00738

* calculated by compiler using a molar volume of nitrous oxide of 0.02430 m^3 at 298.15 K.

\+ partial pressure of nitrous oxide.

AUXILIARY INFORMATION

METHOD APPARATUS/PROCEDURE:

Volumetric method. The apparatus of Bodor, Bor, Mohai, and Sipos (1) was used.

SOURCE AND PURITY OF MATERIALS:

No details given.

ESTIMATED ERROR:

$\delta x_{N_2O} = \pm 3\%$

REFERENCES:

1. Bodor, E.; Bor, Gy.; Mohai, B.; Sipos, G.
Veszpremi Vegyip. Egy. Kozl. 1957, *1*, 55.
Chem. Abstr. 1961, *55*, 3175h

COMPONENTS:

1. Nitrous oxide; N_2O; [10024-97-2]
2. 1-Propanol; C_3H_8O; [71-23-8]

ORIGINAL MEASUREMENTS:

Makranczy, J.; Rusz, L.; Balog-Megyery, K.

Hung. J. Ind. Chem. 1979, *7*, 41-6.

VARIABLES:

PREPARED BY:

C.L. Young

EXPERIMENTAL VALUES:

T/K	P^{+}/kPa	Ostwald coefficient	Mole fraction of nitrous oxide*, x_{N_2O}
298.15	101.3	2.740	0.008404

* calculated by compiler using a molar volume of nitrous oxide of 0.02430 m^3 at 298.15 K.

\+ partial pressure of nitrous oxide.

AUXILIARY INFORMATION

METHOD/APPARATUS/PROCEDURE:

Volumetric method. The apparatus of Bodor, Bor, Mohai and Sipos (1) was used.

SOURCE AND PURITY OF MATERIALS:

No details given.

ESTIMATED ERROR:

$\delta x_{N_2O} = \pm 3\%$

REFERENCES:

1. Bodor, E.; Bor, Gy.; Mohai, B.; Sipos, G. *Veszpremi Vegyip. Egy. Kozl.* 1957, *1*, 55. *Chem. Abstr.* 1961, *55*, 3175h

COMPONENTS:

1. Nitrous oxide; N_2O; [10024-97-2]

2. 2-Propanone (Acetone); C_3H_6O; [67-64-1]

ORIGINAL MEASUREMENTS:

Kunerth, W.

Phys. Rev. 1922, *19*, 512-524.

VARIABLES:

Temperature

PREPARED BY:

W. Gerrard/C. L. Young

EXPERIMENTAL VALUES:

T/K	Ostwald coefficient, L	Temperature coefficient, dL/Ldt	Mole Fraction, x_{N_2O}*
291.16	6.30	0.0150	0.0191
293.16	6.03	0.0196	0.0182
295.16	5.78	0.0244	0.0174
297.16	5.50	0.0294	0.0165
299.16	5.21	0.0342	0.0156
301.16	4.84	0.0393	0.0145
303.16	4.46	0.0440	0.0133
305.16	4.07	0.0492	0.0121
307.16	3.66	0.0540	0.0110
309.16	3.23	0.0590	0.0095

* The x_{N_2O} values were calculated by the compiler, the molar volume of N_2O being taken as 22257 × (T/273.16) cm^3, based on the standard density of 1.9775 g dm^{-3} at 273.16 K and 101.325 kPa (ref. 1). The L values were for a total pressure equal to the prevailing barometric pressure. Kunerth's value for x_{N_2O} at 293.16 K was 0.0185.

(cont.)

AUXILIARY INFORMATION

METHOD /APPARATUS/PROCEDURE:

The volume of gas absorbed by a measured volume of degassed liquid was measured at a total pressure equal to the prevailing barometric pressure. Based on the technique of McDaniel (ref. 2). Apparatus consisted of a gas buret attached to leveling tube containing mercury, and to pipet. A layer of liquid was held over the mercury layer in the buret to saturate the gas with vapor of the liquid. Buret and pipet were kept at a common temperature by electric coils.

SOURCE AND PURITY OF MATERIALS:

1. Gas (S.S.White Dental Co.) of 99.7 per cent purity was frozen to remove volatile gases before being passed over P_2O_5. Density was found to be 1.968 g dm^{-3} at 273.16 K and 101.325 kPa.

2. Attested by b.p. and density.

ESTIMATED ERROR:

$\delta T/K = \pm 0.1$; $\delta x_{N_2O} = \pm 2\%$

(estimated by compiler).

REFERENCES:

1. Kaye, G. W. C. and Laby, T. H.; *Tables of Physical and Chemical Constants*, Longmans, London, 1966.

2. McDaniel, A. S.; *J. Phys. Chem.* 1911, *15*, 587.

COMPONENTS:	ORIGINAL MEASUREMENTS:
1. Nitrous oxide; N_2O; [10024-97-2] 2. 2-Propanone (Acetone); C_3H_6O; [67-64-1]	Kunerth, W. *Phys. Rev.* 1922, *19*, 512-524.

EXPERIMENTAL VALUES:

Smoothed Data

$$\Delta G° = -RT \ln x_{N_2O} = -(299647 - 2055.7(T/K) + 3.639(T/K)^2) \text{ J mol}^{-1}$$

(Std. deviation = 23.5 J mol^{-1})

T/K	Mole fraction of nitrous oxide, x_{N_2O}
291.16	0.0189
293.16	0.0183
295.16	0.0175
297.16	0.0166
299.16	0.0156
301.16	0.0146
303.16	0.0133
305.16	0.0121
307.16	0.0108
309.16	0.0096

COMPONENTS:	ORIGINAL MEASUREMENTS:
1. Nitrous oxide; N_2O; [10024-97-2] 2. 2-Propanone (Acetone); C_3H_6O; [67-64-1]	Horiuti, J. *Sci. Pap. Inst. Phys. Chem. Res. (Japan)*† 1931/32, *17*, 125-256.
VARIABLES:	PREPARED BY:
Temperature	M. E. Derrick

EXPERIMENTAL VALUES: Total pressure is 1 atm for original measurements.

T/K	$10^2\, x_{N_2O}$	Ostwald coefficient, L *	Bunsen coefficient, α
313.15	1.371	4.73	4.13
298.15	1.765	5.95	5.45
283.15	2.325	7.64	7.37
271.74	2.888	9.30	9.35

* original data

Smoothed Data

$$\Delta G° = -RT \ln x_{N_2O} = (-12769.1 + 76.4162T/K)\ \mathrm{J\ mol^{-1}}$$

Std. dev. $\Delta G°$ = 11.95 J mol^{-1}. Coef. Corr. = 1.00

$\Delta H°/\mathrm{J\ mol^{-1}}$ = -12769.1; $\Delta S°/\mathrm{J\ mol^{-1}\ K^{-1}}$ = -76.4162

T/K	$\Delta G°/\mathrm{J\ mol^{-1}}$	x_{N_2O}
268.15	7721.94	0.03132
273.15	8104.02	0.02820
278.15	8486.10	0.02549
283.15	8868.18	0.02312
288.15	9250.27	0.02104
293.15	9632.34	0.01922
298.15	10014.4	0.01760
303.15	10396.5	0.01617
308.15	10778.6	0.01489
313.15	11160.7	0.01375

AUXILIARY INFORMATION

METHOD/APPARATUS/PROCEDURE:

Composed of a gas buret, a solvent reservoir, and an absorption pipet. Volume of pipet is determined at various heights of the meniscus using a weighed quantity of water, measuring the height of the meniscus using a cathetometer. Dry gas is introduced into degassed solvent. System is mixed using a magnetic stirrer until saturation occurs. Care is taken to prevent solvent vapor from mixing with gas in the gas buret. Volume of gas determined from gas buret reading; volume of liquid determined from height of meniscus in the absorption pipet.

SOURCE AND PURITY OF MATERIALS:

1. Nitrous oxide prepared from hydroxylamine sulfate and sodium nitrite both from Kahlbaum (extra pure grade). Gas washed to remove nitric oxide and dried. Fractionated.
2. C_3H_6O No. 1, extra pure from Nippon Pure Chem. Co. or Merck, extra pure, recrystallized with sodium sulfite used. Stored over calcium chloride then fractionated. B.P. at normal P, 56.09 °C.

ESTIMATED ERROR:

$\delta T/K = \pm 0.1$.

REFERENCES:

† Also reported in: Horiuti, J. *Bull. Inst. Phys. Chem. Res., Tokyo* 1928, *7(2)*, 119-172.

COMPONENTS:	ORIGINAL MEASUREMENTS:
1. Nitrous oxide; N_2O; [10024-97-2] 2. Aliphatic compounds containing oxygen.	Hsu, H.; Campbell, D. *Aerosol Age.* 1964, *December*, 34.
VARIABLES:	PREPARED BY: W. Gerrard/C.L. Young

EXPERIMENTAL VALUES:

SOLVENT	T/K	Ostwald coefficient, L	Mole fraction* of nitrous oxide in liquid, x_{N_2O}
2-Propanone, (Acetone); C_3H_6O; [67-64-1]	294.3	5.92	0.0178
1,1´-Oxybisethane, (diethyl ether); $C_4H_{10}O$; [60-29-7]	294.3	7.65	0.0321

* calculated by compiler as for a partial pressure of gas of 101.325 kPa. Molar volume of gas at 294.24K taken as 23967 cm^3 based on the density (1) 1.9775 g dm^{-3} at 101.325 kPa and 273.15K.

AUXILIARY INFORMATION

METHOD/APPARATUS/PROCEDURE:

Copper equilibrium cell fitted with Bourdon gauge and thermometer. Total amount of gas needed to attain given, but not stated, pressure measured. Ostwald coefficient calculated from knowledge of volume of liquid and container.

SOURCE AND PURITY OF MATERIALS:

No details given.

ESTIMATED ERROR:

$\delta T/K = \pm 0.5$; $\delta x_{N_2O} = \pm 3\%$.

(estimated by compiler).

REFERENCES:

1. Kaye, G.W.C.; Laby, T.H. *Tables of Physical and Chemical Constants*, Longmans, London, *1966*.

COMPONENTS:	ORIGINAL MEASUREMENTS:
1. Nitrous oxide; N_2O; [10024-97-2] 2. 1-Butanol; $C_4H_{10}O$; [71-36-3]	Makranczy, J.; Rusz, L.; Balog-Megyery, K. *Hung. J. Ind. Chem.* 1979, *7*, 41-6.
VARIABLES:	PREPARED BY: C.L. Young

EXPERIMENTAL VALUES:

T/K	P^+/kPa	Ostwald coefficient	Mole fraction of nitrous oxide*, x_{N_2O}
298.15	101.3	2.532	0.009494

* calculated by compiler using a molar volume of nitrous oxide of 0.02430 m^3 at 298.15 K.

\+ partial pressure of nitrous oxide.

AUXILIARY INFORMATION

METHOD/APPARATUS/PROCEDURE:

Volumetric method. The apparatus of Bodor, Bor, Mohai, and Sipos (1) was used.

SOURCE AND PURITY OF MATERIALS:

No details given.

ESTIMATED ERROR:

$\delta x_{N_2O} = \pm 3\%$

REFERENCES:

1. Bodor, E.; Bor, Gy.; Mohai, B.; Sipos, G. *Veszpremi Vegyip. Egy. Kozl.* 1957, *1*, 55. *Chem. Abstr.* 1961, *55*, 3175h

COMPONENTS:

1. Nitrous oxide; N_2O; [10024-97-2]
2. 3-Methyl-1-butanol, (*iso*-amyl alcohol); $C_5H_{12}O$; [123-51-3]

ORIGINAL MEASUREMENTS:

Kunerth, W.

Phys. Rev. 1922, *19*, 512-524.

VARIABLES:

Temperature

PREPARED BY:

W. Gerrard/C. L. Young

EXPERIMENTAL VALUES:

T/K	Ostwald coefficient, L	Temperature coefficient, dL/Ldt	Mole fraction, x_{N_2O}*
291.16	2.52	0.010	0.0114
293.16	2.47	0.0098	0.0111
295.16	2.43	0.0097	0.0109
297.16	2.37	0.0096	0.0106
299.16	2.32	0.0096	0.0103
301.16	2.27	0.0095	0.0100
303.16	2.24	0.0095	0.00987
305.16	2.19	0.0094	0.00970
307.16	2.16	0.0094	0.00961
309.16	2.12	0.0093	0.00939

* The x_{N_2O} values were calculated by the compiler, the molar volume of N_2O being taken as 22257 × (T/273.16) cm^3, based on the standard density of 1.9775 g dm^{-3} at 273.16 K and 101.325 kPa (ref. 1). The L values were for a total pressure equal to the prevailing barometric pressure. Kunerth's value for x_{N_2O} at 293.16 K was 0.0111.

(cont.)

AUXILIARY INFORMATION

METHOD /APPARATUS/PROCEDURE:

The volume of gas absorbed by a measured volume of degassed liquid was measured at a total pressure equal to the prevailing barometric pressure. Based on the technique of McDaniel (ref. 2). Apparatus consisted of a gas buret attached to leveling tube containing mercury, and to pipet. A layer of liquid was held over the mercury layer in the buret to saturate the gas with vapor of the liquid. Buret and pipet were kept at a common temperature by electric coils.

SOURCE AND PURITY OF MATERIALS:

1. Gas (S.S.White Dental Co.) of 99.7 per cent purity was frozen to remove volatile gases before being passed over P_2O_5. Density was found to be 1.968 g dm^{-3} at 273.16 K and 101.325 kPa.
2. Attested by b.p. and density.

ESTIMATED ERROR:

$\delta T/K = \pm 0.1$; $\delta x_{N_2O} = \pm 2\%$ (estimated by compiler).

REFERENCES:

1. Kaye, G. W. C. and Laby, T. H.; *Tables of Physical and Chemical Constants*, Longmans, London, 1966.
2. McDaniel, A. S.; *J. Phys. Chem.* 1911, *15*, 587.

COMPONENTS:	ORIGINAL MEASUREMENTS:
1. Nitrous oxide; N_2O; [10024-97-2] 2. 3-Methyl-1-butanol, (*iso*-amyl alcohol); $C_5H_{12}O$; [123-51-3]	Kunerth, W. *Phys. Rev.* 1922, *19*, 512-524.

EXPERIMENTAL VALUES:

Smoothed Data

$$\Delta G° = -RT \ln x_{N_2O} = (48775.2 + 336.14\ (T/K) - 0.4515\ (T/K)^2)\ J\ mol^{-1}$$

(Std. deviation = 12.1 J mol^{-1})

T/K	Mole fraction of nitrous oxide, x_{N_2O}
291.16	0.0114
293.16	0.0111
295.16	0.0108
297.16	0.0106
299.16	0.0103
301.16	0.0101
303.16	0.00989
305.16	0.00971
307.16	0.00955
309.16	0.00941

COMPONENTS:	ORIGINAL MEASUREMENTS:
1. Nitrous oxide; N_2O; [10024-97-2] 2. 1-Pentanol; $C_5H_{12}O$; [71-41-0] or 1-Hexanol; $C_6H_{14}O$; [111-27-3]	Makranczy, J.; Rusz, L.; Balog-Megyery, K. *Hung. J. Ind. Chem.* 1979, *7*, 41-6
VARIABLES:	PREPARED BY: C.L. Young

EXPERIMENTAL VALUES:

T/K	P^+/kPa	Ostwald coefficient	Mole fraction of nitrous oxide *, x_{N_2O}
		1-Pentanol	
298.15	101.3	2.369	0.01048
		1-Hexanol	
298.15	101.3	2.270	0.01157

* calculated by compiler using a molar volume of nitrous oxide of 0.02430 m^3 at 298.15 K.

\+ partial pressure of nitrous oxide.

AUXILIARY INFORMATION

METHOD/APPARATUS/PROCEDURE:

Volumetric method. The apparatus of Bodor, Bor, Mohai and Sipos (1) was used.

SOURCE AND PURITY OF MATERIALS:

No details given

ESTIMATED ERROR:

$\delta x_{N_2O} = \pm 3\%$

REFERENCES:

1. Bodor, E.; Bor, Gy.; Mohai, B.; Sipos, G. *Veszpremi Vegyip Egy. Kozl.* 1957, *1*, 55. *Chem. Abstr.* 1961, *55*, 3175h

COMPONENTS:

1. Nitrous oxide; N_2O; [10024-97-2]
2. Cyclohexanone; $C_6H_{10}O$; [108-94-1]

ORIGINAL MEASUREMENTS:

Rusz, L.; Makranczy, J.; Balog-Megyery, K.; Patyi, L.

Hung. J. Ind. Chem. 1977, *5*, 225-232.

VARIABLES:

Pressure

PREPARED BY:

C.L. Young

EXPERIMENTAL VALUES:

T/K	p/atm	p/MPa	Bunsen coefficient, α	Volume[+] coefficient,	Mole fraction of nitrous oxide§, x_{N_2O}
293.15	10.0	1.01	59.1	1.048	0.214
	20.0	2.03	141.1	1.226	0.394
	29.5	2.99	253.4	1.431	0.538
	37.5	3.80	498.0	1.900	0.696
	47.7	4.83	1179.8	3.400	0.845

+ $V = \dfrac{\text{Volume of gas-saturated liquid}}{\text{Volume of pure solvent}}$

§ calculated by compiler assuming ideal gas molar volume for nitrous oxide.

AUXILIARY INFORMATION

METHOD/APPARATUS/PROCEDURE:

Static glass cell fitted with Teflon-coated magnetic stirrer. Pressure measured by dead-weight gauge and null detector. Volume of saturated liquid measured.

SOURCE AND PURITY OF MATERIALS:

No details given.

ESTIMATED ERROR:

$\delta T/K = \pm 0.1$; $\delta\alpha = \pm 4.0\%$

(estimated by compiler).

REFERENCES:

COMPONENTS:	ORIGINAL MEASUREMENTS:
1. Nitrous oxide; N_2O; [10024-97-2] 2. Cyclohexanol, $C_6H_{12}O$; [108-93-0]	Cauquil, G. *J. Chim. Phys.*, 1927, *24*, 53-55.
VARIABLES:	PREPARED BY: W. Gerrard

EXPERIMENTAL VALUES:

T/K	Ostwald coefficient, L	Mole fraction* of nitrous oxide x_{N_2O}
299.15	0.741	0.00322

Pressure = 101.325 kPa.

* calculated by compiler on basis of molar volume of nitrous oxide at 273.15 K and 101.325 kPa being 22254 cm^3 and density of cyclohexanol being 0.9446 at 299.15 K.

AUXILIARY INFORMATION

METHOD/APPARATUS/PROCEDURE:

Measurement of the initial and final volumes of gas in contact with liquid at 299.15 K and known pressure, nearly 101.325 kPa. Vapor pressure of liquid was ignored. Apparatus appears to be of the Bunsen type, with one absorption vessel.

SOURCE AND PURITY OF MATERIALS:

1. Purity of N_2O not stated.
2. Cyclohexanol, b.p. 334.0 K at 102.125 kPa, degassed and tested to be air free.

ESTIMATED ERROR:

REFERENCES:

COMPONENTS:

1. Nitrous oxide; N_2O; [10024-97-2]

2. 1-Heptanol; $C_7H_{16}O$; [111-70-6]
   or
   1-Octanol; $C_8H_{18}O$; [111-87-5]

ORIGINAL MEASUREMENTS:

Makranczy, J.; Rusz, L.; Balog-Megyery, K.

Hung. J. Ind. Chem. 1979, *7*, 41-6.

VARIABLES:

PREPARED BY:

C.L. Young

EXPERIMENTAL VALUES:

T/K	P^+/kPa	Ostwald coefficient	Mole fraction of nitrous oxide *, x_{N_2O}
		1-Heptanol	
298.15	101.3	2.194	0 0.01265
		1-Octanol	
298.15	101.3	2.139	0.01376

* calculated by compiler using a molar volume of nitrous oxide of 0.02430 m^3 at 298.15 K

\+ partial pressure of nitrous oxide

AUXILIARY INFORMATION

METHOD/APPARATUS/PROCEDURE:

Volumetric method. The apparatus of Bodor, Bor, Mohai, and Sipos (1) was used.

SOURCE AND PURITY OF MATERIALS:

No details given

ESTIMATED ERROR:

$\delta x_{N_2O} = \pm 3\%$

REFERENCES:

1. Bodor, E.; Bor, Gy.; Mohai, B.; Sipos. G.
Veszpremi Vegyip. Egy. Kozl. 1957, *1*, 55.
Chem. Abstr. 1961, *55*, 3175h

COMPONENTS:	ORIGINAL MEASUREMENTS:
1. Nitrous oxide; N_2O; [10024-97-2] 2. 1-Nonanol; $C_9H_{20}O$; [143-08-8] or 1-Decanol; $C_{10}H_{22}O$; [112-30-1]	Makranczy, J.; Rusz, L.; Balog-Megyery, K. *Hung. J. Ind. Chem.* 1979, *7*, 41-6.
VARIABLES:	PREPARED BY: C.L. Young

EXPERIMENTAL VALUES:

T/K	P^+/kPa	Ostwald coefficient	Mole fraction of nitrous oxide*, x_{N_2O}
1-Nonanol; $C_9H_{20}O$; [143-08-8]			
298.15	101.3	2.096	0.01487
1-Decanol; $C_{10}H_{22}O$; [112-30-1]			
298.15	101.3	2.052	0.01591

* calculated by compiler using a molar volume of nitrous oxide of 0.02430 m^3 at 298.15 K.

\+ partial pressure of nitrous oxide.

AUXILIARY INFORMATION

METHOD/APPARATUS/PROCEDURE:

Volumetric method. The apparatus of Bodor, Bor, Mohai and Sipos (1) was used.

SOURCE AND PURITY OF MATERIALS:

No details given.

ESTIMATED ERROR:

$\delta x_{N_2O} = \pm 3\%$

REFERENCES:

1. Bodor, E.; Bor, Gy.; Mohai, B.; Sipos. G.
Veszpremi Vegyip. Egy. Kozl. 1957, *1*, 55.
Chem. Abstr. 1961, *55*, 3175h

COMPONENTS:	ORIGINAL MEASUREMENTS:
1. Nitrous oxide; N_2O; [10024-97-2] 2. 1-Dodecanol; $C_{12}H_{26}O$; [112-53-8] or 1-Undecanol; $C_{11}H_{24}O$; [112-42-5]	Makranczy, J.; Rusz, L.; Balog-Megyery, K. *Hung. J. Ind. Chem.* 1979, *7*, 41-6.
VARIABLES:	PREPARED BY: C.L. Young

EXPERIMENTAL VALUES:

T/K	P^+/kPa	Ostwald coefficient	Mole fraction of nitrous oxide *, x_{N_2O}
	1-Dodecanol; $C_{12}H_{26}O$; [112-53-8]		
298.15	101.3	1.987	0.0180
	1-Undecanol; $C_{11}H_{24}O$; [112-42-5]		
298.15	101.3	2.019	0.0170

* calculated by compiler using a molar volume of nitrous oxide of 0.02430 m^3 at 298.15 K.

\+ partial pressure of nitrous oxide.

AUXILIARY INFORMATION

METHOD/APPARATUS/PROCEDURE:

Volumetric method. The apparatus of Bodor, Bor, Mohai and Sipos (1) was used.

SOURCE AND PURITY OF MATERIALS:

No details given.

ESTIMATED ERROR:

$\delta x_{N_2O} = \pm 3\%$

REFERENCES:

1. Bodor, E.; Bor, Gy.; Mohai, B.; Sipos. G.
Veszpremi Vegyip. Egy. Kozl. 1957, *1*, 55.
Chem. Abstr. 1961, *55*, 3175h

COMPONENTS:

1. Nitrous oxide; N_2O; [10024-97-2]

2. Acetic acid; CH_3CO_2H; [64-19-7]

ORIGINAL MEASUREMENTS:

Kunerth, W.

Phys. Rev. 1922, *19*, 512-524.

VARIABLES:

Temperature

PREPARED BY:

W. Gerrard/C. L. Young

EXPERIMENTAL VALUES:

T/K	Ostwald coefficient, L	Temperature coefficient, dL/Ldt	Mole fraction, x_{N_2O}*
291.16	5.00	0.0160	0.0119
293.16	4.85	0.0162	0.0115
295.16	4.70	0.0163	0.0111
297.16	4.55	0.0163	0.0107
299.16	4.39	0.0164	0.0103
301.16	4.25	0.0164	0.0099
303.16	4.11	0.0165	0.0095
305.16	3.98	0.0165	0.0092
307.16	3.84	0.0165	0.0088
309.16	3.75	0.0166	0.0086

* The x_{N_2O} values were calculated by the compiler, the molar volume of N_2O being taken as 22257 × (T/273.16) cm^3, based on the standard density of 1.9775 g dm^{-3} at 273.16 K and 101.325 kPa (ref. 1). The L values were for a total pressure equal to the prevailing barometric pressure. Kunerth's value for x_{N_2O} at 293.16 K was 0.0115.

(cont.)

AUXILIARY INFORMATION

METHOD/APPARATUS/PROCEDURE:

The volume of gas absorbed by a measured volume of degassed liquid was measured at a total pressure equal to the prevailing barometric pressure. Based on the technique of McDaniel (ref. 2). Apparatus consisted of a gas buret attached to leveling tube containing mercury, and to pipet. A layer of liquid was held over the mercury layer in the buret to saturate the gas with vapor of the liquid, Buret and pipet were kept at a common temperature by electric coils.

SOURCE AND PURITY OF MATERIALS:

1. Gas (S.S.White Dental Co.) of 99.7 per cent purity was frozen to remove volatile gases before being passed over P_2O_5. Density was found to be 1.968 g dm^{-3} at 273.16 K and 101.325 kPa.
2. Attested by b.p. and density.

ESTIMATED ERROR:

$\delta T/K = \pm 0.1$; $\delta x_{N_2O} = \pm 2\%$ (estimated by compiler).

REFERENCES:

1. Kaye, G. W. D. and Laby, T. H.; *Tables of Physical and Chemical Constants*, Longmans, London, 1966.
2. McDaniel, A. S.; *J. Phys. Chem.* 1911, *15*, 587.

COMPONENTS:	ORIGINAL MEASUREMENTS:
1. Nitrous oxide; N_2O; [10024-97-2]	Kunerth, W.
2. Acetic acid; CH_3CO_2H; [64-19-7]	*Phys. Rev.* 1922, *19*, 512-524.

EXPERIMENTAL VALUES:

Smoothed Data

$$\Delta G° = -RT \ln x_{N_2O} = (-13911.9 + 84.578(T/K)) \text{ J mol}^{-1}$$

(Std. deviation = 10.8 J mol^{-1})

T/K	Mole fraction of nitrous oxide, x_{N_2O}
291.16	0.0120
293.16	0.0115
295.16	0.0111
297.16	0.0107
299.16	0.0103
301.16	0.0099
303.16	0.0095
305.16	0.0092
307.16	0.0089
309.16	0.0086

COMPONENTS:	ORIGINAL MEASUREMENTS:
1. Nitrous oxide; N_2O; [10024-97-2] 2. Acetic acid and pentyl ester	Hsu, H.; Campbell, D. *Aerosol Age*, 1964, *December*, 34.
VARIABLES:	**PREPARED BY:** W. Gerrard / C.L. Young

EXPERIMENTAL VALUES:

Solvent	T/K	Ostwald coefficient, L	Mole fraction* of nitrous oxide in liquid, x_{N_2O}
Acetic acid; $C_2H_4O_2$; [64-19-7]	294.3	4.77	0.0113
Acetic acid, pentyl ester, (amyl acetate); $C_7H_{14}O_2$; [628-63-7]	294.3	5.1	0.0305

* calculated by compiler as for a partial pressure of gas of 101.325 kPa. Molar volume of gas at 294.24K taken as 23967 cm^3 based on the density (1) 1.9775 g dm^{-3} at 101.325 kPa and 273.15K.

AUXILIARY INFORMATION

METHOD/APPARATUS/PROCEDURE:

Copper equilibrium cell fitted with Bourdon gauge and thermometer. Total amount of gas needed to attain given, but not stated, pressure measured. Ostwald coefficient calculated from knowledge of volume of liquid and container.

SOURCE AND PURITY OF MATERIALS:

No details given.

ESTIMATED ERROR:

$\delta T/K = \pm 0.5$; $\delta x_{N_2O} = \pm 3\%$.

(estimated by compiler).

REFERENCES:

1. Kaye, G.W.C.; Laby, T.H. *Tables of Physical and Chemical Constants*, Longmans, London, *1966*,

COMPONENTS:	ORIGINAL MEASUREMENTS:
1. Nitrous oxide; N_2O; [10024-97-2] 2. Acetic acid, Methyl ester (Methyl acetate); $C_3H_6O_2$; [79-20-9]	Horiuti, J. *Sci. Pap. Inst. Phys. Chem. Res. (Japan)*† 1931/32, *17*, 125-256.
VARIABLES:	PREPARED BY:
Temperature	M. E. Derrick

EXPERIMENTAL VALUES: Total pressure is 1 atm for original measurements.

T/K	$10^2\ x_{N_2O}$	Ostwald coefficient, L *	Bunsen coefficient, α
313.15	1.547	4.95	4.32
298.15	2.004	6.27	5.74
283.15	2.635	8.035	7.751

* original data

Smoothed Data

$$\Delta G° = -RT \ln x_{N_2O} = (-13101.1 + 76.4835T/K)\ J\ mol^{-1}$$

Std. dev. $\Delta G°$ = 8.52 J mol^{-1}. Coef. Corr = 1.00

$\Delta H°/J\ mol^{-1} = -13101.1$; $\Delta S°/J\ mol^{-1}\ K^{-1} = -76.4835$

T/K	$\Delta G°/J\ mol^{-1}$	x_{N_2O}
283.15	8555.17	0.02641
288.15	8937.59	0.02398
293.15	9320.00	0.02184
298.15	9702.42	0.01996
303.15	10084.8	0.01829
308.15	10467.3	0.01681
313.15	10849.7	0.01550

AUXILIARY INFORMATION

METHOD/APPARATUS/PROCEDURE:

Composed of a gas buret, a solvent reservoir, and an absorption pipet. Volume of pipet is determined at various heights of the meniscus using a weighed quantity of water, measuring the height of the meniscus using a cathetometer. Dry gas is introduced into degassed solvent. System is mixed using a magnetic stirrer until saturation occurs. Care is taken to prevent solvent vapor from mixing with gas in the gas buret. Volume of gas determined from gas buret reading; volume of liquid determined from height of meniscus in the absorption pipet.

SOURCE AND PURITY OF MATERIALS:

1. Nitrous oxide prepared from hydroxylamine sulfate and sodium nitrite both from Kahlbaum (extra pure grade). Gas washed to remove nitric oxide and dried. Fractionated.
2. $C_3H_6O_2$ Merck, extra pure used. Treated with phosphorus pentoxide several times and distilled several times. B.P. at normal *P*, 57.12 °C.

ESTIMATED ERROR:

$\delta T/K = \pm 0.1$.

REFERENCES:

† Also reported in: Horiuti, J. *Bull. Inst. Phys. Chem. Res., Tokyo* 1928, *7(2)*, 119-172.

COMPONENTS:	ORIGINAL MEASUREMENTS:
1. Nitrous oxide; N_2O; [10024-97-2] 2. Acetic acid, pentyl ester (Amyl acetate); $C_7H_{14}O_2$; [628-63-7]	Kunerth, W. *Phys. Rev.* 1922, *19*, 512-524.
VARIABLES:	PREPARED BY:
Temperature	W. Gerrard/C. L. Young

EXPERIMENTAL VALUES:

T/K	Ostwald coefficient, L	Temperature coefficient, dL/Ldt	Mole fraction, x_{N_2O}*
291.16	5.24	0.0103	0.0313
293.16	5.14	0.0104	0.0306
295.16	5.05	0.0105	0.0299
297.16	4.93	0.0106	0.0291
299.16	4.83	0.0107	0.0285
301.16	4.71	0.0109	0.0277
303.16	4.60	0.0110	0.0269
305.16	4.49	0.0111	0.0262
307.16	4.39	0.0112	0.0255
309.16	4.30	0.0113	0.0249

* The x_{N_2O} values were calculated by the compiler, the molar volume of N_2O being taken as 22257 × (T/273.16) cm^3, based on the standard density of 1.9775 g dm^{-3} at 273.16 K and 101.325 kPa (ref. 1). The L values were for a total pressure equal to the prevailing barometric pressure. Kunerth's value for x_{N_2O} at 293.16 K was 0.0312.

(cont.)

AUXILIARY INFORMATION

METHOD /APPARATUS/PROCEDURE:

The volume of gas absorbed by a measured volume of degassed liquid was measured at a total pressure equal to the prevailing barometric pressure. Based on the technique of McDaniel (ref. 2). Apparatus consisted of a gas buret attached to leveling tube containing mercury, and to pipet. A layer of liquid was held over the mercury layer in the buret to saturate the gas with vapor of the liquid. Buret and pipet were kept at a common temperature by electric coils.

SOURCE AND PURITY OF MATERIALS:

1. Gas (S.S.White Dental Co.) of 99.7 per cent purity was frozen to remove volatile gases before being passed over P_2O_5. Density was found to be 1.968 g dm^{-3} at 273.16 K and 101.325 kPa.
2. Attested by b.p. and density.

ESTIMATED ERROR:

$\delta T/K = \pm 0.1$; $\delta x_{N_2O} = \pm 2\%$ (estimated by compiler).

REFERENCES:

1. Kaye, G. W. C. and Laby, T. H.; *Tables of Physical and Chemical Constants*, Longmans, London, 1966.
2. McDaniel, A. S.; *J. Phys. Chem.* 1911, *15*, 587.

COMPONENTS:	ORIGINAL MEASUREMENTS:
1. Nitrous oxide; N_2O; [10024-97-2]	Kunerth, W.
2. Acetic acid, pentyl ester (Amyl acetate); $C_7H_{14}O_2$; [628-63-7]	*Phys. Rev.* 1922, *19*, 512-524.

EXPERIMENTAL DATA:

Smoothed Data

$$\Delta G^\circ = -RT \ln x_{N_2O} = (-9670.1 + 61.962(T/K)) \text{ J mol}^{-1}$$

(Std. deviation = 9.40 J mol^{-1})

T/K	Mole fraction of nitrous oxide, x_{N_2O}
291.16	0.0315
293.16	0.0307
295.16	0.0298
297.16	0.0291
299.16	0.0283
301.16	0.0276
303.16	0.0269
305.16	0.0262
307.16	0.0256
309.16	0.0250

COMPONENTS:

1. Nitrous oxide; N_2O; [10024-97-2]
2. Benzaldehyde; C_7H_6O; [100-52-7]

ORIGINAL MEASUREMENTS:

Kunerth, W.

Phys. Rev. 1922, *19*, 512-524.

VARIABLES:

Temperature

PREPARED BY:

W. Gerrard/C. L. Young

EXPERIMENTAL VALUES:

T/K	Ostwald coefficient, L	Temperature coefficient, dL/Ldt	Mole fraction, x_{N_2O}*
291.16	3.23	0.0126	0.0136
293.16	3.15	0.0126	0.0132
295.16	3.07	0.0125	0.0128
297.16	3.00	0.0125	0.0125
299.16	2.93	0.0125	0.0122
301.16	2.85	0.0126	0.0118
303.16	2.78	0.0126	0.0114
305.16	2.72	0.0126	0.0111
307.16	2.65	0.0126	0.0108
309.16	2.59	0.0126	0.0105

* The x_{N_2O} values were calculated by the compiler, the molar volume of N_2O being taken as 22257 × (T/273.16) cm^3, based on the standard density of 1.9775 g dm^{-3} at 273.16 K and 101.325 kPa (ref. 1). The L values were for a total pressure equal to the prevailing barometric pressure. Kunerth's value for x_{N_2O} at 293.16 K was 0.0134.

(cont.)

AUXILIARY INFORMATION

METHOD/APPARATUS/PROCEDURE:

The volume of gas absorbed by a measured volume of degassed liquid was measured at a total pressure equal to the prevailing barometric pressure. Based on the technique of McDaniel (ref. 2). Apparatus consisted of a gas buret attached to leveling tube containing mercury, and to pipet. A layer of liquid was held over the mercury layer in the buret to saturate the gas with vapor of the liquid. Buret and pipet were kept at a common temperature by electric coils.

SOURCE AND PURITY OF MATERIALS:

1. Gas (S.S.White Dental Co.) of 99.7 per cent purity was frozen to remove volatile gases before being passed over P_2O_5. Density was found to be 1.968 g dm^{-3} at 273.16 K and 101.325 kPa.
2. Attested by b.p. and density.

ESTIMATED ERROR:

$\delta T/K = \pm 0.1$; $\delta x_{N_2O} = \pm 2\%$ (estimated by compiler).

REFERENCES:

1. Kaye, G. W. C. and Laby, T. H.; *Tables of Physical and Chemical Constants*, Longmans, London, 1966.
2. McDaniel, A. S.; *J. Phys. Chem.* 1911, *15*, 587.

COMPONENTS:	ORIGINAL MEASUREMENTS:
1. Nitrous oxide; N_2O; [10024-97-2]	Kunerth, W.
2. Benzaldehyde; C_7H_6O; [100-52-7]	*Phys. Rev.* 1922, *19*, 512-524.

EXPERIMENTAL VALUES:

Smoothed Data

$$\Delta G° = -RT \ln x_{N_2O} = (-10791.5 + 72.774(T/K)) \text{ J mol}^{-1}$$

(Std. deviation = 8.9 J mol^{-1})

T/K	Mole fraction of nitrous oxide, x_{N_2O}
291.16	0.0136
293.16	0.0132
295.16	0.0128
297.16	0.0125
299.16	0.0121
301.16	0.0118
303.16	0.0114
305.16	0.0111
307.16	0.0108
309.16	0.0105

COMPONENTS:

1. Nitrous oxide; N_2O; [10024-97-2]
2. Aniline, (Benzenamine); $C_6H_5NH_2$; [62-53-3]

ORIGINAL MEASUREMENTS:

Kunerth, W.

Phys. Rev. 1922, *19*, 512-524.

VARIABLES:

Temperature

PREPARED BY:

W. Gerrard/C. L. Young

EXPERIMENTAL VALUES:

T/K	Ostwald coefficient, L	Temperature coefficient, dL/Ldt	Mole fraction, x_{N_2O}*
291.16	1.50	0.0083	0.00573
293.16	1.48	0.0084	0.00562
295.16	1.45	0.0085	0.00548
297.16	1.42	0.0087	0.00534
299.16	1.40	0.0089	0.00523
301.16	1.37	0.0090	0.00510
303.16	1.35	0.0091	0.00500
305.16	1.32	0.0093	0.00487
307.16	1.31	0.0094	0.00480
309.16	1.28	0.0096	0.00467

* The x_{N_2O} values were calculated by the compiler, the molar volume of N_2O being taken as $22257 \times (T/273.16)$ cm^3, based on the standard density of 1.9775 g dm^{-3} at 273.16 K and 101.325 kPa (ref. 1). The L values were for a total pressure equal to the prevailing barometric pressure. Kunerth's value for x_{N_2O} at 293.16 K was 0.0053.

(cont.)

AUXILIARY INFORMATION

METHOD /APPARATUS/PROCEDURE:

The volume of gas absorbed by a measured volume of degassed liquid was measured at a total pressure equal to the prevailing barometric pressure. Based on the technique of McDaniel (ref. 2). Apparatus consisted of a gas buret attached to leveling tube containing mercury, and to pipet. A layer of liquid was held over the mercury layer in the buret to saturate the gas with vapor of the liquid. Buret and pipet were kept at a common temperature by electric coils.

SOURCE AND PURITY OF MATERIALS:

1. Gas (S.S.White Dental Co.) of 99.7 per cent purity was frozen to remove volatile gases before being passed over P_2O_5. Density was found to be 1.968 g dm^{-3} at 273.16 K and 101.325 kPa.
2. Attested by b.p. and density.

ESTIMATED ERROR:

$\delta T/K = \pm 0.1$; $\delta x_{N_2O} = \pm 2\%$ (estimated by compiler).

REFERENCES:

1. Kaye, G. W. C. and Laby, T. H.; *Tables of Physical and Chemical Constants*, Longmans, London, 1966.
2. McDaniel, A. S.; *J. Phys. Chem.* 1911, *15*, 587.

COMPONENTS:	ORIGINAL MEASUREMENTS:
1. Nitrous oxide; N_2O; [10024-97-2]	Kunerth, W.
2. Aniline, (Benzenamine); $C_6H_5NH_2$; [62-53-3]	*Phys. Rev.* 1922, *19*, 512-524.

EXPERIMENTAL VALUES:

Smoothed Data

$$\Delta G° = -RT \ln x_{N_2O} = (-8524.16 + 72.178(T/K)) \text{ J mol}^{-1}$$

(Std. deviation = 5.9 J mol^{-1})

T/K	Mole fraction of nitrous oxide, x_{N_2O}
291.16	0.00574
293.16	0.00561
295.16	0.00574
297.16	0.00535
299.16	0.00523
301.16	0.00511
303.16	0.00500
305.16	0.00489
307.16	0.00478
309.16	0.00468

COMPONENTS:	ORIGINAL MEASUREMENTS:
1. Nitrous oxide; N_2O; [10024-97-2] 2. Pyridine; C_5H_5N; [110-86-1]	Kunerth, W. *Phys. Rev.* 1922, *19*, 512-524.
VARIABLES: Temperature	PREPARED BY: W. Gerrard/C. L. Young

EXPERIMENTAL VALUES:

T/K	Ostwald coefficient, L	Temperature coefficient, dL/Ldt	Mole fraction, x_{N_2O}*
293.16	3.58	0.0111	0.0119
295.16	3.50	0.0114	0.0116
297.16	3.45	0.0117	0.0114
299.16	3.34	0.0120	0.0110
301.16	3.25	0.0123	0.0106
303.16	3.17	0.0126	0.0103
305.16	3.10	0.0129	0.0101
307.16	3.02	0.0132	0.00975
309.16	2.94	0.0136	0.0095

* The x_{N_2O} values were calculated by the compiler, the molar volume of N_2O being taken as 22257 × (T/273.16) cm^3, based on the standard density of 1.9775 g dm^{-3} at 273.16 K and 101.325 kPa (ref. 1).

The L values were for a total pressure equal to the prevailing barometric pressure. Kunerth's value for x_{N_2O} at 293.16 K was 0.0120.

(cont.)

AUXILIARY INFORMATION

METHOD/APPARATUS/PROCEDURE:

The volume of gas absorbed by a measured volume of degassed liquid was measured at a total pressure equal to the prevailing barometric pressure. Based on the technique of McDaniel (ref. 2). Apparatus consisted of a gas buret attached to leveling tube containing mercury, and to pipet. A layer of liquid was held over the mercury layer in the buret to saturate the gas with vapor of the liquid. Buret and pipet were kept at a common temperature by electric coils.

SOURCE AND PURITY OF MATERIALS:

1. Gas (S.S.White Dental Co.) of 99.7 per cent purity was frozen to remove volatile gases before being passed over P_2O_5. Density was found to be 1.968 g dm^{-3} at 273.16 K and 101.325 kPa.
2. Attested by b.p. and density.

ESTIMATED ERROR:

$\delta T/K = \pm 0.1$; $\delta x_{N_2O} = \pm 2\%$ (estimated by compiler).

REFERENCES:

1. Kaye, G. W. D. and Laby, T. H.; *Tables of Physical and Chemical Constants*, Longmans, London, 1966.
2. McDaniel, A. S.; *J. Phys. Chem.* 1911, *15*, 587.

COMPONENTS:	ORIGINAL MEASUREMENTS:
1. Nitrous oxide; N_2O; [10024-97-2]	Kunerth, W.
2. Pyridine; C_5H_5N; [110-86-1]	*Phys. Rev.* <u>1922</u>, *19*, 512-524.

EXPERIMENTAL VALUES:

<u>Smoothed Data</u>

$$\Delta G° = -RT \ln x_{N_2O} = (10870.5 + 73.869(T/K)) \text{ J mol}^{-1}$$

(Std. deviation = 13.5 J mol^{-1})

T/K	Mole fraction of nitrous oxide, x_{N_2O}
293.16	0.0120
295.16	0.0116
297.16	0.0113
299.16	0.0110
301.16	0.0106
303.16	0.0103
305.16	0.0101
307.16	0.00978
309.16	0.00951

COMPONENTS:	ORIGINAL MEASUREMENTS:
1. Nitrous oxide; N_2O; [10024-97-2] 2. Tetrachloromethane; (Carbon tetrachloride); CCl_4; [56-23-5]	Horiuti, J. *Sci. Pap. Inst. Phys. Chem. Res. (Japan)* † 1931/32, *17*, 125-256.
VARIABLES:	PREPARED BY:
Temperature	M. E. Derrick

EXPERIMENTAL VALUES:

Total pressure is 1 atm for original measurements.

T/K	$10^2\ x_{N_2O}$	Ostwald coefficient, L *	Bunsen coefficient, α
313.15	1.354	3.565	3.110
308.15	1.446	3.775	3.346
303.15	1.549	4.005	3.609
298.15	1.672	4.285	3.926
293.15	1.801	4.57	4.26
288.15	1.947	4.89	4.64
283.15	2.111	5.26	5.07

* original data

Smoothed Data

$$\Delta G^\circ = -RT \ln x_{N_2O} = (-10953.3 + 70.7618T/K)\ \mathrm{J\ mol^{-1}}$$

Std. dev. ΔG° = 3.43 J mol^{-1}. Coef. Corr. = 1.00

$\Delta H^\circ/\mathrm{J\ mol^{-1}}$ = -10953.3; $\Delta S^\circ/\mathrm{J\ mol^{-1}\ K^{-1}}$ = -70.7618

T/K	$\Delta G^\circ/\mathrm{J\ mol^{-1}}$	x_{N_2O}
283.15	9082.88	0.02111
288.15	9436.69	0.01947
293.15	9790.50	0.01801
298.15	10144.3	0.01670
303.15	10498.1	0.01553
308.15	10851.9	0.01447
313.15	11205.7	0.01352

AUXILIARY INFORMATION

METHOD/APPARATUS/PROCEDURE:

Composed of a gas buret, a solvent reservoir, and an absorption pipet. Volume of pipet is determined at various heights of the meniscus using a weighed quantity of water, measuring the height of the meniscus using a cathetometer. Dry gas is introduced into degassed solvent. System is mixed using a magnetic stirrer until saturation occurs. Care is taken to prevent solvent vapor from mixing with gas in the gas buret. Volume of gas determined from gas buret reading; volume of liquid determined from height of meniscus in the absorption pipet.

SOURCE AND PURITY OF MATERIALS:

1. Nitrous oxide prepared from hydroxylamine sulfate and sodium nitrite both from Kahlbaum (extra pure grade). Gas washed to remove nitric oxide and dried. Fractionated.
2. CCl_4 sample from Kahlbaum, dried and distilled. B.P. at normal *P*, 76.74 °C.

ESTIMATED ERROR:

$\delta T/K = \pm 0.1$.

REFERENCES:

† Also reported in: Horiuti, J. *Bull. Inst. Phys. Chem. Res., Tokyo* 1928, *7(2)*, 119-172.

COMPONENTS:

1. Nitrous oxide; N_2O; [10024-97-2]
2. Tetrachloromethane; (Carbon tetrachloride); CCl_4; [56-23-5]

ORIGINAL MEASUREMENTS:

Yen, L. C.; McKetta, J. J., Jr.
J. Chem. Eng. Data 1962, *7*, 288-289.

VARIABLES:

Temperature

PREPARED BY:

W. Gerrard/C. L. Young

EXPERIMENTAL VALUES:

T/K	Bunsen coefficient	Mole fraction of nitrous oxide, x_{N_2O}
273.16	6.071	0.02506
283.16	5.096	0.02137
293.16	4.262	0.01815
303.16	3.634	0.01570
313.16	3.111	0.01364

The partial pressure of the gas was 101.325 kPa.

Smoothed Data

$$\Delta G° = -RT \ln x_{N_2O} = (-10858 + 70.366 \times T/K) \text{ J mol}^{-1}$$

(Std. deviation = 8.4 J mol^{-1})

T/K	Mole fraction of nitrous oxide, x_{N_2O}
273.16	0.02516
283.16	0.02125
293.16	0.01816
303.16	0.01568
313.16	0.01366

AUXILIARY INFORMATION

METHOD /APPARATUS/PROCEDURE:

Measurement of volume of gas absorbed by a known volume of degassed liquid at a partial pressure of 101.325 kPa. The vapor pressure of the liquid was allowed for. Gas buret and absorption pipet. Modified form of apparatus and technique used by Markham and Kobe (1).

SOURCE AND PURITY OF MATERIALS:

1. Gas of 98 per cent purity was purified at 193.16 K. Mass spectrograph then showed purity of 99.5 per cent.
2. J. T. Baker Chemical Co., Baker analyzed grade freshly fractionated.

ESTIMATED ERROR:

$\delta T/K = \pm 0.1$; $\delta x_{N_2O} = \pm 2\%$ (estimated by compiler).

REFERENCES:

(1) Markham, A. E.; Kobe, K. A.; *J. Am. Chem. Soc.* 1941, *63*, 449.

COMPONENTS:	ORIGINAL MEASUREMENTS:
1. Nitrous oxide; N_2O; [10024-97-2] 2. Halogenated hydrocarbons	Hsu, H.; Campbell, D. *Aerosol Age*. 1964, *December*, 34.
VARIABLES:	PREPARED BY: W. Gerrard / C.L. Young

EXPERIMENTAL VALUES:

Solvent	T/K	Ostwald coefficient, L	Mole fraction* of nitrous oxide in liquid, x_{N_2O}
Tetrachloromethane; (Carbon tetrachloride); CCl_4, [56-23-5]	294.3	2.5	0.00997
Trichloromethane; (Chloroform); $CHCl_3$; [67-66-3]	294.3	5.54	0.0182
1,2-Dichloroethane; $C_2H_4Cl_2$; [107-06-2]	294.3	3.2	0.0104
1,2-Dibromoethane; $C_2H_4Br_2$; [106-93-4]	294.3	2.7	0.00962
"Chlorothene"	294.3	4.96	-

* calculated by compiler as for a partial pressure of gas of 101.325 kPa. Molar volume of gas at 294.26 K taken as 23967 cm^3 based on the density (1) 1.9775 g dm^{-3} at 101.325 kPa and 273.15K.

AUXILIARY INFORMATION

METHOD/APPARATUS/PROCEDURE:

Copper equilibrium cell fitted with Bourdon gauge and thermometer. Total amount of gas needed to attain given, but not stated, pressure measured. Ostwald coefficient calculated from knowledge of volume of liquid and container.

SOURCE AND PURITY OF MATERIALS:

No details given.

ESTIMATED ERROR:

$\delta T/K = \pm 0.5$; $\delta x_{N_2O} = \pm 3\%$.

(estimated by compiler).

REFERENCES:

1. Kaye, G.W.C.; Laby, T.H. *Tables of Physical and Chemical Constants*, Longmans, London, *1966*.

COMPONENTS:	ORIGINAL MEASUREMENTS:
1. Nitrous oxide; N_2O; [10024-97-2] 2. Trichloromethane; (Chloroform); $CHCl_3$; [67-66-3]	Kunerth, W. *Phys. Rev.* 1922, *19*, 512-524.
VARIABLES:	**PREPARED BY:**
Temperature	W. Gerrard/C. L. Young

EXPERIMENTAL VALUES:

T/K	Ostwald coefficient, L	Temperature coefficient, dL/Ldt	Mole fraction, x_{N_2O}*
291.16	5.70	0.0066	0.0189
293.16	5.60	0.0103	0.0184
295.16	5.51	0.0140	0.0181
297.16	5.26	0.0180	0.0172
299.16	5.07	0.0220	0.0165
301.16	4.83	0.0258	0.0157
303.16	4.57	0.0298	0.0148
305.16	4.29	0.0335	0.0138
307.16	4.03	0.0370	0.0130
309.16	3.70	0.0407	0.0119

* The x_{N_2O} values were calculated by the compiler, the molar volume of N_2O being taken as 22257 × (T/273.16) cm^3, based on the standard density of 1.9775 g dm^{-3} at 273.16 K and 101.325 kPa (ref. 1). The L values were for a total pressure equal to the prevailing barometric pressure. Kunerth's value for x_{N_2O} at 293.16 K was 0.0182.

(cont.)

AUXILIARY INFORMATION

METHOD /APPARATUS/PROCEDURE:

The volume of gas absorbed by a measured volume of degassed liquid was measured at a total pressure equal to the prevailing barometric pressure. Based on the technique of McDaniel (ref. 2). Apparatus consisted of gas buret attached to leveling tube containing mercury, and to pipet. A layer of liquid was held over the mercury layer in the buret to saturate the gas with vapor of the liquid. Buret and pipet were kept at a common temperature by electric coils.

SOURCE AND PURITY OF MATERIALS:

1. Gas (S.S.White Dental Co.) of 99.7 per cent purity was frozen to remove volatile gases before being passed over P_2O_5. Density was found to be 1.968 g dm^{-3} at 283.16 K and 101.325 kPa.
2. Attested by b.p. and density.

ESTIMATED ERROR:

$\delta T/K = \pm 0.1$; $\delta x_{N_2O} = \pm 2\%$

(estimated by compiler).

REFERENCES:

1. Kaye, G. W. C. and Laby, T. H.; *Tables of Physical and Chemical Constants*, Longmans, London, 1966.
2. McDaniel, A. S.; *J. Phys. Chem.* 1911, *15*, 587.

COMPONENTS:	ORIGINAL MEASUREMENTS:
1. Nitrous oxide; N_2O; [10024-97-2]	Kunerth, W.
2. Trichloromethane; (Chloroform); $CHCl_3$; [67-66-3]	*Phys. Rev.* 1922, *19*, 512-524.

EXPERIMENTAL VALUES:

Smoothed Data

$$\Delta G^\circ = -RT \ln x_{N_2O} = -(195338 - 1331.7(T/K) + 2.3827(T/K)^2) \text{ J mol}^{-1}$$

(Std. deviation = 12.0 J mol^{-1})

T/K	Mole fraction of nitrous oxide, x_{N_2O}
291.16	0.0189
293.16	0.0184
295.16	0.0179
297.16	0.0173
299.16	0.0165
301.16	0.0157
303.16	0.0148
305.16	0.0139
307.16	0.0129
309.16	0.0119

COMPONENTS:	ORIGINAL MEASUREMENTS:
1. Nitrous oxide; N_2O; [10024-97-2] 2. 1,2-Dibromoethane; CH_2BrCH_2Br; [106-93-4]	Kunerth, W. *Phys. Rev.* 1922, *19*, 512-524.
VARIABLES:	**PREPARED BY:**
Temperature	W. Gerrard/C. L. Young

EXPERIMENTAL VALUES:

T/K	Ostwald coefficient, L	Temperature coefficient, dL/Ldt	Mole fraction, x_{N_2O} *
291.16	2.87	0.0106	0.0103
293.16	2.81	0.0106	0.0100
295.16	2.75	0.0107	0.00977
297.16	2.69	0.0108	0.00952
299.16	2.64	0.0109	0.00930
301.16	2.58	0.0110	0.00904
303.16	2.52	0.0111	0.00879
305.16	2.46	0.0112	0.00855
307.16	2.42	0.0113	0.00837
309.16	2.37	0.0114	0.00816

* The x_{N_2O} values were calculated by the compiler, the molar volume of N_2O being taken as 22257 × (T/273.16) cm^3, based on the standard density of 1.9775 g dm^{-3} at 273.16 K and 101.325 kPa (ref. 1). The L values were for a total pressure equal to the prevailing barometric pressure. Kunerth's value for x_{N_2O} at 293.16 K was 0.0100.

(cont.)

AUXILIARY INFORMATION

METHOD /APPARATUS/PROCEDURE:

The volume of gas absorbed by a measured volume of degassed liquid was measured at a total pressure equal to the prevailing barometric pressure. Based on the technique of McDaniel (ref. 2). Apparatus consisted of a gas buret attached to leveling tube containing mercury, and to pipet. A layer of liquid was held over the mercury layer in the buret to saturate the gas with vapor of the liquid. Buret and pipet were kept at a common temperature by electric coils.

SOURCE AND PURITY OF MATERIALS:

1. Gas (S.S.White Dental Co.) of 99.7 per cent purity was frozen to remove volatile gases before being passed over P_2O_5. Density was found to be 1.968 g dm^{-3} at 273.16 K and 101.325 kPa.
2. Attested by b.p. and density.

ESTIMATED ERROR:

$\delta T/K = \pm 0.1$; $\delta x_{N_2O} = \pm 2\%$ (estimated by compiler).

REFERENCES:

1. Kaye, G. W. C. and Laby, T. H.; *Tables of Physical and Chemical Constants*, Longmans, London, 1966.
2. McDaniel, A. S.; *J. Phys. Chem.* 1911, *15*, 587.

COMPONENTS:	ORIGINAL MEASUREMENTS:
1. Nitrous oxide; N_2O; [10024-97-2] 2. 1,2-Dibromoethane; CH_2BrCH_2Br; [106-93-4]	Kunerth, W. *Phys. Rev.* 1922, *19*, 512-524.

EXPERIMENTAL VALUES:

Smoothed Data

$$\Delta G° = -RT \ln x_{N_2O} = (-9702.7 + 71.360(T/K)) \text{ J mol}^{-1}$$

(Std. deviation = 5.5 J mol^{-1})

T/K	Mole fraction of nitrous oxide, x_{N_2O}
291.16	0.0103
293.16	0.0100
295.16	0.00976
297.16	0.00951
299.16	0.00926
301.16	0.00902
303.16	0.00880
305.16	0.00858
307.16	0.00837
309.16	0.00816

COMPONENTS:	ORIGINAL MEASUREMENTS:
1. Nitrous oxide; N_2O; [10024-97-2] 2. Chlorobenzene; C_6H_5Cl; [108-90-7]	Horiuti, J. *Sci. Pap. Inst. Phys. Chem. Res. (Jpn)*†, 1931/32, *17*, 125-256.
VARIABLES:	PREPARED BY:
Temperature	M. E. Derrick

EXPERIMENTAL VALUES: Total pressure is 1 atm for original measurements.

T/K	$10^2 x_{N_2O}$	Ostwald Coefficient, L*	Bunsen Coefficient, α
328.15	0.8837	2.279	1.897
323.15	0.9399	2.400	2.029
318.15	0.9969	2.520	2.164
313.15	1.059	2.650	2.311
308.15	1.131	2.801	2.483
303.15	1.216	2.981	2.686
298.15	1.309	3.174	2.908
293.15	1.410	3.382	3.151
288.15	1.533	3.636	3.447
283.15	1.659	3.891	3.754

* original data

Smoothed Data

$\Delta G°/J\ mol^{-1} = -RT \ln x_{N_2O} = -10821.7 + 72.3315T/K$

Std. dev. $\Delta G°/J\ mol^{-1} = \pm 10.73$ $\Delta H°/J\ mol^{-1} = -10821.7$

Coeff. Corr. = 1.00 $\Delta S°/J\ mol^{-1}\ K^{-1} = -72.3315$

T/K	$\Delta G°/J\ mol^{-1}$	x_{N_2O}	T/K	$\Delta G°/J\ mol^{-1}$	x_{N_2O}
283.15	9658.91	0.01653	308.15	11467.2	0.01138
288.15	10020.6	0.01526	313.15	11828.9	0.01064
293.15	10382.2	0.01413	318.15	12190.5	0.009966
298.15	10743.9	0.01311	323.15	12552.2	0.009355
303.15	11105.5	0.01220	328.15	12913.8	0.008798

AUXILIARY INFORMATION

METHOD/APPARATUS/PROCEDURE:

Composed of a gas buret, a solvent reservoir, and an absorption pipet. Volume of pipet is determined at various heights of the meniscus using a weighed quantity of water, measuring the height of the meniscus using a cathetometer. Dry gas is introduced into degassed solvent. System is mixed using a magnetic stirrer until saturation occurs. Care is taken to prevent solvent vapor from mixing with gas in the gas buret. Volume gas detn. from gas buret reading: volume of liquid determined from height of meniscus in the adsorption pipet.

SOURCE AND PURITY OF MATERIALS:

N_2O: Prepared from hydroxylamine sulfate and sodium nitrite both from Kahlbaum (extra pure grade). Gas washed to remove nitric oxide and dried. Fractionated.

C_6H_5Cl: Sample from Kahlbaum, dried dried, and distilled. B.P. at normal *P*, 131.96 °C.

ESTIMATED ERROR:

$\delta T/K = \pm 0.1$.

REFERENCES:

† also reported in: Horiuti, J.; *Bull. Inst. Phys. Chem. Res. Tokyo*, 1928, *7(2)*, 119-172.

COMPONENTS:	EVALUATOR:
1. Nitrous oxide; N_2O; [10024-97-2] 2. Biological fluids.	Colin L. Young School of Chemistry University of Melbourne Parkville, Victoria, 3052 AUSTRALIA: June, 1981

CRITICAL EVALUATION:

In study of gas solubility in biological fluids it is evitable that some variation in solubility arises from the source of the sample and its treatment before being used. Therefore careful characterisation of the sample is necessary. As has been pointed out by Kozam *et al* (1) and may be inferred from the early work of Siebeck (2) it appears that some of the discrepancies between the results of various workers for the solubility of nitrous oxide in blood are probably due to the fraction of red cells in the sample. Thorough degassing of the sample is rarely achieved with biological fluids and there is some indication that this also gives rise to a significant additional variation in the reported solubilities

Many gas solubilities in biological fluids have been determined using a Van Slyke apparatus and are therefore not of the highest accuracy. See reference 3 for a discussion of the reliability of this technique. The gas chromatographic (GC) method used by Kozam *et al* (1) appears to be capable of fairly accurate results although, at present, not widely used Jay *et al* (4) have carried out a comparison between GC and the Van Slyke method. The GC method can be used to study the solubility of mixtures of gases. The Scholander microgasometric method (5) has been applied to solubilities in biological fluids by Saidman *et al* (6) and the results appear to be more reliable than those determined by the Van Slyke method.

In general, when more attention is paid to a detailed characterisation of the sample it appears that it will be desirable to further develop techniques which are more accurate than the Van Slyke method. The method is quick but, unless carried out with great care, is not as accurate as the Scholander thechnique.

It is very difficult to establish which solubility measurements are more reliable in the present context. The measurements of Saidman *et al* (6) are probably the most accurate for human blood.

In addition to measurements given in the compiled tables the solubility of nitrous oxide in bovine serum albumen and bovine globulin have been reported by Muehlbaecher *et al* (7). The results were presented in graphical form and little information was given as to the method employed. This workers also quoted Bunsen coefficient for water and blood but it was not clear whether these coefficients were "average" values from the literature or new experimental values. The results are summarized in the table below.

$T/°C = 37$

SOLVENT	BUNSEN COEFFICIENT
Water	0.440
Blood	0.466
Bovine serum albumen in sodium phosphate buffer $pH = 5.6\text{-}6.3$	0.42
Bovine β -globulin in sodium phosphate buffer $pH = 6.3\text{-}6.6$	0.41
Bovine ∂ -globulin in sodium phosphate buffer $pH = 6.3\text{-}6.6$	0.42
Bovine hemoglobin in sodium phosphate buffer $pH = 6.3\text{-}6.6$	0.43

continued....

COMPONENTS:	EVALUATOR:
1. Nitrous oxide; N_2O; [10024-97-2] 2. Biological fluids.	Colin L. Young School of Chemistry University of Melbourne Parkville, Victoria, 3052 AUSTRALIA June, 1981

CRITICAL EVALUATION:

REFERENCES:

1. Kozam, R.L.; Landau, S,M.; Cubina, J.M.; Lukas, D.S. *J. Appl. Physiol.* 1970, *29*, 593.

2. Siebeck, R.; *Skand. Arch. Physiol.* 1909, *21*, 368.

3. Markham, A.E.; Kobe, K.A. *Chem. Revs.* 1941, *28*, 519.

4. Jay, B.E.; Wilson, R.H.; Doty, V.; Pingree, H.; Hargis, B. *Anal. Chem.* 1962, *34*, 414.

5. Douglas, E.; *J. Phys. Chem.* 1964, *68*, 169.

6. Saidman, L.J.; Eger, E.I.; Manson, E.S.; Severinghaus, J.W. *Anesthesiology*, 1966, *27*, 180.

7. Muehlbaecher, C.; DeBon, F.L.; Featherstone, R.M.; *Intern. Anesth. Clinics.* 1963, *1*, 937.

COMPONENTS:

1. Nitrous oxide; N_2O; [10024-97-2]

2. Ox blood and red blood cells

ORIGINAL MEASUREMENTS:

Siebeck, R.

Skand. Arch. Physiol.

1909, *21*, 368-382.

VARIABLES:

Temperature, pressure

PREPARED BY:

C. L. Young

EXPERIMENTAL VALUES:

T/°C	T/K	Partial pressure of nitrous oxide p/mmHg	p/kPa	Ratio of volume of absorbed gas (reduced to 273.15 K and 101.3 kPa to volume of solution	Bunsen coefficient, α
		Ox Blood			
23.1	296.3	707.28	94.296	0.5486	0.5895
23.2	296.4	714.22	95.221	0.5532	0.5887
22.5	295.7	543.90	72.514	0.4212	0.5885
23.1	296.3	174.10	23.211	0.1253	0.5470
		Red blood cells "solution"			
22.9*	296.1	663.50	88.459	0.5520	0.6323
23.2§	296.4	706.60	94.206	0.5727	0.6160
37.9†	311.1	500.44	66.720	0.2705	0.4108
38.0†	311.2	645.70	86.086	0.3630	0.4273
37.7†	310.9	581.50	77.527	0.3159	0.4129
38	311.2	202.65	27.018	0.1023	0.3837
38	311.2	357.60	47.676	0.1911	0.4061
38	311.2	702.45	93.652	0.3741	0.4047

* sample contained 66.0 mg of Fe per 100 g solution.

§ sample contained 59.7 mg of Fe per 100 g solution.

† sample contained 39.7 mg of Fe per 100 g solution.

AUXILIARY INFORMATION

METHOD/APPARATUS/PROCEDURE:

Solvent placed in a 3 dm^3 glass cylinder. Cylinder flushed with nitrous oxide to remove air and then solvent equilibrated with gas at known pressure. Samples of solvent removed and gas extracted under reduced pressure and estimated volumetrically.

SOURCE AND PURITY OF MATERIALS:

No details given.

ESTIMATED ERROR:

$\delta\alpha/\alpha = \pm 4\%$ (estimated by compiler).

REFERENCES:

COMPONENTS:

1. Nitrous oxide; N_2O; [10024-97-2]
2. Ox blood and ox serum

ORIGINAL MEASUREMENTS:

Findlay, A.; Creighton, H. J. M.

Biochem. J.

1910, *5*, 294-305.

VARIABLES:

Pressure

PREPARED BY:

C. L. Young

EXPERIMENTAL VALUES:

T/K	Density of soln. /g cm^{-3}	p_{N_2O}/mmHg	p_{N_2O}/MPa	Solubility, $S^{\dagger}$
		Blood		
298.15	1.065	745	0.099	0.521
		854	0.114	0.530
		1012	0.135	0.539
		1152	0.154	0.544
		1277	0.170	0.547
		1408	0.188	0.548
		Serum		
298.15	1.025	737	0.098	0.517
		846	0.113	0.509
		925	0.123	0.513
		1036	0.138	0.519
		1169	0.156	0.524
		1402	0.187	0.528

† Solubility, S, given as $\dfrac{\text{Concentration of gas in the liquid phase}}{\text{Concentration of gas in the gas phase}}$.

AUXILIARY INFORMATION

METHOD/APPARATUS/PROCEDURE:

Gas buret and adsorption pipet similar to that of Geffcken (1) except that the manometer tube was longer to give the higher pressures.

SOURCE AND PURITY OF MATERIALS:

1. Obtained by heating pure ammonium nitrate, purified by passing through solutions of potassium hydroxide and ferrous sulfate.
2. As obtained from slaughter house: not degassed.

ESTIMATED ERROR:

$\delta T/K = \pm 0.1$; $\delta S/S = \pm 2\%$

(estimated by compiler)

REFERENCES:

1. Geffcken, G.
Z. Phys. Chem.
1904, *49*, 257.

COMPONENTS:	ORIGINAL MEASUREMENTS:
1. Nitrous oxide; N_2O; [10024-97-2] 2. Blood.	Orcutt, F.S.; Seevers, M.H., *J. Biol.Chem.* 1937, *117*,509-15.
VARIABLES:	PREPARED BY: W. Gerrard

EXPERIMENTAL VALUES:

T/K	Bunsen coefficient, α	Ostwald coefficient, L
310.65	0.416	0.454

AUXILIARY INFORMATION

METHOD /APPARATUS/PROCEDURE:

Van Slyke-Neill (1) manometric apparatus. Gas extracted from the solution. Pressure of extracted gas was measured, allowance being made for the gas not extracted.

SOURCE AND PURITY OF MATERIALS:

Not specified.

ESTIMATED ERROR:

REFERENCES:

1. Van Slyke, D.D., Neill, J.M. *J. Biol. Chem.* 1924, *61*,523.

COMPONENTS:

1. Nitrous oxide; N_2O; [10024-97-2]
2. Human blood

ORIGINAL MEASUREMENTS:

Cullen, S. C.; Cook, E. V.

J. Biol. Chem.

1943, *147*, 23-26.

VARIABLES:

Pressure

PREPARED BY:

C. L. Young

EXPERIMENTAL VALUES: T/°C = 37.5; T/K = 310.7

Sample No.	Equilibrated gas, mole per cent composition CO_2	O_2	N_2	N_2O	Total pressure /mmHg	Partial pressure of N_2O, p_{N_2O} /mmHg	Solubility†
1	5.95	9.48	0	84.5	740	583	30.9
2	6.15	4.21	0	89.6	737	616	33.2
3	6.30	1.04	4.16	88.5	738	609	34.0
4	5.75	17.40	0	76.8	740	531	28.7
5	6.10	5.33	21.30	66.3	740	458	25.2
6	5.30	46.80	0	47.9	738	330	17.8
7	5.60	9.50	38.00	46.9	738	323	17.7
8	5.80	1.66	0	92.5	740	639	34.0
9	1.23	0.60	0	98.1	740	678	37.4
10	5.80	33.70	0	60.5	738	417	22.0
11	5.70	7.50	30.00	56.8	737	391	21.8
12	5.25	75.80	0	19.0	736	131	7.4
13	5.38	15.60	62.40	16.6	736	114	6.6
14	0.42	0.40	0	99.1	736	681	37.6
15	5.45	16.25	65.00	13.3	738	92	4.8
16	5.33	63.80	0	30.8	742	213	11.7
17	0.57	0.28	0	99.1	745	690	37.9
18	5.53	53.60	0	40.8	738	281	15.1
19	6.44	11.00	44.00	38.5	738	265	14.2
20	5.71	39.60	0	54.7	737	376	19.9
21	5.63	8.12	32.48	53.7	736	369	20.9

† Solubility is the volume (reduced to 273.2 K and 1 atmosphere pressure) dissolved by 100 cm^3 of blood.

Bunsen coefficient = 0.415.

Ostwald coefficient at 1 atmosphere pressure = 0.472.

AUXILIARY INFORMATION

METHOD/APPARATUS/PROCEDURE:

Blood equilibrated with gas of stated composition.
Samples analyzed by the Van Slyke manometric procedure.
Details in source and ref. (1).

SOURCE AND PURITY OF MATERIALS:

1. No details given.
2. Oxalated venous blood.

ESTIMATED ERROR:

$\delta T/K = \pm 0.1$; $\delta\alpha/\alpha = \pm 3\%$ (estimated by compiler).

REFERENCES:

1. Cullen, S. C.; Cook, E. V. *Am. J. Physiol.* 1942, *137*, 238.

COMPONENTS:	ORIGINAL MEASUREMENTS:
1. Nitrous oxide; N_2O; [10024-97-2] 2. Blood.	Kety, S.S.; Harmel, M.H., Broomell, H.T., Rhode, C.B. *J. Biol. Chem.* 1948, *173*, 487-496.
VARIABLES:	PREPARED BY: W. Gerrard

EXPERIMENTAL VALUES:

	Dog, 6 animals α*	Man Red blood cell hematocrit	Man α*
	0.419	28.8	0.400
	0.419	34.2	0.408
	0.433	41.0	0.410
	0.421	44.0	0.414
	0.435	51.5	0.425
	0.421	-	-
Mean:	0.425		0.412
Standard error:	0.003		0.004

α* Bunsen coefficient for 310.15 K defined as cm^3 of N_2O (reduced to 273.15 K and 101.325 kPa) dissolved by 1 cm^3 of blood when equilibrated at a nitrous oxide pressure of 101.325 kPa.

AUXILIARY INFORMATION

METHOD/APPARATUS/PROCEDURE:	SOURCE AND PURITY OF MATERIALS:
Blood is equilibrated with nitrous oxide in a 50 cm^3 glass syringe, closed and shaken at 310.15 K. Gas was removed and the nitrous oxide content of the liquid was determined using a Van Slyke-Neill manometric apparatus.	1. Not stated. 2. Freshly shed heparinised whole blood.
	ESTIMATED ERROR:
	REFERENCES:

COMPONENTS:	ORIGINAL MEASUREMENTS:
1. Nitrous oxide; N_2O; [10024-97-2] 2. Blood	Hattox, J. S.; Saari, J. M.; Faulconer, A. *Anesthesiology* 1953, *14*, 584-590.
VARIABLES:	PREPARED BY: C. L. Young

EXPERIMENTAL VALUES:

T = 37 °C

Sample No.	Distribution coefficient,[†] D	Concentration[§]/mg cm^{-3} (a)	(b)
1	0.453	0.713	0.373
2	0.451	0.706	0.354
3	0.454	0.689	0.364
4	0.460	0.716	0.367
5	-	0.684	0.372
Mean	0.455	0.702	0.366
Standard deviation		0.0145	0.0075

† Ostwald coefficient.

§ Concentration calculated assuming Ostwald coefficient independent of pressure in milligrams of N_2O per cubic centimetre of blood
(a) partial pressure of N_2O = 1 atmosphere
(b) partial pressure of N_2O = 0.5 atmosphere.

AUXILIARY INFORMATION

METHOD/APPARATUS/PROCEDURE:

Mass spectrometric method based on the relative height of argon and nitrous oxide peaks at 40 and 30 respectively. Solubility of argon in blood assumed to be same as in water.

SOURCE AND PURITY OF MATERIALS:

No details given.

ESTIMATED ERROR:

$\delta T/K = \pm 0.5$; $\delta D = \pm 2\%$.

REFERENCES:

COMPONENTS:	ORIGINAL MEASUREMENTS:
1. Nitrous oxide; N_2O; [10024-97-2] 2. Human blood	Jay, B.E.; Wilson, R.H.; Doty, V.; Pingree, H.; Hargis, B. *Anal. Chem.* 1962, *34*, 414-418
VARIABLES:	PREPARED BY: C.L. Young

EXPERIMENTAL VALUES:

T/K	Ostwald coefficient,[a] L
309.2	0.405 ± 0.004

a Average of 14 values. Incorrectly called Bunsen coefficient in source.

AUXILIARY INFORMATION

METHOD/APPARATUS/PROCEDURE:

Saturated blood samples were analysed using GC with an activated charcoal or activated silica gel stationary phase. Thermal conductivity detector used and helium was used as carrier gas. Results checked against Van Slyke method.

SOURCE AND PURITY OF MATERIALS:

1. Matheson Gas Co. sample.
2. Hemoglobin content 14g per 100 cm^3 of solution.

ESTIMATED ERROR:

$\delta T/K = \pm 0.5$

REFERENCES:

COMPONENTS:	ORIGINAL MEASUREMENTS:
1. Nitrous oxide; N_2O; [10024-97-2] 2. Dog blood (*in vivo*)	Sy, W. P.; Hasbrouck, J. D. *Anesthesiology* 1964, *25*, 59-63.
VARIABLES:	**PREPARED BY:**
	C. L. Young

EXPERIMENTAL VALUES:

T/K	Partial pressure[†] of nitrous oxide P/mmHg	Ostwald coefficient
310.65	499.1	0.400
		0.395
		0.396
		0.392
		0.393
		0.395

Mean Ostwald coefficient = 0.395

Standard deviation = 0.002

[†] calculated by subtracting vapor pressure of water, partial pressure of oxygen and partial pressure of carbon dioxide from atmospheric pressure.

AUXILIARY INFORMATION

METHOD/APPARATUS/PROCEDURE:

Healthy dog anesthetized with sodium pentobarbital, trachea intubated and lungs mechanically ventilated with a mixture of 80 per cent nitrous oxide and 20 per cent oxygen. End expiratory partial pressure of carbon dioxide maintained at 39.5-40.5 mmHg. After 75 minutes samples of arterial blood taken and analysed with van Slyke apparatus. Details in source.

SOURCE AND PURITY OF MATERIALS:

No details given.

ESTIMATED ERROR:

$\delta T/K = \pm 0.1$; $\delta P/\text{mmHg} = \pm 0.1$

REFERENCES:

COMPONENTS:

1. Nitrous oxide; N_2O; [10024-97-2]

2. Human blood

ORIGINAL MEASUREMENTS:

Saidman, L. J.; Eger, E. I.; Munson, E. S.; Severinghaus, J. W. *Anesthesiology* 1966, *27*, 180-184.

VARIABLES:

PREPARED BY:

C. L. Young

EXPERIMENTAL VALUES:

T/°C	T/K	No. of samples	Ostwald coefficient Mean	Ostwald coefficient Standard deviation
37	310.2	5	0.462	0.007
25	298.2	5	0.602	0.007

AUXILIARY INFORMATION

METHOD/APPARATUS/PROCEDURE:

Modified Scholander apparatus used. Known amount of blood equilibrated with a known volume of gas and change in volume used to estimate Ostwald coefficient. Details of apparatus in source and ref. 1. Correction made for nitrogen in sample.

SOURCE AND PURITY OF MATERIALS:

1. No details given.

2. Nitrogen bubbled through sample to remove other gases. Mean hemoglobin 14 per cent.

ESTIMATED ERROR:

$\delta T/K = \pm 0.1$ (estimated by compiler).

REFERENCES:

1. Douglas, E. *J. Phys. Chem.* 1964, *68*, 169.

COMPONENTS:

1. Nitrous oxide; N_2O; [10024-97-2]

2. Dog blood

ORIGINAL MEASUREMENTS:

Saidman, L. J.; Eger, E. I.; Manson, E. S.; Severinghaus, J. W. *Anesthesiology* 1966, *27*, 180-184.

VARIABLES:

PREPARED BY:

C. L. Young

EXPERIMENTAL VALUES:

T/°C	T/K	No. of samples	Ostwald coefficient Mean	Ostwald coefficient Standard deviation
37	310.2	5	0.472	0.009
25	298.2	3	0.618	0.015

AUXILIARY INFORMATION

METHOD/APPARATUS/PROCEDURE:

Modified Scholander apparatus used. Known amount of blood equilibrated with a known volume of gas and change in volume used to estimate Ostwald coefficient. Details of apparatus in source and ref. 1.

SOURCE AND PURITY OF MATERIALS:

1. No details given.

2. Nitrogen bubbled through sample to remove other gases.

ESTIMATED ERROR:

$\delta T/K = \pm 0.1$ (estimated by compiler).

REFERENCES:

1. Douglas, E. *J. Phys. Chem.* 1964, *68*, 169.

COMPONENTS:

1. Nitrous oxide; N_2O; [10024-97-2]

2. Human blood

ORIGINAL MEASUREMENTS:

Ostiguy, G. L.; Becklake, M. R.
J. Appl. Physiol. 1966, *21*, 1397-1399.

VARIABLES:

PREPARED BY:

C. L. Young

EXPERIMENTAL VALUES:

$T/°C = 37$

	No. of samples	Cholesterol range 10^3 g m^{-3}	Total lipids 10^3 g m^{-3}	Ostwald coefficient Mean	SD*
Normal subjects (both sexes)	40			0.465	0.0095
Thyrotoxic subjects	4	1.04-2.84		0.461	0.0093
Subjects with hyperlipidemia	4	2.41-4.54	10.75-17.95	0.472	0.0200

* standard deviation

AUXILIARY INFORMATION

METHOD /APPARATUS/PROCEDURE:

Venous blood taken in heparinized syringes from 40 normal healthy subjects and from 4 thyrotoxic subjects with hyperlipidemia. Blood equilibrated with gas mixture (8% N_2O, 20% O_2 and 72% He) in flow system. Samples then analysed by extracting in stream of oxygen and using infrared analyser. Details in source and ref. 1.

SOURCE AND PURITY OF MATERIALS:

See method.

ESTIMATED ERROR:

$\delta T/K = \pm 0.1$.

REFERENCES:

1. Lawther, P. J.; Bates, D. V. *Clin. Sci.* 1953, *12*, 91.

COMPONENTS:	ORIGINAL MEASUREMENTS:
1. Nitrous oxide; N_2O; [10024-97-2] 2. Human blood	Ostiguy, G. L.; Becklake, M. R. *J. Appl. Physiol.* 1966, *21*, 1397-1399.
VARIABLES:	PREPARED BY: C. L. Young

EXPERIMENTAL VALUES: T/°C = 37

Men

Age/yr	No. of samples	Ostwald coefficient Mean	Range
Under 20	3	0.472*	0.461 - 0.491
20-29	4	0.456	0.450 - 0.460
30-39	6	0.461	0.458 - 0.466
40-49	3	0.465	0.460 - 0.468
50-59	3	0.464	0.448 - 0.473
60-69	2	0.465	0.462 - 0.468
70 and over	3	0.461	0.459 - 0.462
Total	24	0.463 ± 0.0083	(standard deviation)

Women

Age/yr	No. of samples	Ostwald coefficient Mean	Range
20-29	3	0.463	0.457 - 0.468
30-39	3	0.458	0.449 - 0.463
40-49	3	0.468	0.463 - 0.473
50-59	4	0.475†	0.455 - 0.490
60-69	3	0.476†	0.473 - 0.477
Total	16	0.468 ± 0.0105	(standard deviation)

* Authors claim this high mean was due to one high value. No obvious cause for this high value was discovered.

† Authors claim values for women aged 50-69 yr were significantly different from those for women aged 20-49 yr.

AUXILIARY INFORMATION

METHOD /APPARATUS/PROCEDURE:	SOURCE AND PURITY OF MATERIALS:
Venous blood taken in heparinized syringes from normal healthy men and women. Blood equilibrated with gas mixture (8% N_2O, 20% O_2 and 72% He) in flow system. Samples then analysed by extracting nitrous oxide in stream of oxygen and using infrared analyser. Details in source and ref. 1.	See method.
	ESTIMATED ERROR: $\delta T/K = \pm 0.1$
	REFERENCES: 1. Lawther, P. J.; Bates, D. V. *Clin. Sci.* 1953, *12*, 91.

COMPONENTS:

1. Nitrous oxide; N_2O; [10024-97-2]

2. Rabbit Tissue and Blood

ORIGINAL MEASUREMENTS:

Mapleson, W. W.; Evans, D. E.; Flook, V.

Brit. J. Anaesth. 1970, *42*, 1033-1041.

VARIABLES:

PREPARED BY:

C. L. Young

EXPERIMENTAL VALUES:

T = 37.5 °C

Ostwald coefficients

Sample No.	Blood *vivo*	Blood *vitro*	Brain *vivo*	Brain *vitro*	Heart *vivo*	Heart *vitro*
1	0.51	0.48	0.44	0.46	0.34	-
2	0.47	0.42	0.48	0.46	0.44	0.48
3	0.47	0.44	0.34	0.44	0.39	0.35
4	0.45	0.46	0.45	0.48	0.42	0.31
5	0.47	0.47	0.37	0.44	0.41	0.43
6	0.46	0.46	0.49	0.50	0.43	0.48

Sample No.	Kidney *vivo*	Kidney *vitro*	Liver *vivo*	Liver *vitro*	Muscle *vivo*	Muscle *vitro*
1	0.45	0.52	0.39	0.48	0.42	0.40
2	0.38	0.40	0.46	0.51	0.27	0.35
3	0.36	0.40	0.35	0.39	0.31	0.22
4	0.39	0.40	0.37	0.37	0.31	0.38
5	0.41	0.38	0.43	0.45	0.43	0.44
6	0.39	0.42	0.41	0.45	0.38	0.40

AUXILIARY INFORMATION

METHOD/APPARATUS/PROCEDURE:

Samples taken from rabbits of a variety of strains. Details in source. Nitrous oxide extracted from samples using van Slyke apparatus. Loss of nitrous oxide during handling of *in vivo* samples estimated at about 3 per cent.

SOURCE AND PURITY OF MATERIALS:

1. No details given.

2. Details of sample preparation given in source.

ESTIMATED ERROR:

Detailed analysis of error given in paper.

REFERENCES:

COMPONENTS:

1. Nitrous oxide; N_2O; [10024-97-2]

2. Human blood components

ORIGINAL MEASUREMENTS:

Kozam, R. L.; Landau, S. M.; Cubina, J. M.; Lukas, D. S.
J. Appl. Physiol. 1970, *29*, 593-597.

VARIABLES:

PREPARED BY:

C. L. Young

EXPERIMENTAL VALUES:

T/°C = 37

Substance	No. of samples	No. of analyses	Bunsen coefficient, α Mean	SD*	SE*
Water	6	19	0.433	0.008	0.002
0.156 M NaCl	5	33	0.398	0.003	0.001
Plasma	13	62	0.400	0.008	0.002
Hyperlipidemic plasma	2	8	0.436	0.002	0.001
Red blood cells	14	75	0.450	0.021	0.002
Red cell membrane	1	8	0.412	0.002	0.001

* SD - standard deviation; SE - standard error

AUXILIARY INFORMATION

METHOD/APPARATUS/PROCEDURE:

Blood samples taken from five normal subjects. Details of sample preparation in source. Samples equilibrated by bubbling nitrous oxide through mixture and then allowing to stand. Nitrous oxide then estimated by gas chromatography.

SOURCE AND PURITY OF MATERIALS:

1. Nitrous oxide-USP Pure sample.

2. Water-distilled: details of other samples in source.

ESTIMATED ERROR:

$\delta T/K = \pm 0.2$.

REFERENCES:

COMPONENTS:	ORIGINAL MEASUREMENTS:
1. Nitrous oxide; N_2O; [10024-97-2]	Kety, S.S.; Harmel, M.H. Broomell, H.T.; and Rhode, C.B.
2. Brain, homogenized in distilled water.	*J. Biol. Chem.* 1948, *173*, 487-496.
VARIABLES:	PREPARED BY: W. Gerrard.

EXPERIMENTAL VALUES:

Dog (7 animals)	α	Man (7 deceased patients)	α
	0.434		0.428
	0.406		0.434
	0.430		0.437
	0.420		0.425
	0.458		0.438
	0.455		0.432
	0.454		0.464
Mean	0.437		0.437
Standard error	0.008		0.005

Bunsen coefficient, α, for 310.15 K defined as cm^3 of N_2O (reduced to 273.15 K and 101.325 kPa) dissolved by 1 gram of brain when equilibrated at 310.15 K and at a nitrous oxide pressure of 101.325 kPa. Given by :

$$\frac{\alpha_h \frac{W_b}{1.05} + V_w - V_w\alpha_w}{W_b}$$

$\alpha_h = \alpha$ for 1 cm^3 of homogenate; α for 1 cm^3 of water; W_b = weight of brain sample; V_w = cm^3 of water; 1.05 = specific gravity of brain.

AUXILIARY INFORMATION

METHOD/APPARATUS/PROCEDURE:

Homogenate was equilibrated with nitrous oxide in a 50 cm^3 glass syringe, capped and shaken at 310.15 K. Undissolved gas removed and the nitrous oxide content of the liquid was determined using a Van Slyke-Neill manometric apparatus.

SOURCE AND PURITY OF MATERIALS:

N_2O: Not stated.

ESTIMATED ERROR:

REFERENCES:

COMPONENTS:	ORIGINAL MEASUREMENTS:
1. Nitrous oxide; N_2O; [10024-97-2] 2. Human lung tissue-blood free homogenate.	Cander, L.; *J. Appl. Physiol.* 1959 *14*, 538-540.
VARIABLES:	PREPARED BY: W. Gerrard.

EXPERIMENTAL VALUES:

From 5 deceased patients.	α^*
1	0.433
2	0.371
	0.390
	0.384
3	0.403
	0.413
4	0.414
5	0.429
	0.423
Mean	0.407 ±7%

* Bunsen coefficient, α, for 310.15 K defined as cm^3 of N_2O (reduced to 273.15 K, and 101.325 kPa) dissolved by 1 cm^3 of lung tissue in equilibrium with N_2O gas assumed to be at 101.325 kPa, and 310.15 K.

AUXILIARY INFORMATION

METHOD/APPARATUS/PROCEDURE:

5 cm^3 of homogenate deaerated and transferred to a 50 cm^3 syringe containing nitrous oxide. After equilibration at 310.15 K (1), the undissolved gas removed and the nitrous oxide content of the liquid phase was determined using a Van Slyke manometric apparatus.

SOURCE AND PURITY OF MATERIALS:

1. 98.0% pure.
2. Lung samples from deceased patients with no history of acute or chronic lung disease.

ESTIMATED ERROR:

$\delta T/K = \pm 0.5$; $\delta\alpha = \pm 2\text{-}9\%$ (estimated by compiler).

REFERENCES:

1. Kety, S.S.; Harmel, M.H., Broomell, H.T.; and Rhode, C.B. *J. Biol. Chem.* 1948, *173*, 487.

COMPONENTS:	ORIGINAL MEASUREMENTS:
1. Nitrous oxide; N_2O; [10024-97-2] 2. Human Fetal and Uterine Tissues	Assali, N. S.; Ross, M. *Soc. Exp. Biol. Med. Proc.* 1959, *100*, 497-498.
VARIABLES:	PREPARED BY: C. L. Young

EXPERIMENTAL VALUES:

T = 37 °C Pressure = 1 atmosphere = 1.01325 bar

Tissue	Kuenen Coefficient[†], S
Spleen	0.589
Skeletal muscle	0.473
Skin	0.458
Liver	0.377
Brain	0.439
Heart	0.496
Lung	0.432
Scalp	0.327
Placenta	0.331
Umbilical cord	0.333
Uterine muscle	0.472
Mean	0.430

[†] Incorrectly called Bunsen coefficient per gram of tissue in original paper.

AUXILIARY INFORMATION

METHOD /APPARATUS/PROCEDURE:

Samples homogenized and equilibrated with nitrous oxide. No details of apparatus given. Five to eight determinations made for each type of tissue.

SOURCE AND PURITY OF MATERIALS:

1. No details given.
2. Samples of fetal material taken from one to five month immature fetus (therapeutic abortion) and one 8 months premature fetus and one full term fetus who died shortly after delivery. Myometrial tissue obtained from patients undergoing cesarean section. Placental tissues collected from spontaneously delivered pregnancies.

ESTIMATED ERROR:

$\delta T/K = \pm 0.5$; $\delta S = \pm 1\%$.

REFERENCES:

COMPONENTS:

1. Nitrous oxide; N_2O; [10024-97-2]

2. Human myocardium

ORIGINAL MEASUREMENTS:

Kozam, R. L.; Landau, S. M.; Cubina, J. M.; Lukas, D. S.

J. Appl. Physiol. 1970, *29*, 593-597.

VARIABLES:

PREPARED BY:

C. L. Young

EXPERIMENTAL VALUES:

$T/°C = 37$

Nature of sample	No. of samples	No. of analyses	Mean density 10^6 g m^{-3}	Bunsen coefficient, α Mean	SD*	SE*
Normal						
Left ventricle	5	30	1.055	0.395	0.016	0.003
Right ventricle	3	22	1.056	0.462	0.038	0.008
Hypertrophy						
Left ventricle	3	20	1.046	0.396	0.029	0.007
Right ventricle	3	19	1.048	0.448	0.023	0.005
ASHD and MI†						
Left ventricle	2	13		0.382	0.013	0.004
Right ventricle	1	8		0.449	0.010	0.004
Cardiomyopathy						
Left ventricle	1	9	1.053	0.409	0.012	0.004
Right ventricle	1	7	1.039	0.468	0.102	0.005

* SD - standard deviation; SE - standard error

† Arteriosclerotic heart disease and myocardial infaction

AUXILIARY INFORMATION

METHOD/APPARATUS/PROCEDURE:

All large vascular structures and segments of connective tissues removed from left and right ventricular myocardium. Density determined by displacement of water. Muscle minced and homogenized. Details of sample preparation in source. Samples equilibrated by bubbling nitrous oxide through mixture. Allowed to stand and then nitrous oxide estimated by gas chromatography.

SOURCE AND PURITY OF MATERIALS:

1. Nitrous oxide - USP Pure sample.

2. Obtained at autopsy from subjects 5-12 hr after death (some samples stored at -15 °C).

ESTIMATED ERROR:

$\delta T/K = \pm 0.2$.

REFERENCES:

COMPONENTS:

1. Nitrous oxide; N_2O; [10024-97-2]

2. Dog myocardium

ORIGINAL MEASUREMENTS:

Kozam, R. L.; Landau, S. M.; Cubina, J. M.; Lukas, D. S.

J. Appl. Physiol. 1970, *29*, 593-597.

VARIABLES:

PREPARED BY:

C. L. Young

EXPERIMENTAL VALUES:

T/°C = 37

Nature of sample	No. of samples	No. of analyses	Mean density 10^6g m^{-3}	Bunsen coefficient, α Mean	SD*	SE*
Left ventricle	7	40	1.037	0.386	0.022	0.008
Right ventricle	1	4	$1.037^{\dagger}$	0.374		

* SD - standard deviation; SE - standard error

† estimated

AUXILIARY INFORMATION

METHOD/APPARATUS/PROCEDURE:

All large vascular structures and segments of connective tissues removed from left and right ventricular myocardium. Density determined by displacement of water. Muscle minced and homogenized. Details of sample preparation in source. Samples equilibrated by bubbling nitrous oxide through mixture. Allowed to stand and then nitrous oxide estimated by gas chromatography.

SOURCE AND PURITY OF MATERIALS:

1. Nitrous oxide-USP Pure sample.

2. Myocardium samples obtained from freshly killed dogs.

ESTIMATED ERROR:

$\delta T/K = \pm 0.2$.

REFERENCES:

COMPONENTS:	ORIGINAL MEASUREMENTS:
1. Nitrous oxide; N_2O; [10024-97-2] 2. Rat abdominal muscle	Campos Carles, A.; Kawashiro, T.; Piiper, J. *Pflugers Arch.* 1975, *359*, 209-218.
VARIABLES: T/K: 310.15	PREPARED BY: A.L. Cramer H.L. Clever

EXPERIMENTAL VALUES:

T/K	Solubility coefficient $\mu mol\ dm^{-3}\ torr^{-1}$	Corrected[2] Solubility coefficient $\mu mol\ dm^{-3}\ torr^{-1}$	Bunsen Coefficient, α
310.15	24.6 ± 0.7	27.7	0.471

[1] Mean value ± standard error of 12 measurements.

[2] Corrected for unextracted gas in the sample, and for gas lost during transfer of the sample.

Another report from this laboratory gives Krogh's diffusion constant, K = $(20.0 \pm 0.4) \times 10^{-9}$ m mol m^{-1} cm^{-1} $torr^{-1}$, and the diffusion coefficient, D = 12.1×10^{-6} cm^2 s^{-1}, for nitrous oxide in rat abdominal muscle at 310.15 K (1).

The sample is a non-homogenised solid, not a liquid.

AUXILIARY INFORMATION

METHOD/APPARATUS/PROCEDURE:

The nitrous oxide gas was pre-saturated with water vapor, then passed through an equilibration chamber containing the muscle sample resting on a screen to expose all sides. The gas was passed through the equilibration chamber for one hour at a rate of 8 cm^3 m^{-1}.

The muscle was transferred to an extraction chamber filled with air for the same length of time as equilibration. The gas in the extraction chamber was then forced into a gas chromatograph by mercury entering the chamber.

SOURCE AND PURITY OF MATERIALS:

1. Nitrous oxide. Source not given. Stated to be 99.9 per cent pure.

2. Rat abdominal muscle. Flat abdominal wall muscle layer of about 1.6 g, 1.4 mm thick, and surface area of 10 cm^2 on one side. Sample taken from 250-430g rat.

ESTIMATED ERROR:

REFERENCES:

1. Kawashiro, T.; Campos Carles, A.; Perry, S.F.; Piiper, J.

Pflugers Arch. 1975, *359*, 219.

COMPONENTS:	ORIGINAL MEASUREMENTS:
1. Nitrous oxide; N_2O; [10024-97-2] 2. Water, H_2O; [7732-18-5] 3. Serum albumen (colloidal)	Findlay, A.; Creighton, H.J.M. *J. Chem. Soc.* 1910, *97*, 536 - 561
VARIABLES:	PREPARED BY:
Pressure, concentration	W. Gerrard

EXPERIMENTAL VALUES:

Solubility, *S*, given as $\frac{\text{concentration of the gas in the liquid phase}}{\text{concentration of the gas in the gaseous phase}}$

T/K = 298.16.

Conc. of colloid / 10^{-2} g cm^{-3}	Density of solution	Pressure of gas /kPa	Solubility, *S*
0.32	0.998	99.457	0.583
0.32	0.998	116.388	0.581
0.32	0.998	130.387	0.579
0.32	0.998	150.118	0.586
0.32	0.998	167.850	0.588
0.32	0.998	185.981	0.591
1.40	1.001	99.057	0.537
1.40	1.001	112.255	0.538
1.40	1.001	121.721	0.545
1.40	1.001	139.719	0.550
1.40	1.001	163.717	0.558
1.40	1.001	185.048	0.562

AUXILIARY INFORMATION

METHOD /APPARATUS/PROCEDURE:

Gas buret and absorption pipet similar to that of Geffcken (1), except that the manometer tube was longer to give the higher pressures.

SOURCE AND PURITY OF MATERIALS:

1. Self prepared and purified; not attested.
2. Neutral serum-albumen was obtained from fresh ox-blood by a method described.

ESTIMATED ERROR:

$\delta S/S$ Stated to be $\mp$ 0.25%.

REFERENCES:

1. Geffcken, G. *Z. Phys. Chem.* 1904, *49*, 257.

COMPONENTS:	ORIGINAL MEASUREMENTS:
1. Nitrous oxide; N_2O; [10024-97-2] 2. Water; H_2O; [7732-18-5] 3. Serum albumen (colloidal) solution.	Shkol'nikova, R.I. *Uchenye Zapiski Leningrad. Gosudart*, 1959, No. 18, Part 272, 64-86.
VARIABLES:	PREPARED BY:
Temperature, concentration of colloid	W. Gerrard

EXPERIMENTAL VALUES:

T/K	Concn. of colloid %	α	*L*
283.15	0.575	0.7094	0.7354
288.15		0.6245	0.6588
293.15		0.5520	0.5923
298.15		0.5131	0.5600
303.15		0.4850	0.5379
308.15		0.4031	0.4548
313.15		0.3490	0.3999
283.15	1.15	0.6988	0.7244
288.15		0.6162	0.6500
293.15		0.5440	0.5837
298.15		0.4943	0.5396
303.15		0.4210	0.4669
308.15		0.3681	0.4153
313.15		0.3000	0.3438
283.15	1.68	0.6324	0.6556
288.15		0.5852	0.6173
293.15		0.5390	0.5783
298.15		0.4513	0.4927
303.15		0.4430	0.4580
308.15		0.3340	0.3768
313.15		0.2380	0.2727
283.15	1.99	0.5900	0.6116
288.15		0.5374	0.5669

AUXILIARY INFORMATION

METHOD/APPARATUS/PROCEDURE:

The volume of N_2O absorbed was measured by the use of a gas buret and absorbtion pipet.

Reproducibility stated to be to within ±0.2%.

Heat of solution, *U*, appeared to be based on :

$$\frac{d \log L}{dT} = - \frac{U}{RT^2}$$

SOURCE AND PURITY OF MATERIALS:

(1) Nitrous oxide stated to be of a purity of 99.6 - 99.7%; 0.4 to 0.3% N_2.

(2) Water may be taken as of satisfactory purity.

(3) Obtained by "salting out from the serum of horse blood by ammonium sulfate." Dialized

ESTIMATED ERROR:

REFERENCES:

COMPONENTS:	ORIGINAL MEASUREMENTS:
1. Nitrous oxide; N_2O; [10024-97-2] 2. Water; H_2O; [7732-18-5] 3. Serum albumen (colloidal) solution.	Shkol'nikova, R.I. *Uchenye Zapiski Leningrad Gosudart*, 1959, No. 18, Part 272, 64-86.

EXPERIMENTAL VALUES:

T/K	Concn. of colloid %	α	*L*
293.15	1.99	0.4120	0.4421
298.15		0.4021	0.4390
303.15		0.3920	0.4347
308.15		0.3091	0.3487
313.15		0.2190	0.2510

α = the bunsen absorbtion coefficient.

L = α x T/K /273, but defined as the ratio of the concentration of the gas in the liquid phase to that in the gas phase. Partial pressure of gas taken to be 760 mmHg. 760mmHg = 1 atm = 101.325 kPa.

Heat of solution / cal mol^{-1}

Concn. of colloid	α	*L*
0.575%	4190	3580
1.15 %	5430	4830
1.68 %	7460	6860
1.99 %	5770	5160

COMPONENTS:	ORIGINAL MEASUREMENTS:
1. Nitrous oxide; N_2O; [10024-97-2] 2. Water; H_2O; [7732-18-5] 3. Egg albumen, (colloidal).	Findlay, A.; Creighton, H.J.M. *J. Chem. Soc.* 1910, *97*, 536-561
VARIABLES: Pressure, concentration	PREPARED BY: W. Gerrard

EXPERIMENTAL VALUES:

Solubility, *S*, given as $\frac{\text{concentration of the gas in the liquid phase}}{\text{concentration of the gas in the gaseous phase}}$

T/K = 298.16

Conc. of soln. $/10^{-2}$ g cm^{-3}	Density of solution	Pressure of gas /kPa	Solubility, *S*
0.35	0.998	97.990	0.580
0.35	0.998	110.656	0.578
0.35	0.998	127.187	0.580
0.35	0.998	151.851	0.581
0.35	0.998	166.517	0.580
0.35	0.998	181.715	0.580
0.75	1.000	97.990	0.569
0.75	1.000	109.322	0.562
0.75	1.000	116.255	0.564
0.75	1.000	126.787	0.567
0.75	1.000	147.185	0.573
0.75	1.000	179.182	0.577
1.60	1.005	97.190	0.548
1.60	1.005	108.123	0.535
1.60	1.005	118.122	0.540
1.60	1.005	126.121	0.544
1.60	1.005	159.851	0.553
1.60	1.005	186.515	0.558

AUXILIARY INFORMATION

METHOD: / APPARATUS/PROCEDURE:

Gas buret and absorption pipet similar to that of Geffcken (1), except that the manometer tube was longer to give the higher pressures.

The concentration of the colloid was determined by heating the solution to effect complete coagulation. The coagulate was dried at 373.16K and weighed.

SOURCE AND PURITY OF MATERIALS:

1. Self prepared and purified: not attested.
2. Egg albumen was obtained from fresh eggs by the method described. A small amount of toluene was added to prevent putrifaction.

ESTIMATED ERROR:

$\delta S/S$ Stated to be $\mp$ 0.25%.

REFERENCES:

1. Geffcken, G., *Z. Phys. Chem.* 1904, *49*, 257.

COMPONENTS:

1. Nitrous oxide; N_2O; [10024-97-2]
2. Water; H_2O; [7732-18-5]
3. Egg-albumen (colloid)

ORIGINAL MEASUREMENTS:

Findlay, A.; Howell, O. R.

J. Chem. Soc. 1914, *105*, 291-8.

VARIABLES:

Pressure, concentration

PREPARED BY:

W. Gerrard

EXPERIMENTAL VALUES:

Temperature not stated: presumably 298.16 T/K

Solubility, s, given as $\frac{\text{concentration of the gas in the liquid phase}}{\text{concentration of the gas in the gaseous phase}}$

Conc. of colloid 10^2g cm^{-3} (soln.)	Density of solution	P_{N_2O}† kPa	s	P_{N_2O}† kPa	s	P_{N_2O}† kPa	s
0.38	0.998	33.157	0.572	48.169	0.573	70.753	0.573
		84.485	0.572	100.750	0.570	132.813	0.571
0.62	1.000	33.877	0.568	50.662	0.569	60.461	0.568
		85.925	0.567	101.657	0.565	136.053	0.571
0.38	0.998	34.943	0.572	49.395	0.573	71.633	0.573
		84.592	0.572	97.790	0.571	121.988	0.568

† P_{N_2O} is the pressure of N_2O over the solution.

AUXILIARY INFORMATION

METHOD/APPARATUS/PROCEDURE:

Measurement of volume of N_2O by gas buret and pipet (ref. 1).

SOURCE AND PURITY OF MATERIALS:

1. Nitrous oxide self prepared and purified, see ref. 2.
3. Commercial egg-albumen was treated with water, the solution was filtered and dialysed.

ESTIMATED ERROR:

REFERENCES:

1. Findlay, A.; Williams, T. *J. Chem. Soc.* 1913, *103*, 636.

2. Findlay, A.; Creighton. H. J. M. *J. Chem. Soc.* 1910, *97*, 536.

COMPONENTS:	ORIGINAL MEASUREMENTS:
1. Nitrous oxide; N_2O; [10024-97-2] 2. Water; H_2O; [7732-18-5] 3. Dextrin (colloidal);	Findlay, A.; Howell, O.R. *J. Chem. Soc.* 1914, *105*, 291-8
VARIABLES:	PREPARED BY:
Pressure, concentration	W. Gerrard

EXPERIMENTAL VALUES: Temperature not stated: presumably 298.16 T/K

Solubility, S, given as $\dfrac{\text{concentration of the gas in the liquid phase}}{\text{concentration of the gas in the gaseous phase}}$

Conc. of soln. /10^{-2}g cm^3.	Density of solution	$p_{N_2O}$$^+$ /kPa	S	$p_{N_2O}$$^+$ /kPa	S	$p_{N_2O}$$^+$ /kPa	S
6.82	1.019	37.530	0.557	54.275	0.550	75.339	0.542
6.82	1.019	89.738	0.542	109.202	0.547	133.920	0.554
6.70	1.019	37.903	0.555	54.368	0.550	74.753	0.544
6.70	1.019	88.618	0.544	103.070	0.546	130.734	0.554
12.41	1.037	37.836	0.537	54.355	0.532	76.632	0.526
12.41	1.037	88.058	0.527	104.750	0.526	131.320	0.534
12.50	1.037	37.543	0.535	55.501	0.530	76.939	0.526
12.50	1.037	89.551	0.526	103.190	0.524	129.507	0.532
19.24	1.060	39.063	0.515	56.194	0.510	79.819	0.504
19.24	1.060	92.724	0.501	106.563	0.500	132.988	0.506
19.31	1.060	38.423	0.516	55.114	0.510	74.886	0.504
19.31	1.060	86.125	0.502	103.590	0.500	132.813	0.506

+ p_{N_2O} is the pressure of N_2O over the solution.

AUXILIARY INFORMATION

METHOD:/APPARATUS/PROCEDURE:	SOURCE AND PURITY OF MATERIALS:
Measurement of volume of N_2O by gas buret and pipet (1). The concentration of the dextrin in the solutions was determined by evaporating to dryness, drying the residue in a steam oven and weighing.	1. Self prepared and purified (2). 2. Kahlbaum's purest dextrin was used. ESTIMATED ERROR: REFERENCES: 1. Findlay, A.; Williams, T. *J. Chem. Soc.* 1913, *103*, 636. 2. Findlay, A.; Creighton, H.J.M. *J. Chem. Soc.* 1910, *97*, 536.

COMPONENTS:

1. Nitrous oxide; N_2O; [10024-97-2]
2. Water, H_2O; [7732-18-5]
3. Glycogen, (colloidal).

ORIGINAL MEASUREMENTS:

Findlay, A.; Creighton, H.J.M.

J. Chem. Soc. 1910, *97*, 536-561

VARIABLES:

Pressure, concentration

PREPARED BY:

W. Gerrard

EXPERIMENTAL VALUES:

Solubility, S given as $\dfrac{\text{concentration of the gas in the liquid phase}}{\text{concentration of the gas in the gaseous phase}}$

T/K = 298.16

Conc. of colloid /g cm^{-3} (soln)	Density of solution	Pressure of gas /kPa.	Solubility, S
0.0049	0.999	98.390	0.590
0.0049	0.999	118.521	0.588
0.0049	0.999	130.254	0.591
0.0049	0.999	146.919	0.594
0.0049	0.999	165.183	0.594
0.0049	0.999	184.782	0.594
0.0100	1.002	98.257	0.585
0.0100	1.002	116.122	0.584
0.0100	1.002	132.120	0.589
0.0100	1.002	139.986	0.591
0.0100	1.002	160.117	0.594
0.0100	1.002	181.315	0.596

AUXILIARY INFORMATION

METHOD /APPARATUS/PROCEDURE:

Gas buret and absorption pipet similar to that of Geffcken (1), except that the manometer tube was longer to give the higher pressure.

SOURCE AND PURITY OF MATERIALS:

1. Self prepared and purified; not attested.

2. Kahlbaum's pure glycogen was dialysed. Toluene was added to prevent putrifaction.

ESTIMATED ERROR:

$\delta S/S$ Stated to be $\mp$ 0.25%.

REFERENCES:

1. Geffcken, G.; *Z. Phys. Chem.* 1904, *49*, 257.

COMPONENTS:	ORIGINAL MEASUREMENTS:
1. Nitrous oxide; N_2O; [10024-97-2] 2. Water; H_2O; [7732-18-5] 3. Dextrin, (colloidal);	Findlay, A.; Creighton, H.J.M. *J. Chem. Soc.* 1910, *97*, 536-561
VARIABLES:	PREPARED BY:
Pressure, concentration	W. Gerrard

EXPERIMENTAL VALUES:

Solubility, *S* given as $\dfrac{\text{concentration of the gas in the liquid phase}}{\text{concentration of the gas in the gaseous phase}}$

T/K = 298.16

Conc. of colloid /g cm^{-3} (soln)	Density of solution	Pressure of gas /kPa	Solubility, *S*
0.0698	1.018	98.523	0.549
0.0698	1.018	109.589	0.550
0.0698	1.018	126.521	0.555
0.0698	1.018	145.585	0.560
0.0698	1.018	165.183	0.562
0.0698	1.018	182.382	0.569
0.1301	1.039	97.190	0.529
0.1301	1.039	111.456	0.523
0.1301	1.039	121.854	0.526
0.1301	1.039	136.386	0.533
0.1301	1.039	164.917	0.540
0.1301	1.039	181.049	0.544
0.2030	1.062	98.657	0.503
0.2030	1.062	111.456	0.499
0.2030	1.062	121.454	0.503
0.2030	1.062	153.185	0.509
0.2030	1.062	171.983	0.513
0.2030	1.062	181.315	0.516

AUXILIARY INFORMATION

METHOD /APPARATUS/PROCEDURE:

Gas buret and absorption pipet similar to that of Geffcken (1), except that the manometer tube was longer to give the higher pressures.

SOURCE AND PURITY OF MATERIALS:

1. Self prepared and purified: not attested.
2. Kahlbaum's purest dextrin was used

ESTIMATED ERROR:

$\delta S/S$ Stated to be $\mp$ 0.25%

REFERENCES:

(1) Geffcken, G.; *Z. Phys. Chem.* 1904, *49*, 257.

COMPONENTS:	ORIGINAL MEASUREMENTS:
1. Nitrous oxide; N_2O; [10024-97-2] 2. Olive oil.	Meyer, K.H.; Gottlieb-Billroth, H.; *Z. Phys. Chem.* 1921, *112*, 55.
VARIABLES:	PREPARED BY: W. Gerrard

EXPERIMENTAL VALUES:

T/K — Ostwald coefficient, L, taken as

$$\frac{\text{Concn. of gas in the liquid phase}}{\text{Concn. of gas in the gaseous phase}}$$

(Final pressures, accurately measured, were about 0.79 x 101.325 kPa.)

T/K	L	
310.16	1.44	
	1.47	Mean 1.40 ±0.06
	1.34	
	1.34	

AUXILIARY INFORMATION

METHOD/APPARATUS/PROCEDURE:	SOURCE AND PURITY OF MATERIALS:
Cylindrical absorption pipet attached to buret and levelling tube assembly. Measurement of concentration of gas in the liquid and in the gas phase by measurement of volume and pressure, and by the assumption of the ideal gas laws.	Not specified.
	ESTIMATED ERROR:
	REFERENCES:

COMPONENTS:	ORIGINAL MEASUREMENTS:
1. Nitrous oxide; N_2O; [10024-97-2] 2. Petroleum.	Gniewosz, S.; Walfisz, A. *Z. Physik. Chem.* 1887. *1*,70-72.

VARIABLES:	PREPARED BY:
Temperature	M.E. Derrick.

EXPERIMENTAL VALUES:

No pressures given; however work carried out at atmospheric pressure. Gas is assumed to be ideal and Henry's Law, to be obeyed.

T/K	Ostwald coefficient L	Bunsen coefficient,* α
293.15	2.26	2.11
283.15	2.58	2.49

* original data.

AUXILIARY INFORMATION

METHOD/APPARATUS/PROCEDURE:	SOURCE AND PURITY OF MATERIALS:
Composed of an absorption flask connected to a gas buret using a flexible lead capillary. System is thermostated in a large water bath. Volume of gas absorbed in a volume of degassed liquid measured using a gas buret.	Gas: no information given. Petroleum: Russian petroleum used. Cleaned by boiling in a large copper flask.
	ESTIMATED ERROR:
	REFERENCES:

COMPONENTS:	ORIGINAL MEASUREMENTS:
1. Nitrous oxide; N_2O; [10024-97-2] 2. Nitrogen dioxide; NO_2; [10102-44-0]	Rocker, A.W. *Anal. Chem.* 1952, *24*, 1322-1324.
VARIABLES:	PREPARED BY:
Temperature, pressure	W. Gerrard

EXPERIMENTAL VALUES:

T/K	Partial pressure p_{N_2O}/kPa	Solubility weight N_2O,%	Mole fraction* x_1 based on NO_2 observed	Mole fraction* x_1 based on NO_2 101.325 kPa	Mole fraction* x_1 based on N_2O_4 observed	Mole fraction* x_1 based on N_2O_4 for 101.325kPa
263.0	82.66	1.28	0.0134	0.0164	0.0268	0.0328
263.1	83.99	1.33	0.0139	0.0167	0.0278	0.0334
263.1	83.99	1.52	0.0159	0.0191	0.0318	0.0382
268.	75.99	1.21	0.0126	0.0167	0.0252	0.0334
268	75.99	1.27	0.0133	0.0177	0.0266	0.0354
273.2	67.99	1.11	0.0116	0.0172	0.0232	0.0344
273.0	67.99	1.04	0.0109	0.0162	0.0218	0.0324
278.2	58.66	0.81	0.00853	0.0146	0.0169	0.0292
278.4	57.33	0.82	0.00857	0.0152	0.0171	0.0304
283.0	49.33	0.65	0.00679	0.0140	0.0136	0.0280
283.1	46.66	0.59	0.00617	0.0136	0.0123	0.0272
283.1	45.33	0.63	0.00658	0.0146	0.0132	0.0292

* Calculated by compiler

The mole fractions, x_1, for p_{N_2O} = 101.325 kPa are based on the assumption that x_1 changes linearly with p_{N_2O} up to 101.325 kPa.

AUXILIARY INFORMATION

METHOD /APPARATUS/PROCEDURE:

Used a modified Ostwald apparatus. The gas buret was a Fisher precision model 100 cm^3, graduated in 0.1 cm^3. The absorption tube was graduated in 0.1 cm^3 to 50 cm^3. Connections were by plastic tube. Glass joints were lubricated by fluorogrease. The gas buret was fitted with a levelling tube containing mercury. A slight positive difference of gas pressure was maintained on the buret side to hinder the flowback of nitrogen dioxide.

SOURCE AND PURITY OF MATERIALS:

1. The nitrous oxide was of commercial grade, 99.2% purity, or better.

2. The nitrogen dioxide was commercial grade, 97.8-99% pure.

ESTIMATED ERROR:

REFERENCES:

COMPONENTS:	ORIGINAL MEASUREMENTS:
1. Nitrous oxide; N_2O; [10024-97-2] 2. Carbon disulfide; CS_2; [75-15-0]	Yen, L. C.; McKetta, J. J., Jr. *J. Chem. Eng. Data* 1962, *7*, 288-289.
VARIABLES:	PREPARED BY:
Temperature	W. Gerrard/ C. L. Young

EXPERIMENTAL VALUES:

T/K	Bunsen coefficient	Mole fraction of nitrous oxide, x_{N_2O}
263.16	3.409	0.008842
273.16	2.801	0.007359
283.16	2.349	0.006250
293.16	2.124	0.005721

The partial pressure of the gas was 101.325 kPa.

Smoothed Data

$$\Delta G^\circ = -RT \ln x_{N_2O} = (-9394 + 75.146 \times T/K)\ \text{J mol}^{-1}$$

(Std. deviation = 53 J mol^{-1})

T/K	Mole fraction of nitrous oxide, x_{N_2O}
263.16	0.008697
273.16	0.007432
283.16	0.006422
293.16	0.005605

AUXILIARY INFORMATION

METHOD/APPARATUS/PROCEDURE:

Measurement of volume of gas absorbed by a known volume of degassed liquid at a partial pressure of 101.325 kPa. The vapor pressure of the liquid was allowed for. Gas buret and absorption pipet. Modified form of apparatus and technique used by Markham and Kobe (1).

SOURCE AND PURITY OF MATERIALS:

1. Gas of 98 per cent purity was purified at 193.16 K. Mas spectrograph then showed purity of 99.5 per cent.
2. Allied Chemical and Dye Corp. ACS grade, freshly fractionated.

ESTIMATED ERROR:

$\delta T/K = \pm 0.1$; $\delta x_{N_2O} = \pm 2\%$ (estimated by compiler).

REFERENCES:

(1) Markham, A. E.; Kobe, K. A. *J. Am. Chem. Soc.* 1941, *63*, 449.

COMPONENTS:	EVALUATOR
1. Nitric oxide; NO; [10102-43-9] 2. Water; H_2O; [7732-18-5]	Rubin Battino, Department of Chemistry, Wright State University, Dayton, Ohio, 45431, U.S.A.

CRITICAL EVALUATION:

Only the data determined by Winkler (1) was considered to be of sufficient accuracy to use in the smoothing equation. We used all nine of his data points. The standard deviation of the fit was 0.76%. The fitting equation is :

$$\ln x_1 = 62.8086 + 82.3420/(T/100K) + 22.8155 \ln (T/100K) \qquad (1)$$

where x_1 is the mole fraction solubility at 101.325 kPa (1 atm) partial pressure of gas. Table 1 gives smoothed values of the mole fraction (at 101.325 kPa) and the Ostwald coefficient at 5K intervals. Table 1 also gives values of the thermodynamic functions $\Delta\overline{G}_1^\circ$, $\Delta\overline{H}_1^\circ$, $\Delta\overline{S}_1^\circ$ and $\Delta\overline{C}_{p_1}^\circ$ for the transfer of the gas from the vapor phase at 101.325 kPa partial gas pressure to the (hypothetical) solution phase of unit mole fraction.

The earlier paper by Winkler (2) contains an identical set of data. Although Usher's datum point was quite close to the smoothed values it was not used in the fitting equation so that the data from one consistent set could be used alone (3). Armor's single value was considerably further off and this may be due to the chemical method used for analysis (4).

REFERENCES:

1. Winkler, L.W.; *Ber.* 1901, *34*,1408-22.

2. Winkler, L.W., *Z.Physik.Chem.* 1892, *9*, 171-5.

3. Usher, F.L., *Z. Physik.Chem.* 1908, *62*, 622-5.

4. Armor, J.N., *J.Chem. Eng. Data.* 1974, *19*, 82-4.

(Table 1 on next page)

Table 1. Smoothed values of nitric oxide solubility in water and thermodynamic functions[a] using equation (1) at 101.325 kPa (1 atm) partial pressures of gas.

T/K	$x_1 \times 10^5$ [b]	L x 10^2 [c]	ΔG_1° [d]	ΔH_1° [d]	ΔS_1° [e]
273.15	5.905	7.346	22.11	-16.65	-141.9
278.15	5.196	6.583	22.81	-15.70	-138.6
283.15	4.625	5.964	23.50	-14.75	-135.1
288.15	4.163	5.460	24.17	-13.80	-131.8
293.15	3.786	5.047	24.82	-12.85	-128.5
298.15	3.477	4.708	25.45	-11.90	-125.3
303.15	3.222	4.430	26.07	-10.96	-122.1
308.15	3.012	4.203	26.67	-10.01	-119.0
313.15	2.838	4.017	27.26	- 9.06	-116.0
318.15	2.695	3.867	27.83	- 8.11	-113.0
323.15	2.577	3.748	28.39	- 7.16	-110.0
328.15	2.481	3.656	28.93	- 6.21	-107.1
333.15	2.404	3.587	29.46	- 5.26	-104.3
338.15	2.343	3.539	29.97	- 4.32	-101.4
343.15	2.297	3.511	30.47	- 3.37	- 98.6
348.15	2.264	3.500	30.96	- 2.42	- 95.9
353.15	2.242	3.506	31.43	- 1.47	- 93.2
358.15	2.232	3.527	31.89	- 0.52	- 90.5

a. $\Delta \bar{C}^\circ_{p_1}$ was independent of temperature and had the value of 190 J K^{-1} mol^{-1}.

b. Mole fraction solubility at 101.325 kPa partial pressure of gas.

c. Ostwald coefficient.

d. Units are J mol^{-1}. cal_{th} = 4.184 J.

e. Units are J K^{-1} mol^{-1}.

COMPONENTS:

1. Nitric oxide; NO; [10102-43-9]
2. Water; H_2O; [7732-18-5]

ORIGINAL MEASUREMENTS:

Winkler, L.W.,

Ber., 1901, *34*, 1408-22.

VARIABLES:

T/K: 273-353

PREPARED BY:

R. Battino

EXPERIMENTAL VALUES:

T/K [a]	$x_1 \times 10^5$ [b]	$L \times 10^2$ [c]	$\alpha \times 10^2$ [d]
273.22	5.922	7.369	7.367
283.17	4.587	5.915	5.706
293.17	3.787	5.049	4.704
303.16	3.231	4.443	4.003
313.11	2.842	4.022	3.509
323.19	2.562	3.727	3.151
333.09	2.414	3.601	2.954
343.20	2.311	3.532	2.809
353.00	2.231	3.487	2.698

a. Temperature reported to 0.01^0C.

b. Mole fraction solubility at 101.325 Pa (1 atm) partial pressure of gas. Calculated by compiler.

c. Ostwald coefficient calculated by compiler.

d. Bunsen coefficient.

AUXILIARY INFORMATION

METHOD/APPARATUS/PROCEDURE:

Using his "absorptionmeter" method.

SOURCE AND PURITY OF MATERIALS:

1. Nitric oxide - prepared and purified chemically by author.

ESTIMATED ERROR:

$\delta\alpha/\alpha = 0.01$ (compiler's estimate)

REFERENCES:

COMPONENTS:

1. Nitric oxide; NO; [10102-43-9]

2. Water; H_2O; [7732-18-5]

ORIGINAL MEASUREMENTS:

Usher, F.L.

Z. Physik. Chem., 1908, *62*, 622-5.

VARIABLES:

PREPARED BY:

R. Battino

EXPERIMENTAL VALUES:

T/K [a]	$\alpha \times 10^2$ [b]	T/K	$x_1 \times 10^5$ [c]	$L \times 10^2$ [d]
293.15	4.51	293.15	3.74	4.98
	4.49			
	4.48			
	4.68			
	4.71			
	4.73			
	4.87			
Avg. =	4.64 ± 0.15			

a. Temperature reported to 1^0C.

b. Bunsen coefficient.

c. Mole fraction solubility at 101.325 Pa (1 atm) partial) pressure of gas calculated by compiler.

d. Ostwald coefficient calculated by compiler.

AUXILIARY INFORMATION

METHOD/APPARATUS/PROCEDURE:

Used the "Ostwald" apparatus.

SOURCE AND PURITY OF MATERIALS:

1. Nitric oxide - chemically prepared and purified.

2. Water - no comment by author.

ESTIMATED ERROR:

$\delta\alpha/\alpha = 0.03$ (compiler's estimate)

REFERENCES:

COMPONENTS:

1. Nitric oxide; NO; [10102-43-9]

2. Water; H_2O; [7732-18-5]

ORIGINAL MEASUREMENTS:

Armor, J.N.

J. Chem. Eng. Data, 1974, *19*, 82-4.

VARIABLES:

PREPARED BY:

R. Battino

EXPERIMENTAL VALUES:

T/K [a]	$x_1 \times 10^5$ [b]	$L \times 10^2$ [c]	$M \times 10^3$ [d]
298.15	3.523	4.770	1.95

a. Temperature reported to 0.1 K.

b. Mole fraction solubility at 101.325 Pa (1 atm) partial pressure of gas. Calculated by compiler.

c. Ostwald coefficient. Calculated by compiler.

d. M is the solubility in molarity or moles per litre of solution.

e. The author also reports solubilities in buffered solutions at various pHs, and salt solutions, all at 25°C.

AUXILIARY INFORMATION

METHOD/APPARATUS/PROCEDURE:

Pure water was saturated for at least 30 min in a reaction vessel. A 5 cm^3 aliquot was removed and injected into 80 cm^3 of oxygen saturated water. This solution was analysed spectrophotometrically for NO_2^-.

SOURCE AND PURITY OF MATERIALS:

1. Nitric oxide - from Matheson Gas Products. "Vigorously" scrubbed to remove NO_2.

2. Water - distilled water redistilled from alkaline permanganate.

ESTIMATED ERROR:

$\delta M/M$ = 0.03 (author's estimate)
δT = 0.1 K (author's estimate)

REFERENCES:

COMPONENTS:	EVALUATOR:
1. Nitric oxide; [10102-43-9] 2. Solutions of metallic salts	W. Gerrard The Polytechnic of North London Holloway, London, N7 8DB UK March 1980

CRITICAL EVALUATION:

Ferrous salts:

Caution should be exercised in looking at all the published data from the aspect of the solubility of nitric oxide in aqueous and nonaqueous solutions of ferrous salts; because the main, if not the only purpose, of the measurements has been to show that the equilibrium

$$Fe^{2+} + NO \rightleftharpoons Fe^{2+}NO$$

occurs up to a maximum ratio of one NO to one Fe^{2+}. Thus Manchot *et al's* [1,2] technique was to measure the volume of nitric oxide absorbed by the iron solution in excess of that volume, not reported, required to saturate the liquid before the addition of the ferrous salt. The maximum volume of nitric oxide absorbed for one mole of ferrous salt was taken to be approximately 22.4 dm^3, achieved at a sufficiently low T/K and high pressure. The evaluator deems it inadvisable to attempt a detailed evaluation, especially as there is a wide variation in concentration of the ferrous iron in the different sets of data. This means that the volume of nitric oxide due to the presence of liquid, water or organic liquid, which is holding the ferrous salt is ignored. When looking at a set of data, readers should ask "what volume of nitric oxide would the stated volume of solution of ferrous salt absorb if the iron salt were not present?"

Gay's [3] measurements appear to have been carefully conducted on a comprehensive scale. Thomas's work [4,5,6] is deemed to be less satisfactory. The work of Manchot *et al.* appears to be of satisfactory accuracy for the purpose of the measurements [i.e. to fix the constitution of the 1:1 complex of $Fe^{2+}NO$]. It must be kept in mind, however, that the reported volume of nitric oxide appears to be that in excess of the volume required to saturate the iron-free liquid. Although the data of Kohschutter and Kutscheroff [7] should be compared with those of Manchot *et al.*, the presentation of these is deemed to be unsatisfactory for the present purpose. Ganz and Mamon [8] were concerned with the absorption of nitric oxide from gaseous mixtures containing that gas. From the aspect of the solubility of nitric oxide their presentation of data is also deemed unsatisfactory. A similarly conclusion was reached by Battino and Clever [9]. Data by Pozin *et al.* [10] on aqueous solutions of ferrous salts should be read alongside those of Manchot *et al.* The Pozin *et al.* data were for several concentrations of ferrous salt and for several temperatures. However a table was given showing the relative absorbing powers of the aqueous solutions of the following salts at 293 K: ferrous sulfate, ferrous chloride, sodium sulfite plus sodium hydroxide, cuprous chloride plus ammonia, nickel sulfate, manganese sulfate, cobalt sulfate, copper sulfate, cupric chloride, and phosphoric acid. The main concern of Pozin *et al.* appears to have been the absorption of nitric oxide in technical operations;

COMPONENTS:	EVALUATOR:
1. Nitric oxide; [10102-43-9] 2. Solutions of metallic salts	W. Gerrard

CRITICAL EVALUATION: Continued

and reference only is now given to the two papers which followed [11,12]. It may, however, be stated that those authors concluded that aqueous solutions of cuprous ammonium chloride, sodium sulfite, and the ferrous salts named in the foregoing were the most efficient absorbing media.

Ferric salts

The data of Manchot [13] on ferric sulfate in aqueous solutions containing each of several concentrations of sulfuric acid should be accepted with caution as gas-liquid solubilities. Thus 1 dm^3 of solution containing 0.01 gram-ion of Fe^{3+}, also contained 18.3 moles of H_2SO_4, and absorbed 44.7 dm^3 of nitric oxide "per gram-atom of iron". This means that 100 dm^3 of solution containing 1 gram-atom of iron and 1832 moles of H_2SO_4 absorbs 44.7 dm^3 [about 2 moles] of nitric oxide. A similar remark applies to the data of Griffith, Lewis, and Wilkinson [14] on an ethanolic solution of ferric chloride. For a molarity of 0.017 of ferric chloride, 21.3 dm^3 of nitric oxide were absorbed "per mole" presumably of ferric chloride. That is, 58.8 dm^3 of solution absorbs 21.3 dm^3 of nitric oxide; whereas 58.8 dm^3 of ethanol itself would absorb about 16.8 dm^3 of the gas.

Cupric salts

The data on cupric salts in ethanol furnished by Manchot [15] and by Griffith *et al.* [14] should be looked at from the aspect of the volume of solution and not from the aspect of the volume of nitric oxide per mole of copper salt. This adjustment of attitude is required also in the scrutiny of the data on the solution of cupric salts in several different organic liquids as given by Kohlschutter and Kutscheroff [16].

Salts of nickel, cobalt and manganese

Hüfner [17] declared that aqueous solutions of these salts absorb "notable quantities" of nitric oxide; but Usher [18] disagreed with Hüfner, and declared that these salts diminished the solubility in water. The evaluator is dubious about the form of presentation of the Hüfner results; and the short paper by Usher is of no consequence. It must be left with readers to draw their own conclusions from the data sheets.

Sodium salts

The Ostwald coefficients based on the data by Kohlschutter and Kutscheroff [7] are less than that for water itself under the same conditions, and are probably approximately of the acceptable magnitude.

COMPONENTS:

1. Nitric oxide; [10102-43-9]

2. Solutions of metallic salts

EVALUATOR:

W. Gerrard

CRITICAL EVALUATION: Continued

REFERENCES:

1. Manchot, W.; Zechentmayer, K. *Annalen,* 1906, *350,* 368
2. Manchot, W.; Huttner, F. *Annalen,* 1910, *372,* 153
3. Gay, J. *Ann. Chim. Phys.* 1885, [6], *5,* 145
4. Thomas, V. *Compt. rend.* 1896, *123,* 943
5. Thomas, V. *Compt. rend.* 1897, *124,* 366
6. Thomas, V. *Bull. Soc. Chim. (France)* 1898, [3], *19,* 343, 419
7. Kohlschutter, V.; Kutscheroff, M. *Ber.* 1907, *40,* 873
8. Ganz, S.N.; Mamon, L.I. *J. Applied Chem.* [USSR], 1953, *26,* 927
9. Battino, R.; Clever, H.L. *Chem. Rev.* 1966, *66,* 395
10. Pozin, M.E.; Zubov, V.V.; Tereshchenko, L. Ya.; Tarat, E. Ya.; Panomarev, Yu. L. *Izv. Vysshikh. Uchebn. Zavedenii, Khim. i Khim. Tekhnol.* 1963, *6* [4], 608
11. Pozin, M.E.; Tarat, E.Ya.; Zubov, V.V.; Tereshchenko, L.Ya. *Izv. Vysshikh. Uchebn. Zavedenii, Khim. i. Khim.* 1963 *6* [6], 974
12. Pozin, M.E.; Tarat, E.Ya.; Tereshchenko, L.Ya.; Zubov, V.V.; Treushchenko, N.N. *Izv. Vyashikh Uchebn. Zavedenii, Khim. i. Khim. Tekhnol.* 1965, *8* [4], 628
13. Manchot, W. *Annalen,* 1910, *372,* 179
14. Griffith, W.P.; Lewis, J.; Wilkinson, G. *J. Chem. Soc.* 1958, 3993
15. Manchot, W. *Ber.,* 1914. *47,* 1601
16. Kohlschutter, V.; Kutcheroff, M. *Ber.* 1904, *37,* 3044
17. Hüfner, G. *Z. Phys. Chem.* 1907, *59,* 416
18. Usher, F.L. *Z. Phys. Chem.* 1908, *62,* 622

COMPONENTS:

1. Nitric oxide; NO; [10102-43-9]
2. Water; H_2O; [7732-18-5]
3. Iron chloride, (ferrous chloride); $FeCl_2$; [7758-94-3]

ORIGINAL MEASUREMENTS:

Gay, J., *Ann. Chim. Phys.* 1885, (6), *5*, 145.

VARIABLES:

Temperature, pressure and concentration.

PREPARED BY:

W. Gerrard.

EXPERIMENTAL VALUES:

T/K	p_{NO}/kPa	V_{NO}/cm^3	Bunsen coefficient α	Conc* /g l^{-1}
277.66	89.724	993.8	44.87	150.04
276.96	79.432	961.6	49.06	
276.76	64.260	905.7	57.12	
276.46	60.407	881.1	59.10	
276.66	55.861	831.2	60.28	
277.46	53.355	812.0	61.68	
277.66	46.622	789.4	68.61	
280.26	45.329	697.9	62.29	
281.56	44.076	653.9	60.10	
281.56	38.489	612.9	63.05	
281.46	34.783	571.9	66.63	
281.46	30.530	530.1	70.34	
281.76	27.197	486.7	72.50	
281.56	24.544	446.9	73.78	
282.46	22.131	400.3	73.29	
283.16	17.772	305.8	69.73	
282.36	14.079	266.4	76.67	
285.46	14.132	225.5	64.66	
285.56	11.199	187.2	67.73	
285.16	8.079	142.7	71.56	
285.16	5.533	101.9	74.63	
285.16	3.373	57.3	68.82	
285.36	1.680	13.8	33.28	
283.56	64.260	446.4	29.32	90.84

cont.

AUXILIARY INFORMATION

METHOD /APPARATUS/PROCEDURE:

p_{NO} is the pressure of NO over the solution. V_{NO} is the volume of NO (adjusted to 101.325 kPa and 273.16K) absorbed by the stated volume of aqueous solution of salt.

$\alpha = 101.325\ V_{NO}/p_{NO}$ x (vol. of soln. of iron salt). The V_{NO} value was determined by extraction of NO in a manometer assembly; diagram given.

SOURCE AND PURITY OF MATERIALS:

NO: Not stated.

Water was distilled. The concentration of salt was checked by titration with permanganate.

DATA CLASS:

ESTIMATED ERROR:

REFERENCES:

COMPONENTS	ORIGINAL MEASUREMENTS:
1. Nitric oxide; NO; [10102-43-9] 2. Water; H_2O; [7732-18-5] 3. Iron chloride, (ferrous chloride); $FeCl_2$; [7758-94-3]	Gay, J., *Ann. Chim. Phys.* 1885, (6) *5*, 145.

EXPERIMENTAL VALUES:

T/K	p_{NO}/kPa	V_{NO}/cm^3	Bunsen coefficient α	Conc* /gl^{-1}
283.76	51.662	420.9	34.39	90.84
283.76	45.422	391.2	36.35	
281.96	36.423	371.6	43.06	
282.16	32.983	350.1	44.81	
282.16	28.051	312.4	46.99	
282.66	24.358	277.2	48.04	
281.96	19.491	239.0	51.74	
282.16	16.598	204.3	51.96	
281.36	10.812	155.0	60.51	
280.16	7.439	118.0	66.94	
280.56	5.039	78.4	65.68	
280.86	2.866	39.2	57.73	
281.16	1.467	14.5	41.74	
285.56	62.567	410.9	33.25	73.76
284.96	52.395	391.9	37.89	
284.96	47.302	362.6	38.83	
284.96	43.689	326.9	37.89	
284.96	37.383	293.0	39.71	
284.96	30.930	257.8	42.22	
284.96	25.197	222.5	44.73	
284.46	18.891	184.6	49.47	
284.56	13.572	143.1	53.39	
284.66	9.026	101.5	56.96	
284.66	4.840	60.0	62.77	
273.66	1.653	14.5	44.38	

* g of Fe in 1.0 ℓ of soln.

COMPONENTS:

1. Nitric oxide; NO; [10102-43-9]
2. Water; H_2O; [7732-18-5]
3. Iron(II) sulfate, (ferrous sulfate); $FeSO_4$; [7720-78-7]

ORIGINAL MEASUREMENTS:

Gay, J., *Ann. Chim. Phys.* 1885, (6), *5*, 145.

VARIABLES:

Temperature, concentration

PREPARED BY:

W. Gerrard.

EXPERIMENTAL VALUES:

T/K	g Fe in 100 cm^3 solution	g NO per 28 g Fe	Mole No* per mole Fe salt	T/K	g Fe in 100 cm^3 solution	g NO per 28 g Fe	Mole NO* per mole Fe salt
Gravimetric determination				Volumetric determination			
273.16	8.7	10.7	0.713	281.16	5.2	10.6	0.707
279.16	5.0	9.5	0.633	281.16	5.2	10.4	0.693
283.16	3.61	10.8	0.720	281.16	5.2	9.5	0.633
283.16	7.57	9.2	0.613	285.66	5.2	10.6	0.707
285.16	8.48	7.34	0.489	286.66	8.1	7.6	0.507
288.16	4.22	7.6	0.507	287.16	7.4	7.3	0.487
288.16	2.23	7.9	0.527	298.66	9.45	4.72	0.315
289.16	4.54	7.6	0.507	298.66	9.45	4.74	0.316
290.16	3.22	7.4	0.493				
298.16	9.45	4.6	0.307				
299.16	8.73	4.7	0.313				
299.16	8.73	5.4	0.360				
299.16	8.73	5.1	0.340				
299.16	8.73	4.9	0.327				
299.65	8.7	5.8	0.387				
299.65	2.0	6.2	0.413				
299.65	4.0	5.6	0.373				
300.16	10.0	4.0	0.267				
300.96	10.0	4.5	0.300				
300.96	8.8	4.3	0.287				

* Calculated by compiler.

AUXILIARY INFORMATION

METHOD /APPARATUS/PROCEDURE:

(a) Determined weight of NO absorbed by known weight of solution containing the ferrous salt. Water entrained by passage of NO was absorbed by anhydrous $CaCl_2$. Pressure appeared to be 101.325 kPa.

(b) Volume of NO absorbed was determined by "extraction" in a manometer assembly. The volume was adjusted to 273.16K and 101.325 kPa.

SOURCE AND PURITY OF MATERIALS:

NO, not stated.

The water was distilled, and the concentration was checked by titration with permanganate.

ESTIMATED ERROR:

REFERENCES:

COMPONENTS:

1. Nitric oxide; NO; [10102-43-9]
2. Water; H_2O; [7732-18-5]
3. Iron chloride, (ferrous chloride); $FeCl_2$; [7758-94-3]

ORIGINAL MEASUREMENTS:

Gay, J., *Ann. Chim. Phys.* 1885, (6) *5*, 145.

VARIABLES:

Temperature, concentration

PREPARED BY:

W. Gerrard.

EXPERIMENTAL VALUES:

Gravimetric determination

T/K	g Fe per 100 cm^3 of solution.	g NO per 28 g Fe
294.66	23.9	7.03
294.66	23.9	6.85
294.66	23.9	7.26
294.66	23.9	6.95
298.16	11.0	5.95
298.16	11.0	6.01
298.16	11.0	5.80
298.16	11.0	6.33
298.16	15.6	5.96
298.16	15.6	6.04
298.16	15.6	6.02
298.16	15.6	5.88

Volumetric determination

T/K	g Fe per 100 cm^3 of solution	g NO per 28 g Fe
277.16	15.0	9.96
285.16	6.56	9.52
289.16	21.4	7.3
289.16	21.4	7.7
289.16	21.4	7.6
289.16	26.5	7.45
292.66	18.2	7.4
292.66	18.2	7.6
295.16	15.6	7.3
298.16	2.6	5.7
298.16	2.6	5.1
298.16	2.6	5.1

AUXILIARY INFORMATION

METHOD /APPARATUS/PROCEDURE:

(a) Determined weight of NO absorbed by known weight of solution containing the ferrous salt. Water entrained by passage of NO was absorbed by anhydrous $CaCl_2$. Pressure appeared to be atmospheric.

(b) Volume of NO absorbed was determined by "extraction" in a manometer assembly (diagram given) over several days. The volume was adjusted to that at 273.16K and 101.325 kPa.

SOURCE AND PURITY OF MATERIALS:

NO, not stated.
Distilled water was used, and the concentration of salt was checked by titration with permangate.

ESTIMATED ERROR:

REFERENCES:

COMPONENTS:	ORIGINAL MEASUREMENTS:
1. Nitric oxide; NO; [10102-43-9] 2. Water; H_2O; [7732-18-5] 3. Iron(II) ammonium sulfate; $Fe(NH_4)_2(SO_4)_2$; [10045-89-3]	Gay, J. *Ann. Chim. Phys.* 1885, *5* (6), 145.
VARIABLES:	PREPARED BY:
Temperature, pressure, concentration	W. Gerrard

EXPERIMENTAL VALUES:

T/K	g Fe/100 cm^3 soln.	g NO/28 g Fe
Gravimetric determination		
273.16	1.15	10.2
273.16	5.7	9.8
279.16	3.5	11.1
288.16	1.15	7.45
299.66	5.7	5.9
299.66	2.8	5.6
Volumetric determination		
295.66	4.7	4.8
298.66	4.5	5.7

(cont.)

AUXILIARY INFORMATION

METHOD/APPARATUS/PROCEDURE:

P_{NO} is the pressure of NO over the solution.

V_{NO} is the volume of NO (adjusted to 101.325 kPa and 273.16 K) absorbed by the stated volume of aqueous solution of ferrous salt containing the stated weight of ferrous iron (Fe).

The volume of NO was determined by extraction during several days.

The weight of NO was that absorbed at the atmospheric pressure.

Entrained water was absorbed in anhydrous $CaCl_2$.

SOURCE AND PURITY OF MATERIALS:

1. NO: not stated.
2. Water was distilled.
3. The salt was pure crystalline standard; the solution was titrated by permanganate.

ESTIMATED ERROR:

REFERENCES:

COMPONENTS:	ORIGINAL MEASUREMENTS:
1. Nitric oxide; NO; [10102-43-9] 2. Water; H_2O; [7732-18-5] 3. Iron(II) ammonium sulfate; $Fe(NH_4)_2(SO_4)_2$; [10045-89-3]	Gay, J. *Ann. Chim. Phys.* 1885, *5* (6), 145.

EXPERIMENTAL VALUES:

Solution	T/K	P_{NO}/kPa	V_{NO}/cm^3	α
0.440 g Fe in 20 cm^3 solution	273.16	91.324	109.1	6.05
		65.820	104.9	8.07
		46.182	90.7	9.95
		29.450	75.0	12.90
		14.000	50.3	18.20
		5.506	18.0	16.56
1.250 g Fe in 25 cm^3 solution	285.16	65.380	229.2	11.28
	284.96	55.634	211.1	15.38
		47.195	181.9	15.62
	285.16	38.943	157.2	16.35
	284.77	29.557	126.1	17.30
	284.66	20.971	95.8	18.51
		12.759	62.1	19.73
		4.653	24.1	21.00
	284.56	2.786	5.0	7.26
0.685 g Fe in 25 cm^3 solution	284.16	74.939	137.8	7.45
	284.66	60.061	119.0	8.02
	284.56	43.862	98.6	9.12
		29.597	79.6	10.88
		21.131	61.5	11.79
	283.56	13.599	42.0	12.49
	283.96	3.040	7.8	10.40

$\alpha = V_{NO}/(\text{Volume of solution} \times \text{partial pressure})$

COMPONENTS:

1. Nitric oxide; NO; [10102-43-9]
2. Water; H_2O; [7732-18-5]
3. Iron bromide (ferrous bromide); $FeBr_2$; [7789-46-0]

ORIGINAL MEASUREMENTS:

Thomas, V. *Bull. Soc. Chim.* 1898, *19*, (3), 343.

See also Ref. (1), (3).

VARIABLES:

Temperature, concentration

PREPARED BY:

W. Gerrard

EXPERIMENTAL VALUES:

Weight of ferrous Fe in 1 cm^3 of solution g.	Volume of solution $/cm^3$	T/K	Volume of NO absorbed $/cm^3$	Mole of NO per mole of salt*
0.05978	2	279.16	30	0.628
0.05978	9.1	283.16	97	0.446
0.05978	2.8	268.16	53	0.792
0.05978	2.8	279.16	45	0.673
0.05978	3.5	284.16	38.5	0.460
0.05978	3.7	281.16	47	0.532
0.05978	2.3	281.16	25	0.455
0.05978	4.1	275.16	65	0.663
0.05978	3.5	281.16	38.8	0.465
0.05978	2.9	283.16	29.9	0.431
0.05978	3	283.16	29.5	0.412
0.05978	2	273.16	36.5	0.764
0.59978	1.5	277.16	25.4	0.0706
0.59978	3.5	293.66	22.7	0.0205
0.59978	3.5	277.16	50.7	0.0604
0.59978	2.5	280.16	41	0.0684

* Calculated by compiler.

AUXILIARY INFORMATION

METHOD/APPARATUS/PROCEDURE:

The liquid was contained in a simple glass vessel furnished with an inlet tube leading into the liquid, and an exit tube. Nitric oxide was slowly passed into the liquid, and entrained water vapor was collected and weighed. It was simply stated that the quantity of absorbed gas was determined. The pressure was not mentioned.

In the calculation of the mole ratio, NO/Fe, the compiler has assumed the correction to 273.16 K, and the pressure, p_{NO}, to be 101.325 kPa.

SOURCE AND PURITY OF MATERIALS:

NO: Not mentioned; but presumably prepared from mercury and nitric acid as in ref.(2)

DATA CLASS:

ESTIMATED ERROR:

REFERENCES:

(1) Thomas, V., *Compt.rend.*, 1896, *123*, 943.

(2) Thomas, V., *Bull Soc. Chim.*, 1898, (3), *19*, 419.

(3) Thomas, V., *Ann. Chim. Phys.* 1898, *13*, (7), 145.

COMPONENTS:

1. Nitric oxide; NO; [10102-43-9]
2. Water; H_2O; [7732-18-5]
3. Iron iodide, (ferrous iodide); FeI_2; [7783-86-0]

ORIGINAL MEASUREMENTS:

Thomas, V.; *Bull, Soc. Chim.*, 1898, *19*, (3), 343.

Thomas, V., *Ann. Chim. Phys.* 1898, *13*, (7), 145.

VARIABLES:

Temperature, concentration

PREPARED BY:

W. Gerrard

EXPERIMENTAL VALUES:

Weight of ferrous iron, Fe, g, in cm^3 of solution.	Volume of solution of salt $/cm^3$	T/K	Volume of NO, absorbed $/cm^3$	Mole NO/56 g Fe*
3.171	12.2	289.16	58	0.375
3.171	9.6	286.16	45	0.367
2.926	8.5	270.16	36	0.362
2.926	5.8	291.16	19	0.280
2.926	8.3	286.16	37	0.381

* The calculated mole ratio calculated by the compiler, NO/Fe, is based on the assumption that the volume of NO observed was adjusted to that at 273.16 K, and that the pressure, p_{NO} was 101.325 kPa.

AUXILIARY INFORMATION

METHOD: /APPARATUS/PROCEDURE:

The liquid was contained in a simple glass vessel furnished with an inlet tube passing right into the liquid, and an exit tube. Nitric oxide was slowly passed into the liquid, the entrained water vapor being collected and weighed. It was stated that the quantity of absorbed gas was determined. The pressure was not mentioned.

SOURCE AND PURITY OF MATERIALS:

NO: Not mentioned; but presumably it was prepared from mercury and nitric acid (1).

ESTIMATED ERROR:

REFERENCES:

1. Thomas. V., *Bull. Soc. Chim.* 1898, *19*, (3), 419.

COMPONENTS:	ORIGINAL MEASUREMENTS:
1. Nitric oxide; NO; [10102-45-9] 2. Water; H_2O; [7732-18-5] 3. Iron(II) sulfate (ferrous sulfate); $FeSO_4$; [7780-78-7] 4. Sodium sulfate; Na_2SO_4; [7757-82-6] or Ammonium sulfate; $(NH_4)_2SO_4$; [10043-02-4]	Manchot, W.; Zechentmayer, K. *Annalen* 1906, *350*, 368-389.
VARIABLES:	PREPARED BY: W. Gerrard

EXPERIMENTAL VALUES:

T/K = 289.35

1 cm^3 of ferrous sulfate solution contained 0.1013 g of ferrous iron.

The solution was made up of:	Molarity of ferrous sulfate /mol dm^{-3}	Pressure of nitric oxide /kPa	Volume, V_1, of nitric oxide for 1 mole of ferrous salt /dm^3
40 cm^3 of a saturated solution of sodium sulfate, 10 cm^3 of water, and 20 cm^3 of ferrous iron.	0.0257	98.390	10.5
	0.0257	107.989	10.6
	0.0257	116.122	10.7
20 cm^3 of a saturated solution of ammonium sulfate, 5 cm^3 of water, and 40 cm^3 of ferrous sulfate solution.	0.0402	89.058	8.9
	0.0402	95.990	9.3
	0.0402	113.055	10.3

AUXILIARY INFORMATION

METHOD/APPARATUS/PROCEDURE:

Volumetric apparatus. A three-way tube was connected to a gas buret and absorption pipet; the third port was used for exhausting the apparatus and for adding gas. The pressure and volume of gas in the buret could be measured using a levelling tube containing mercury.

SOURCE AND PURITY OF MATERIALS:

1. Nitric oxide was prepared from sodium nitrite and dilute sulfuric acid and washed with alkali. Stated to be 100% pure.

3 and 4. Appear to be of acceptable purity.

ESTIMATED ERROR:

REFERENCES:

COMPONENTS:

1. Nitric oxide; NO; [10102-45-9]
2. Water; H_2O; [7732-18-5]
3. Iron(II) ammonium sulfate; $FeSO_4(NH_4)_2SO_4$; $6H_2O$; [10045-89-3]

ORIGINAL MEASUREMENTS:

Manchot, W.; Zechentmayer, K.

Annalen

1906, *350*, 368-389.

VARIABLES:

PREPARED BY:

W. Gerrard

EXPERIMENTAL VALUES:

T/K = 272.65

SERIES A

A known weight of salt was dissolved in 50 cm^3 of water plus 2 cm^3 of sulfuric acid (17%). The molarity is stated in the table. The volume of nitric oxide absorbed refers to the volume of liquid containing 1 mole of ferrous salt.

Molarity of salt /mol dm^{-3}	Pressure of nitric oxide /kPa	Volume, V_1, of nitric oxide for 1 mole salt /dm^3
0.0242	6.426	3.8
0.0222	11.839	5.3
0.0222	13.572	5.8
0.0328	25.704	11.6
0.0363	26.000	11.7
0.0337	41.729	13.6
0.033	61.994	15.5
0.033	76.792	16.2
0.033	81.325	17.0
0.033	89.724	17.4
0.033	97.537	18.1
0.033	109.322	18.8
0.033	128.920	19.2
0.033	140.386	19.5
0.033	164.384	20.4
0.033	177.982	21.5

(cont.)

AUXILIARY INFORMATION

METHOD/APPARATUS/PROCEDURE:

Volumetric apparatus. A three-way tube was connected to a gas buret and absorption pipet; the third port was used for exhausting the apparatus and for adding gas. The pressure and volume of the gas in the buret could be measured using a levelling tube containing mercury.

SOURCE AND PURITY OF MATERIALS:

1. Nitric oxide was prepared from sodium nitrite and dilute sulfuric acid and washed with alkali. Stated to be 100% pure.

2 and 3. Appear to be of acceptable purity.

ESTIMATED ERROR:

REFERENCES:

COMPONENTS:	ORIGINAL MEASUREMENTS:
1. Nitric oxide; NO; [10102-45-9]	Manchot, W.; Zechentmayer, K.
2. Water; H_2O; [7732-18-5]	*Annalen*
3. Iron(II) ammonium sulfate; $FeSO_4(NH_4)_2SO_4$; $6H_2O$; [10045-89-3]	1906, *350*, 368-389.

EXPERIMENTAL VALUES:

SERIES A (cont.)

Molarity of salt /mol dm^{-3}	Pressure of nitric oxide /kPa	Volume, V_1, of nitric oxide for 1 mole salt /dm^3
0.033	159.584	20.2
0.033	177.316	21.0
0.033	189.581	21.2
0.033	202.113	21.2
(; *)	258.774	21.5
0.034	258.507	20.9
0.0298	205.046	21.0
0.0329	236.910	21.8
0.0332	253.975	21.7
0.0332	257.574	22.3
0.0382	298.370	21.8

* Was given in error as 0.0033

SERIES B

T/K	Weight of salt /g	Volume of water /cm^3	Volume of acid* /cm^3	Molarity of salt mol dm^{-8}	Pressure, nitric oxide /kPa	Volume, V_1, of nitric oxide for 1 mole of salt /dm^3
270.65	0.6680	100	4	0.0164	246.775	20.9
270.65	0.6868	100	10	0.0159	246.375	21.5
270.65	0.6868	100	10	0.0159	265.973	21.5
268.15	0.1321	100	100	-	239.976	20.6
268.15	0.8015	40	60	0.0204	273.573	20.7
272.65	0.3255	100	-	0.0083	258.774	21.2

* 17% sulfuric acid.

SERIES C

The stated weight of salt was dissolved in 50 cm^3 of water plus 2 cm^3 sulfuric acid.

T/K	Weight of salt /g	Pressure of nitric oxide /kPa	Volume, V_1, of nitric oxide for 1 mole of salt /dm^3
272.65	0.6748	97.537	18.15
275.75	0.6766	99.483	17.11
277.96	0.6820	98.670	16.26
281.65	0.6530	100.497	15.15
285.05	0.6851	98.857	14.27
287.05	0.6704	98.657	13.39
290.05	0.6604	97.617	12.01
291.76	0.7763	97.230	11.40
292.65	0.6899	97.244	10.98
292.75	0.6770	97.977	10.88
298.65	0.6873	96.257	8.64
304.55	0.6728	93.417	6.22
317.45	0.6842	88.698	2.34

In all series of measurements V_1 was reduced to 273.15 K and 101.3 kPa.

COMPONENTS:

1. Nitric oxide; NO; [10102-43-9]
2. Water; H_2O; [7732-18-5]
3. Iron II sulfate, (ferrous sulfate); $FeSO_4$; [7720-78-7]

ORIGINAL MEASUREMENTS:

Manchot, W.; Zechentmayer, K.

Annalen, 1906, 350, 368-389.

VARIABLES:

Temperature

PREPARED BY:

W. Gerrard

EXPERIMENTAL VALUES:

T/K	Molarity of salt / mol dm^{-3}	Pressure of gas /kPa	Volume of nitric oxide absorbed by volume of solution containing 1 mole of salt /dm^3 (101.325 kPa, 273.15K)
289.25-289.45	0.0257	92.257	12.4
	0.0257	96.790	12.8
	0.0257	109.06	13.4
	0.0257	116.26	14.0
	0.0257	128.25	14.4
	0.0257	143.32	15.6
275.25	0.0257	99.19	17.3
	0.0257	115.32	18.0
	0.0257	129.72	18.3
	0.0257	143.85	18.9
	0.0257	161.72	19.4

AUXILIARY INFORMATION

METHOD /APPARATUS/PROCEDURE:

Volumetric apparatus. A three-way tube was connected to a gas buret and an absorption pipet; the third member of the connection was used for exhausting the apparatus, and for passing in the gas. By means of a levelling tube containing mercury, the pressure and volume of gas in the buret could be regulated. Water, and a small tube containing the salt were put into the absorption pipet. The volume of gas absorbed by the water was first determined; the small tube was broken, and the absorption for the solution was measured.

Drops of dilute sulfuric acid were added to the ferrous sulfate solution.

SOURCE AND PURITY OF MATERIALS:

1. Nitric oxide was prepared from sodium nitrate and diluted sulfuric acid, and washed with alkali. Stated to be 100% pure.
2. and 3. Appeared to be of acceptable purity.

ESTIMATED ERROR:

REFERENCES:

COMPONENTS:

1. Nitric oxide; NO; [10102-43-9]
2. Water; H_2O; [7732-18-5]
3. Iron chloride (Ferrous chloride); $FeCl_2$; [7758-94-3]

ORIGINAL MEASUREMENTS:

Manchot, W.; Zechentmayer, K.

Annalen, 1906, *350*, 368-389.

VARIABLES:

PREPARED BY:

W. Gerrard

EXPERIMENTAL VALUES:

T/K = 273.65. The solution of ferrous chloride was made up by adding the stated weight of ferrous chloride, $FeCl_2$, to 50 cm^3 of water plus 2 cm^3 of hydrochloric acid (14% HCl)

Weight of $FeCl_2$ /g	Molarity of $FeCl_2$ / mol dm^{-3}	Pressure of nitric oxide. /kPa	Volume, V_1, of nitric oxide absorbed per 1 mole of $FeCl_2$, /dm^3 (101.325 kPa)
0.2528	0.0383	34.530	12.1
*0.229	0.034	98.977	17.3
*0.1795	0.027	100.310	17.1
*0.1639	0.0248	232.377	21.2
0.2542	(0.0385)**	267.440	21.8

*Stated to contain 40.9% Fe.

**Incorrectly given in the original table as 0.0538.

For a solution made up by adding 0.1974 g of $FeCl_2$ to 52 cm^3 of hydrochloric acid (14% HCl), V_1 was 17.1 dm^3 at T/K = 273.65 and for 100.097 kPa.

For a solution made up by adding 0.1954 g of $FeCl_2$ to 50 cm^3 of water, V_1 was 17.7 dm^3 at T/K = 273.65 and for 100.097 kPa.

AUXILIARY INFORMATION

METHOD /APPARATUS/PROCEDURE:

Volumetric apparatus. A three-way tube was connected to a gas buret and an absorption pipet; the third member was used for exhausting the apparatus, and for passing in the gas. By means of a levelling tube containing mercury, the pressure and volume of gas in the buret could be regulated.

SOURCE AND PURITY OF MATERIALS:

1. Nitric oxide was prepared from sodium nitrite and dilute sulfuric acid, and washed with alkali. Stated to be 100% pure.
3. Prepared by action of hydrogen chloride on heated sheet iron. Contained 43.96-44.06% Fe.

DATA CLASS:

ESTIMATED ERROR:

REFERENCES:

COMPONENTS:	ORIGINAL MEASUREMENTS:
1. Nitric oxide; NO; [10102-43-9] 2. Water; H_2O; [7732-18-5] 3. Iron II sulfate, (ferrous sulfate); $FeSO_4$; [7720-78-7]	Hufner, G. *Z. Phys. Chem.* 1907, *59*, 416-423.
VARIABLES:	PREPARED BY:
Concentration of salt	W. Gerrard

EXPERIMENTAL VALUES:

The form in which the data were given requires a detailed analysis by the compiler.

The volume of the solution of ferrous salt was the same ($205.69\ cm^3$) for each series.

W = weight of ferrous iron in $205.69\ cm^3$ of solution.

P = pressure of nitric oxide.

V_1= volume of nitric oxide, cm^3, absorbed by $205.69\ cm^3$ of solution at the stated pressure.

By a certain argument, the author derived the following equation:

$V_1 = a + bP$, where a is a constant, believed to represent the chemically combined part; and b is a constant representing the "simply" absorbed part. A number referred to as the "absorption coefficient, α," was obtained by the operation: (b/205.69) x 760; but this is not the quantity which is usually understood by the term "absorption coefficient". There is consequent confusion in the reviews:

(cont.)

AUXILIARY INFORMATION

METHOD/APPARATUS/PROCEDURE:	SOURCE AND PURITY OF MATERIALS:
The volume of nitric oxide absorbed by a fixed volume of solution was measured at the observed pressure of gas.	1. Nitric oxide was prepared from hydrogen iodide and nitric acid (Winkler,1). 3. Attested by titration with permanganate.
	ESTIMATED ERROR:
	REFERENCES:
	1. Winkler, L.W, *Ber.* 1901, *34*, 1408.

COMPONENTS:	ORIGINAL MEASUREMENTS:
1. Nitric oxide; NO; [10102-43-9]	Hufner, G.
2. Water; H_2O; [7732-18-5]	*Z. Phys. Chem.* <u>1907</u>, *59*, 416-423.
3. Iron II sulfate, (ferrous sulfate); $FeSO_4$; [7720-78-7]	

EXPERIMENTAL VALUES:

Additional columns in the table are : V_3* = volume, dm^3, of solution containing 56 g of ferrous iron. V_1* = volume of NO/dm^3, absorbed by V_3* of solution, as for 101.325 kPa. V_w(NO) = volume of NO/dm^3, absorbed by V_3* of water itself.

T/K	W/g	*P*/mmHg	V_1/cm^3		
293.25	0.0221	760.0	15.33**	*a*	= 2.8466
293.25		704.9	14.42	*b*	= 0.01642
293.25		683.5	14.10	α	= 0.06067
293.25		668.6	13.80	V_3*/dm^3	= 521.2
292.35		651.9	13.58		
293.20		632.9	13.15	V_1*/dm^3	= 38.85
293.15		613.7	12.98	V_w(NO)/dm^3	= 24.50
	0.0296	760.0	15.57**	*a*	= 4.2407
293.20		677.5	14.30	*b*	= 0.0149
293.20		655.3	14.07	α	= 0.05505
293.19		639.1	13.81	V_3*/dm^3	= 389.1
293.15		620.2	13.39		
293.30		600.5	13.20	V_1*/dm^3	= 29.46
293.29		581.2	12.92	V_w(NO)/dm^3	= 18.29
	0.0409	760.0	18.50**	*a*	= 4.749
293.19		667.6	16.79	*b*	= 0.01809
293.17		650.6	16.65	α	= 0.06684
293.15		613.1	15.71	V_3*/dm^3	= 281.6
293.15		594.6	15.41	V_1*/dm^3	= 25.33
293.25		577.1	15.32	V_w(NO)/dm^3	= 13.23
	0.0513	760.0	21.33**	*a*	= 4.9136
293.25		644.8	18.82	*b*	= 0.02160
293.25		623.8	18.47	α	= 0.07981
293.23		606.4	18.02	V_3*/dm^3	= 224.5
293.25		589.7	17.56		
293.25		571.1	17.19	V_1*/dm^3	= 23.28
293.25		553.1	16.95	V_w(NO)/dm^3	= 10.55
	0.0663	760.0	23.32**	*a*	= 6.7589
293.25		697.3	21.91	*b*	= 0.02181
293.25		678.9	21.60	α	= 0.08059
293.25		660.4	21.18	V_3*/dm^3	= 173.7
293.23		638.2	20.71		
293.19		620.7	20.28	V_1*/dm^3	= 19.70
293.15		602.5	19.87	V_w(NO)/dm^3	= 8.16
	0.099	760.0	37.40**	*a*	= 13.78
293.25		649.9	34.26	*b*	= 0.031558
293.30		631.1	33.82	α	= 0.11661
293.35		618.4	33.26	V_3*/dm^3	= 116.3
293.15		603.3	32.76		
293.00		588.6	32.34	V_1*/dm$_3$	= 21.16
293.00		574.2	31.95	V_w(NO)/dm^3	= 5.47

** Calculated from the equation; the original author's value.

V_3*, V_1* and V_w(NO)/dm^3 were calculated by the compiler.

COMPONENTS:	ORIGINAL MEASUREMENTS:
1. Nitric oxide; NO; [10102-43-9] 2. Water; H_2O; [7732-18-5] 3. Iron chloride (Ferrous chloride); $FeCl_2$; [7758-94-3]	Kohlschutter, V.; Kutscheroff, M. *Ber.* <u>1907</u>, *40*, 873-878.
VARIABLES:	**PREPARED BY:**
Concentration	W. Gerrard

EXPERIMENTAL VALUES:

The pressure of nitric oxide, p_{NO}, was not stated. The compiler has taken p_{NO} to be 101.325 kPa.

T/K	Volume of water, V_s containing one mole of ferrous chloride /dm^3	Volume of nitric oxide, V_1 absorbed by V_s (1 mole of salt) /dm^3 a	b	Ostwald coefficient L, based on b Calculated by compiler
295.15	2.5	3.15	3.30	1.320
	5.18	4.62	4.83	0.932
	10.35	5.9	6.56	0.634
	20.7	7.0	8.32	0.402
	51.8	8.6	11.89	0.230

Column (a) shows the volume of nitric oxide absorbed after allowance for the volume absorbed by V_s dm^3 of pure water, i.e., b - a is the volume of NO which would be absorbed by V_s dm^3 of pure water,taken as 0.0636 dm^3 of NO per dm^3 of water.

AUXILIARY INFORMATION

METHOD/APPARATUS/PROCEDURE:

An Ostwald type gas buret and pipet were connected by a lead capillary.

SOURCE AND PURITY OF MATERIALS:

Ferrous chloride prepared from pure iron wire. No details given for other components.

ESTIMATED ERROR:

REFERENCES:

COMPONENTS:

1. Nitric oxide; NO; [10102-43-9]
2. Water; H_2O; [7732-18-5]
3. Iron II sulfate, (ferrous sulfate); $FeSO_4$; [7720-78-7]

ORIGINAL MEASUREMENTS:

Kohlschutter, V.; Kutscheroff, M.

Ber, 1907, *40*, 873-878.

VARIABLES:

Concentration

PREPARED BY:

W. Gerrard.

EXPERIMENTAL VALUES:

The pressure of nitric oxide, p_{NO}, was not stated. The compiler has taken p_{NO} to be 101.325 kPa.

T/K	Volume, V_s, of water in which 1 mole of ferrous sulphate is dissolved /dm^3	Volume, V_1, of nitric oxide absorbed by V_s /dm^3 (a)	(b)	Ostwald coefficient, *L*, based on V_1 (b) calculated by compiler.
298.15	1.2	1.4	1.47	1.225
	1.8	1.9	2.01	1.117
	2.4	2.4	2.55	1.0625
	4.82	4.1	4.40	0.913
	7.2	5.06	5.52	0.767
	12.0	5.70	6.46	0.538
	18.6	6.9	8.01	0.431
	36.0	8.2	10.4	0.289

V_1 (b) - V_1 (a) refers to the volume of nitric oxide which would be absorbed by the stated volume, V_s, of pure water.

AUXILIARY INFORMATION

METHOD /APPARATUS/PROCEDURE:

An Ostwald type gas buret and pipet were connected by a lead capillary.

SOURCE AND PURITY OF MATERIALS:

No details given.

ESTIMATED ERROR:

REFERENCES:

COMPONENTS:	ORIGINAL MEASUREMENTS:
1. Nitrogen oxide, (Nitric oxide); NO; [10102-43-9] 2. Water; H_2O; [7732-18-5] 3. Iron chloride, (ferrous chloride); $FeCl_2$; [7758-94-3]	Kohlschutter, V.; Kutscheroff, M. *Ber*, 1907, *40*, 873-878.
VARIABLES:	PREPARED BY:
Concentration	W. Gerrard

EXPERIMENTAL VALUES:

The pressure of nitric oxide, p_{NO}, was not stated. The compiler has taken p_{NO} to be 101.325 kPa. Temperature assumed to be 295.15K.

T/K	1 mole of ferrous chloride dissolved in 10.37 dm^3 of the following:	Volume, V_1, of nitric oxide absorbed /dm^3
295.15	Water; H_2O;	6.559
	Sodium chloride; NaCl; (saturated aqueous solution)	6.549
	Ammonium chloride; NH_4Cl; (saturated aqueous solution)	6.549
	Hydrochloric acid (about 30%)	15.64
	(10%)*	6.17

* given as such in original table (probably wt %)

AUXILIARY INFORMATION

METHOD:/APPARATUS/PROCEDURE:	SOURCE AND PURITY OF MATERIALS:
An Ostwald type gas buret and pipet were connected by a lead capillary.	Ferrous chloride solution was prepared from iron wire. No details of source or preparation of other components given.
	ESTIMATED ERROR:
	REFERENCES:

COMPONENTS:	ORIGINAL MEASUREMENTS:
1. Nitric oxide; NO; [10102-43-9] 2. Water; H_2O; [7732-18-5] 3. Iron (II) nitrate (Ferrous nitrate); $Fe(NO_3)_2$; [14013-86-6]	Kohlschutter, V.; Kutscheroff, M. *Ber*, 1907, *40*, 873-878.
VARIABLES:	**PREPARED BY:**
Concentration of salt	W. Gerrard

EXPERIMENTAL VALUES:

The pressure of nitric oxide, p_{NO}, was not stated. The compiler has taken p_{NO} to be 101.325 kPa.

T/K	Volume, V_s, of water in which 1 mole of ferrous nitrate is dissolved /dm^3	Volume, V_1, of nitric oxide absorbed by V_s (1 mole of ferrous nitrate) /dm^3		Ostwald coefficient, L, based on V_1 (b) calculated by compiler
		(a)	(b)	
296.15	3.25	2.56	2.77	0.852
	6.5	3.75	4.16	0.640
	13.0	4.71	5.54	0.426
	26.0	4.90	6.61	0.254

V_1 (b) - V_1 (a) refers to the volume of nitric oxide which would be absorbed by the stated volume, V_s, of pure water.

AUXILIARY INFORMATION

METHOD/APPARATUS/PROCEDURE:	SOURCE AND PURITY OF MATERIALS:
An Ostwald type gas buret and pipet were connected by a lead capillary.	Ferrous nitrate solution was prepared by the interation of ferrous sulfate and barium nitrate.
	ESTIMATED ERROR:
	REFERENCES:

COMPONENTS:	ORIGINAL MEASUREMENTS:
1. Nitric oxide; NO; [10102-43-9] 2. Water; H_2O; [7732-18-5] 3. Iron chloride, (Ferrous chloride); $FeCl_2$; [7758-94-3] 4. Hydrochloric acid; HCl; [7647-01-0]	Kohlschutter, V.; Kutscheroff, M. *Ber*, 1907, *40*,873-878.
VARIABLES:	PREPARED BY:
Concentration	W. Gerrard

EXPERIMENTAL VALUES: The pressure of nitric oxide, p_{NO} was not stated. The compiler has taken p_{NO} to be 101.325 kPa.

The stock solution of ferrous chloride contained 45.365 g/dm^3; 1.234 dm^3 of solution contained 1 mole of $FeCl_2$.

Concentration of "strong hydrochloric acid" was not specified.

The aqueous solutions were made up as specified in the table. The temperature was probably 295.15K.

Ferrous chloride stock solution cm^3	Water cm^3	Hydrochloric acid stock solution cm^3	Volume, V_s, solution containing 1 mole ferrous chloride /dm^3	Volume, V_1 of nitric oxide absorbed by V_s /dm^3
10	40	-	6.16	9.3
10	-	40	6.16	20.48
5	45	-	12.32	10.81
5	15	30	12.32	17.2
5	5	40	12.3	19.2
2.5	47.5	-	24.64	10.42
2.5	37.5	10	24.64	9.165
2.5	27.5	20	24.64	11.62
2.5	7.5	40	24.64	20.62

AUXILIARY INFORMATION

METHOD /APPARATUS/PROCEDURE:

An Ostwald type gas buret and pipet were connected by a lead capillary.

SOURCE AND PURITY OF MATERIALS:

Ferrous chloride solution prepared from iron wire.

No details of other components given.

ESTIMATED ERROR:

REFERENCES:

COMPONENTS:	ORIGINAL MEASUREMENTS:
1. Nitric oxide; NO; [10102-43-9] 2. Water; H_2O; [7732-18-5] 3. Iron(II) sulfate; $FeSO_4$; [7720-78-7]	Manchot, W.; Huttner, F. *Annalen* 1910, *372*, 153-178.
VARIABLES:	PREPARED BY:
Concentration of salt	W. Gerrard

EXPERIMENTAL VALUES:

T/K	Conc. of ferrous sulfate /mol dm^{-3}	Pressure of nitric oxide /kPa	Volume, V, of nitric oxide absorbed of Iron(II) salt (reduced to 273.15 K, 101.325 kPa) /dm^3
273.15	0.0034	98.257	17.8
	0.0429	98.257	17.9
	0.44	98.257	17.2
	1.2140	100.657	15.4

AUXILIARY INFORMATION

METHOD/APPARATUS/PROCEDURE:

The volume of gas absorbed was measured by the Ostwald-type gas buret and absorption pipet. The apparatus and method were stated to be similar to those previously described (1), except that the technique for introducing the ferrous salt was modified.

SOURCE AND PURITY OF MATERIALS:

1. Nitric oxide was probably taken as 100% pure, see ref. (1).

2, 3. Appeared to be of satisfactory purity.

ESTIMATED ERROR:

REFERENCES:

1. Manchot, W.; Zechentmayer, K. *Annalen* 1906, *350*, 368-389.

COMPONENTS:

1. Nitric oxide; NO; [10102-43-9]
2. Water; H_2O; [7732-18-5]
3. Sulfuric acid; H_2SO_4; [7664-93-9]
4. Iron(II) ammonium sulfate; $Fe(NH_4)_2(SO_4)_2$; [10045-89-3]

ORIGINAL MEASUREMENTS:

Manchot, W.; Huttner, F.

Annalen

1910, *372*, 153-178.

VARIABLES:

Concentration of components 3 and 4

PREPARED BY:

W. Gerrard

EXPERIMENTAL VALUES: T/K = 273.15

M /mol dm^{-3}	S /mol dm^{-3}	P /kPa	V /dm^{3}mol^{-1}	M /mol dm^{-3}	S /mol dm^{-3}	P /kPa	V /dm^{3}mol^{-1}
0.0077	18.32 (97.6)	98.124	22.4	0.0396	18.32 (97.6)	151.185	22.6
0.0099	16.92 (90.6)	98.657	22.5	0.0452	18.32 (97.6)	157.184	22.2
0.0105	14.69 (82.0)	98.657	22.1	0.0153	18.32 (97.6)	48.528	22.2
0.0127	1.00 (9.8)	100.657	16.1	0.0156	18.32 (97.6)	48.528	22.8
0.0131	2.25 (19.9)	100.657	16.2				
0.0138	18.32 (97.6)	99.723	23.0	0.0063	0	100.657	17.3
0.0148	18.32 (97.6)	99.323	22.4	0.0120	0	100.657	17.3
0.0152	18.32 (97.6)	99.323	22.7	0.0313	0	100.657	17.4
0.0153	2.25 (19.9)	99.723	16.3	0.0624	0	100.657	17.1
0.0154	6.67 (47.6)	99.723	18.3	0.0974	0	100.657	17.1
0.0155	11.73 (70.8)	99.457	21.0	0.2035	0	100.657	16.6
0.0382	18.32 (97.6)	98.257	22.4				
0.0782	18.32 (97.6)	97.990	22.3	0.0103	18.32 (97.6)	101.723	22.4*

* T/K = 284.15.

M is the number of moles of the sulfate in 1 dm^3 of solution.

S is the number of moles of sulfuric acid in 1 dm^3 of solution; the number () in brackets is the %.

V is the volume in dm^3 (at 273.15 K and 101.325 kPa) of nitric oxide absorbed for 1 mole of ferrous salt.

P is the pressure of nitric oxide.

AUXILIARY INFORMATION

METHOD/APPARATUS/PROCEDURE:

The volume of gas absorbed was measured by the Ostwald-type gas buret and absorption pipet. The apparatus and method were stated to be similar to those previously described (1), except that the technique for introducing the gerrous salt was modified.

SOURCE AND PURITY OF MATERIALS:

1. Nitric oxide probably taken as 100% pure, see ref. (1).

2, 3. Appeared to be of satisfactory purity.

ESTIMATED ERROR:

REFERENCES:

1. Manchot, W.; Zechentmayer, K. *Annalen* 1906, *350*, 368-389.

COMPONENTS:	ORIGINAL MEASUREMENTS:
1. Nitric oxide; NO; [10102-43-9] 2. Water; H_2O; [7732-18-5] 3. Iron(II) ammonium sulfate; $Fe(NH_4)_2(SO_4)_2$; [10045-89-3] 4. Hydrochloric acid; HCl; [7647-01-0]	Manchot, W.; Huttner, F. *Annalen* 1910, *372*, 153-178.
VARIABLES:	PREPARED BY:
Concentration of components 3 and 4	W. Gerrard

EXPERIMENTAL VALUES:

M/mol dm^{-3}	C/mol dm^{-3}	P/kPa	V/dm^3	M/mol dm^{-3}	C/mol dm^{-3}	P/kPa	V/dm^3
0.0122	12.56	100.657	22.1	0.0063	0	100.657	17.3
0.0128	1.6		13.2	0.0120	0		17.3
0.0128	3.4		15.8	0.0313	0		17.4
0.0128	7.76	100.790	19.5	0.0624	0		17.1
0.0128	8.94		21.0	0.0974	0		17.1
0.0130	7.3		19.2	0.2035	0		16.6
0.0131	11.0		22.5	0.4154	0		16.3
0.0302	12.56	100.657	22.5	0.6535	0		15.6
0.1336	12.56	100.123	22.1	1.4730	0		14.6

M is the number of moles of sulfate in 1 dm^3 of solution.

C is the number of moles of hydrogen chloride in 1 dm^3 of solution.

P is the pressure of nitric oxide.

V is the volume in dm^3 (at 273.15 K, 101.325 kPa) of nitric oxide absorbed for 1 mole of iron(II) salt.

AUXILIARY INFORMATION

METHOD/APPARATUS/PROCEDURE:	SOURCE AND PURITY OF MATERIALS:
The volume of gas absorbed was measured by the Ostwald-type gas buret and absorption pipet. The apparatus and method were stated to be similar to those previously described (1), except that the technique for introducing the ferrous salt was modified.	1. Nitric oxide was probably taken as 100% pure, see ref. (1). 2, 3. Appeared to be of satisfactory purity. ESTIMATED ERROR: REFERENCES: 1. Manchot, W.; Zechentmayer, K. *Annalen* 1906, *350*, 368-389.

COMPONENTS:

1. Nitric oxide; NO; [10102-43-9]
2. Water; H_2O; [7732-18-5]
3. Iron(II) chloride; $FeCl_2$; [7758-94-3]
4. Hydrochloric acid; HCl; [7647-01-0]

ORIGINAL MEASUREMENTS:

Manchot, W.; Huttner, F.

Annalen

1910, *372*, 153-178.

VARIABLES:

Concentration of components 3 and 4

PREPARED BY:

W. Gerrard

EXPERIMENTAL VALUES:

T/K	Conc. of Iron salt /mol dm^{-3}	Conc. of HCl /mol dm^{-3} (soln.)	Pressure of NO /kPa	Volume, V_1, of NO, 1 mole of iron salt (101.325 kPa) /dm^3
273.15	0.0054	0	100.123	16.7
	0.0111	12.56	99.457	22.1
	0.0111	0	100.123	16.7
	0.0209	7.3	99.590	19.1
	0.0216	0	100.257	16.2
	0.0226	11.0	99.590	22.6
	0.1078	0	100.123	15.8
	0.2108	0	99.057	15.6
	0.4219	12.56	100.123	20.4
	0.5616	0	99.057	15.1
	0.0219	11.0	165.583	22.0
	0.0226	11.0	159.717	22.6
	0.0201	11.0	48.662	21.3
	0.0208	11.0	48.662	21.3
287.65	0.0306	11.0	98.257	21.8

AUXILIARY INFORMATION

METHOD/APPARATUS/PROCEDURE:

Apparatus and method were similar to those described (1), except that the procedure for introducing the ferrous salt was modified.

The volume of gas absorbed was measured by means of the gas buret and absorption pipet technique.

SOURCE AND PURITY OF MATERIALS:

1. Nitric oxide probably taken as being 100% pure as in ref. (1).

2, 3, 4. Appeared to be of satisfactory purity.

ESTIMATED ERROR:

REFERENCES:

1. Manchot, W.; Zechentmayer, K. *Annalen* 1906, *350*, 368-389.

COMPONENTS:	ORIGINAL MEASUREMENTS:
1. Nitric oxide; NO; [10102-43-9] 2. Water; H_2O; [7732-18-5] 3. Hydrogen bromide; HBr; [10035-10-6] 4. Iron bromide, (Ferrous bromide); $FeBr_2$; [7789-46-0] (see note)	Manchot, W. *Ber.* 1914, *47*,1601-1614
VARIABLES:	PREPARED BY:
Concentration	W. Gerrard

EXPERIMENTAL VALUES:

T/K = 273.15

The pressure of NO was stated to be practically 1 atm (101.325 kPa).

Conc of HBr / mol dm^{-3}	Volume,* V_1, of NO absorbed by 1 mole of ferrous salt / dm^3
0.	17.45
1.918	10.46
2.110	10.06
3.836	12.41
4.54	12.65
6.40	13.55
9.35	17.19
11.52	18.70
12.20	19.60
13.83	21.78
14.33	22.14
14.67	23.61

* appears to be the amount absorbed by ferrous bromide in excess of amount absorbed by hydrogen bromide solution.

AUXILIARY INFORMATION

METHOD: /APPARATUS/PROCEDURE:

The volume of nitric oxide absorbed was observed as previously stated (1). The "Ferrous" molarity was stated to be equivalent to 0.0127 mole of ferrous ammonium sulfate in 1 dm^3 of solution.

SOURCE AND PURITY OF MATERIALS:

3. Hydrogen bromide was prepared from bromine and naththalene, and treated with red phosphorus.

Other components presumably as previously stated in ref. (1).

ESTIMATED ERROR:

REFERENCES:

1. Manchot, W.; Zechentmayer, K. *Annalen*, 1906, *350*,368-389.

See also Manchot, W.;Huttner , F. *Annalen*, 1910,*372*,153-178

COMPONENTS:	ORIGINAL MEASUREMENTS:
1. Nitric oxide; NO; [10102-43-9] 2. Water; H_2O; [7732-18-5] 3. Iron II sulfate, (ferrous sulfate); $FeSO_4$; [7720-78-7]	Manchot, W.; Haunschild, H. *Z. anorg. Chem.* 1924, *140*, 22-36.
VARIABLES:	PREPARED BY: W. Gerrard.

EXPERIMENTAL VALUES:

T/K = 291.15

M is the molarity of ferrous sulfate
V_3 is the volume of solution containing 1 mole of ferrous sulfate.

P is the pressure of nitric oxide
V_1 is the volume of nitric oxide absorbed by the solution containing 1 mole of sulfate;
V_1 is adjusted to (273.15 K, 101.325 kPa).

M/mol dm^{-3}	V_3/dm^3	P/kPa	V_1/dm^3
0.0031	320	90.844	11.4
0.0031	320	90.751	11.4
0.00625	160	90.471	10.9
0.00625	160	90.564	10.9
0.01	100	90.751	10.4
0.01	100	90.938	10.1
0.0125	80	90.658	10.9
0.0125	80	90.844	10.4
0.025	40	90.751	11.2
0.025	40	90.191	11.1
0.03	33.3	90.378	11.0
0.03	33.3	90.751	10.9
0.05	20	90.938	11.4
0.05	20	90.284	10.9
0.064	15.6	90.564	11.0
0.1493	6.69	90.378	10.9
0.363	2.75	90.844	10.6
0.6883	1.45	90.191	9.8
0.911	1.10	91.204	9.9

AUXILIARY INFORMATION

METHOD/APPARATUS/PROCEDURE:

The volume of nitric oxide absorbed was observed as previously stated (1).

SOURCE AND PURITY OF MATERIALS:

Presumably as previously described (1).

ESTIMATED ERROR:

REFERENCES:

1. Manchot, W.; Zechentmayer, K. *Annalen*, 1906, *350*, 368-389.

See also Manchot, W.; Huttner, F. *Annalen*, 1910, *372*, 153-178.

COMPONENTS:	ORIGINAL MEASUREMENTS:
1. Nitric oxide; NO; [10102-43-9] 2. Water; H_2O; [7732-18-5] 3. Ethanol; C_2H_6O; [64-17-5] 4. Iron (II) selenate, (ferrous selenate); $FeSeO_4$; [15857-43-9]	Manchot, W.; Linckh, E. *Z. Anorg. Chem.* 1924, *140*, 37-46.
VARIABLES:	**PREPARED BY:**
Temperature, concentration	W. Gerrard

EXPERIMENTAL VALUES:

Concentration of ethanol was 90%.

T/K	Molarity of ferrous selenate /mol dm^{-3}	Pressure of NO /kPa	Volume of nitric oxide (adjusted to 273.15, 101.325 kPa) absorbed per mole of ferrous salt /dm^3
273.15	0.034	92.031	20.4
273.15	0.0201	93.631	20.6
273.15	0.0142	93.497	20.6
273.15	0.0118	92.031	20.3
273.15	0.01034	93.231	20.2
273.15	0.00981	93.097	21.4
273.15	0.00794	92.564	22.2
267.16	0.0103	92.791	22.2
266.15	0.0103	92.604	22.6
259.15	0.0103	93.164	22.4
258.15	0.0102	93.871	23.2
253.15	0.0101	93.551	22.1

AUXILIARY INFORMATION

METHOD/APPARATUS/PROCEDURE:

The volume of nitric oxide absorbed was measured by means of the apparatus referred to by Manchot and Huttner (1).

The vapor pressure of the alcohol was taken into account.

SOURCE AND PURITY OF MATERIALS:

1. Nitric oxide was stated to be essentially pure.

4. Analytically tested.

ESTIMATED ERROR:

REFERENCES:

1. Manchot, W.; Huttner, F.
Annalen, 1910, *372*, 153.

COMPONENTS:	ORIGINAL MEASUREMENTS:
1. Nitric oxide; NO; [10102-43-9] 2. Water; H_2O; [7732-18-5] 3. Iron (II) selenate; $FeSeO_4$; [15857-43-9] 4. Selenic acid; H_2SeO_4; [7783-08-6]	Manchot, W.; Linckh, E. *Z. Anorg. Chem.* 1924, *140*,37-46.
VARIABLES:	PREPARED BY:
Concentration	W. Gerrard

EXPERIMENTAL VALUES:

T/K is presumably 273.15. The Volume, V_1, of nitric oxide stated to be absorbed for 1 mole of ferrous selenate is for a pressure of nitric oxide equal to 101.325 kPa.

Conc of ferrous selenate /mol dm^{-3}	Concentration of H_2SeO_4 Weight, %	/mol dm^{-3}	V_1, dm^3(NO)
0.0188	29.9	2.62	18.2
0.049	58.8	6.7	20.0
0.0195	67.9	7.15	18.6
0.049	84.8	13.2	11.9
0.0371	91.5	15.1	11.6

AUXILIARY INFORMATION

METHOD/APPARATUS/PROCEDURE:

The volume of nitric oxide absorbed was measured by means of the apparatus referred to by Manchot and Huttner (1).

SOURCE AND PURITY OF MATERIALS:

1. Nitric oxide was stated to be nearly 100% pure.

3,4. Were analytically attested.

ESTIMATED ERROR:

REFERENCES:

1. Manchot, W.; Huttner, F. *Annalen*, 1910, *372*,153.

COMPONENTS:	ORIGINAL MEASUREMENTS:
1. Nitric oxide; NO; [10102-43-9] 2. Water; H_2O; [7732-18-5] 3. Iron (II) selenate; $FeSeO_4$; [15857-43-9]	Manchot, W.; Linckh, E. *Z. Anorg. Chem.* 1924, *140*, 37-46.
VARIABLES: Concentration	PREPARED BY: W. Gerrard

EXPERIMENTAL VALUES:

T/K = 273.15

Conc of ferrous salt /mol dm^{-3}	Pressure of NO /kPa	Volume of NO (adjusted to 101.325 kPa) absorbed per mole of salt / dm^3
0.459	94.44	16.6
0.137	94.44	16.7
0.0755	94.44	17.3
0.0582	95.56	18.0
0.0577	94.31	17.7
0.0403	94.02	17.5
0.029	95.19	17.9
0.027	94.16	19.1
0.027	94.16	19.0
0.021	95.19	18.7
0.0176	94.95	19.9
0.0088	95.56	19.3
0.02	108.42	18.3
0.02	119.04	18.3
0.0202	137.05	20.7
0.0208	146.92	21.5
0.0186	155.58	20.3
0.0165	134.92	20.3
0.01996	180.11	23.9
0.0224	180.92	22.6

AUXILIARY INFORMATION

METHOD/APPARATUS/PROCEDURE:

The volume of nitric oxide absorbed was measured by means of the apparatus described by Manchot and Huttner (1).

SOURCE AND PURITY OF MATERIALS:

1. Nitric oxide stated to be nearly 100% pure.
3. Ferrous selenate was self prepared and analytically attested.

ESTIMATED ERROR:

REFERENCES:

1. Manchot, W.; Huttner, F. *Annalen*, 1910, *372*, 153.

COMPONENTS:
1. Nitric oxide; NO; [10102-43-9]
2. Water; H_2O; [7732-18-5]
3. Iron(II) ammonium sulfate; $Fe(NH_4)_2(SO_4)_2$; [10045-89-3]

ORIGINAL MEASUREMENTS:

Manchot, W.; Haunschild, H.

Z. anorg. Chem. 1924, *140*, 22-36.

VARIABLES:

Temperature, pressure

PREPARED BY:

W. Gerrard.

EXPERIMENTAL VALUES:

The concentration of the ferrous salt was 0.03 moles dm^{-3}. P is the pressure of NO. V_1 is the volume of NO (273.15 K, 101.325 kPa) absorbed by the volume of solution, dm^3, containing 1 mole of ferrous salt, i.e., per mole of ferrous salt. K is the equilibrium constant.
$K = P\,[22.4 - V_1)/V_1]$

T/K	P/kPa	V_1/dm^3		K
273.15	43.662	13.8		27.21
273.15	44.009	13.7		27.95
273.15	61.407	16.4		22.47
273.15	61.247	16.0		24.50
273.15	60.047	16.5		21.47
273.15	83.125	17.3		24.50
273.15	82.618	17.0		26.24
			Mean	24.91
286.15	23.558	6.0		64.39
286.15	23.531	5.7		68.94
286.15	23.518	5.6		70.55
286.15	45.809	9.6		61.08
286.15	44.476	9.4		61.51
286.15	80.125	12.6		62.32
286.15	81.045	12.7		61.90
286.15	79.619	12.5		63.06
			Mean	64.22

AUXILIARY INFORMATION

METHOD/APPARATUS/PROCEDURE:

The volume of nitric oxide absorbed was observed as previously stated (1).

SOURCE AND PURITY OF MATERIALS:

Presumably as previously described (1).

ESTIMATED ERROR:

REFERENCES:

1. Manchot, W.; Zechentmayer, K. *Annalen*, 1906, 350, 368-389.

See also Manchot, W.; Huttner, F. *Annalen*, 1910, 372, 153-178.

COMPONENTS:	ORIGINAL MEASUREMENTS
1. Nitric oxide; NO; [10102-43-9] 2. Water; H_2O; [7732-18-5] 3. Iron(II) ammonium sulfate; $Fe(NH_4)_2(SO_4)_2$; [10045-89-3]	Manchot, W.; Haunschild, H. *Z. anorg. Chem.* <u>1924</u>, *140*, 22-36.

EXPERIMENTAL VALUES:

T/K	P/kPa	V_1/dm^3	K
298.15	46.675	5.3	150.59
298.15	46.449	5.3	149.86
298.15	59.834	6.2	156.34
298.15	60.074	6.2	156.97
298.15	92.004	8.5	150.45
298.15	91.564	8.3	155.55
		Mean	<u>153.29</u>

Heat of reaction was calculated by the van't Hoff formula :

$$\Delta H = \frac{4.571\, T_1 T_2}{T_1 - T_2} (\log K_1 - \log K_2)$$

The following table shows the values given by the original authors, and those calculated by the compiler*.

T/K (1)	T/K (2)	ΔH/kcal mole^{-1}	ΔH/kcal mole^{-1}*	ΔH/kJ mole^{-1}*
286.15	273.15	11.25	11.29	47.24
298.15	273.15	11.74	11.75	49.16
298.15	286.15	12.28	12.29	51.42
		Mean <u>11.76</u>	11.78	49.27

Estimated by a graph, ΔH was given as 11.9 kcal.

Where V_1 is the volume, dm^3, of nitric oxide absorbed for 1 mole of ferrous ammonium sulfate in a 0.03 molar solution, the authors calculated, by means of the equilibrium constant, the pressure of nitric oxide, P, required to give stated value of V_1. The compiler has also given the corresponding kPa values.

V_1, (273.15 K, 101.325 kPa).	Required P_{NO}					
	T/K = 273.15		T/K = 286.15		T/K = 298.15	
	/atm	/kPa	/atm	/kPa	/atm	/kPa
11.2	0.246	24.926	0.633	64.168	1.513	153.321
17.0	0.774	78.407	1.993	201.983	4.76	482.307
17.3	0.834	84.486	2.149	217.715	5.13	519.797
17.8	0.951	96.360	2.450	248.246	5.85	592.751
20.0	2.049	207.582	5.28	534.996	12.61	1277.708
22.2	27.3	2766.	70.3	7123.	168	17022.6

COMPONENTS:

1. Nitric oxide; NO; [10102-43-9]
2. Water; H_2O; [7732-18-5]
3. Iron (II) sulfate; $FeSO_4$; [7720-78-7]

ORIGINAL MEASUREMENTS:

Ganz, S. N.; Mamon, L. I.
Zh. Prikl. Khim. 1953, *26*, 1005-13
and
J. Applied Chem. (USSR) 1953, *26*, 927-935.

VARIABLES:

PREPARED BY:

W. Gerrard

EXPERIMENTAL VALUES:

TABLE 1

Values of the equilibrium pressure, *p*, for solutions of 20% $FeSO_4$ in which a volume, V, of nitric oxide has dissolved.

	V/cm^3	1000	2000	3000	4000	5000	6000	13000
T/K	θ/°C				p_{NO}/mmHg			
293	20	1.26	2.63	5.25	8.8	14.8	21.3	52.5
303	30	3.55	6.9	12.0	20.4	31.6	49	107
313	40	8.5	15.2	24.5	40.6	60.1	87	191
323	50	29	34.7	52.5	79.4	114.0	162	380
333	60	50	74	105	162	224	288	630
343	70	100	135	214	290	380	500	-
353	80	224	295	450	550	603	740	-
363	90	390	500	692	-	-	-	-

V is the volume absorbed adjusted to a pressure of 760 mmHg (101.325 kPa) and a temperature of 273.15 K.

(cont.)

AUXILIARY INFORMATION

METHOD/APPARATUS/PROCEDURE:

The nitric oxide was absorbed in the aqueous solution contained in a fitted scrubber filled with rings and placed in a thermostat. It was simply stated that nitric oxide pressures over the solutions were "investigated" by the static method, and by the dynamic method. The ferrous sulfate content was stated to be "20%". It appears that this is weight per cent of anhydrous salt, *i.e.*, a concentration of 1.317 mole $FeSO_4$ per 1000 g of solution.

SOURCE AND PURITY OF MATERIALS:

No details given.

ESTIMATED ERROR:

No details given.

REFERENCES:

COMPONENTS:	ORIGINAL MEASUREMENTS:
1. Nitric oxide; NO; [10102-43-9] 2. Water; H_2O; [7732-18-5] 3. Iron (II) sulfate; $FeSO_4$; [7720-78-7]	Ganz, S. N.; Mamon, L. I. *Zh. Prikl. Khim.* 1953, *26*, 1005-13 and *J. Applied Chem.* (USSR) 1953, *26*, 927-935.

TABLE 2

Values of the solubility coefficient*, H for solution containing 20% $FeSO_4$.

T/K	293	303	313	323	333	343	353	363
θ/°C	20	30	40	50	60	70	80	90
V/cm^3	Solubility coefficient, H							
1000	9.4	25.0	63.5	149.0	373.0	745.0	1670.0	2910.0
2000	9.8	25.8	56.6	129.0	276.0	505.0	1100.0	2000.0
3000	13.0	30.0	61.0	130.0	288.0	535.0	1120.0	1720.0
4000	16.4	38.2	76.0	148.0	303.0	550.0	1130.0	-
5000	22.0	47.2	90.0	170.0	334.0	568.0	1140.0	-
6000	26.6	61.0	101.0	202.0	359.0	622.0	-	-
13000	30.1	62.0	110.0	218.0	362.0	-	-	-

* defined as pressure of solution containing V cm^3 of NO (measured at 101.325 kPa and 273.15 K) divided by the concentration of nitric oxide in solution as a weight-%.

TABLE 3

Values of the solubility coefficient*, H for solution containing 20% $FeSO_4$ and 2.68 g of nitric oxide + stated % acid. (% acid are probably percentage by weight)

T/K	293	303	313	323	333	343	353	363
θ/°C	20	30	40	50	60	70	80	90
	Solubility coefficient, H							
Without acid	9.8	25.8	56.7	129.0	276.0	505.0	1100	1865
5% H_2SO_4	11.3	29	63.5	139.0	310.0	594.0	1290	2100
5% HNO_3	12.1	31.7	71.0	159.0	373.0	710.0	1590	2580
5% H_2SO_4 + 5% HNO_3	13.1	37.3	93.5	214.0	515.0	1030	2460	-
5% HNO_3 + 8% H_2SO_4	14.5	48.2	121.0	317.0	817.0	1740	-	-
10% H_2SO_4 + 7% HNO_3	16.3	59.0	170.0	526.0	1550.0	-	-	-
15% H_2SO_4 + 10% HNO_3	19.1	85.5	304.0	1072.0	-	-	-	-

(cont.)

COMPONENTS:	ORIGINAL MEASUREMENTS:
1. Nitric oxide; NO; [10102-43-9] 2. Water; H_2O; [7732-18-5] 3. Iron (II) sulfate; $FeSO_4$; [7720-78-7]	Ganz, S. N.; Mamon, L. I. *Zh. Prikl. Khim.* 1953, *26*, 1005-13 and *J. Applied Chem.* (USSR) 1953, *26*, 927-935.

TABLE 4

Effects of additives upon the absorption of nitric oxide in 20% $FeSO_4$ solution at 293 K and 101.3 kPa pressure.

	Solution	Gas: NO %	NO absorbed *cm³	*mol	mol^{-1} $FeSO_4$
$FeSO_4$	pure	76	378	0.017	0.43
	+5% H_2SO_4	79	191	0.0085	0.215
	+ 1% hydroquinone	70	258	0.012	0.301
	+ 1% phenol	72	131	0.0059	0.148
	+ 1% sodium sulfide	72	208	0.0093	0.234
	+ 1% HNO_3	72	202	0.009	0.225
	+ 3% HNO_3	72	184	0.00825	0.208
	+ 5% HNO_3	75	195	0.00875	0.22
	+ 1% HNO_3 + 1% hydroquinone	75	276	0.0123	0.31
	+ 1% hydroquinone + 3% HNO_3	75	276	0.0123	0.31
	+ 1% hydroquinone + 5% HNO_3	75	276	0.0123	0.31
	+ 1% phenol + 3% HNO_3	75	208	0.0093	0.294
	+ 1% phenol + 5% HNO_3	75	212	0.0095	0.24
	+ 1% hydroquinone + 2% H_2SO_4 + 3% HNO_3	75	268	0.012	0.31

* per 30 cm³ of solution.

Note the original paper presents data on solubilities after several regenerations of the solution by boiling.

COMPONENTS:

1. Nitric oxide; NO; [10102-43-9]
2. Water; H_2O; [7732-18-5]
3. Iron (II) sulfate; $FeSO_4$; [7720-78-7]

ORIGINAL MEASUREMENTS:

Pozin, M.E.; Zubov, V.V.; Tereshchenko, L. Ya.; Tarat, E. Ya.; Ponomarev, Yu.L.

Izv. Vysshikh Uchebn. Zavedenii, Khim. i. Khim. Tecknol. 1963, *6*, (4), 608-616.

VARIABLES:

Temperature, concentration of salt

PREPARED BY:

W. Gerrard.

EXPERIMENTAL VALUES:

1 atm = 101.325 kPa.

Concn. of salt / mol dm^{-3}	T/K	Vol. NO. abs. per mole salt /dm^3*	p_{NO} /atm
1.54	283.15	13.5	0.93
	288.15	11.14	0.93
	288.15	11.08	0.93
	293.15	9.35	0.92
	293.15	9.65	0.93
	303.15	5.87	0.77
	303.15	6.62	0.90
	318.15	2.90	0.79
	318.15	3.50	0.86
	328.15	2.12	0.81
	333.15	1.60	0.77
	338.15	1.22	0.68
	338.15	1.175	0.73
	348.15	0.381	0.60
	348.15	0.76	0.57
0.705	278.15	15.05	0.93
	283.15	13.4	0.93
	288.15	11.6	0.93
	293.15	9.81	0.92
	293.15	10.00	0.93
	308.15	5.5	0.945
	323.15	2.8	0.89
	338.15	1.2	0.77

(cont.

AUXILIARY INFORMATION

METHOD/APPARATUS/PROCEDURE:

Volume of gas absorbed was measured by the change in volume in a measuring buret, water saturated with gas being used as a liquid seal. Purified nitrogen was used as a gas diluent to obtain results for various partial pressures of nitric oxide. Vapor pressure of water was taken into account. Absorption was first effected at 363 K, and the temperature was decreased, and then increased again to give 2 values for each temperature. These were stated to agree well. Desorption measurements were also mentioned.

SOURCE AND PURITY OF MATERIALS:

1. Dry gas stated to be 95-98%. Source not stated.
2. Water may be assumed to be of satisfactory purity.
3. Ferrous sulfate may be assumed to be of satisfactory purity.

ESTIMATED ERROR:

REFERENCES:

COMPONENTS:

1. Nitric oxide; NO; [10102-43-9]
2. Water; H_2O; [7732-18-5]
3. Iron (II) sulfate; $FeSO_4$; [7720-78-7]

ORIGINAL MEASUREMENTS:

Pozin, M.E.; Zubov, V.V.; Tereshchenko, L. Ya.; Tarat, E.Ya. Ponomarev, Yu.I.

Izv. Vysshikh Uchebn. Zavedenii, Khim. i. Khim. Tecknol. 1963, *6*, (4), 608-616.

EXPERIMENTAL VALUES: 1 atm = 101.325 kPa.

Concn. of salt / mol dm^{-3}	T/K	Vol. NO. abs. per mole salt /dm^3*	p_{NO} /atm
0.705	353.15	0.3	0.55
0.262	278.75	17.3	0.94
	283.15	15.2	0.935
	283.15	15.3	0.93
	288.15	13.3	0.93
	293.15	10.85	0.924
	293.15	8.9	0.91
	303.15	7.7	0.90
	313.15	4.9	0.87
0.065	283.15	16.2	0.935
	283.15	16.2	0.92
	288.15	13.7	0.93
	293.15	11.3	0.92
	293.15	11.3	0.91
	298.15	8.8	0.92
	303.15	7.3	0.90
	313.15	4.7	0.88
	313.15	4.6	0.88
	328.15	1.8	0.80
	343.15	1.3	0.66
	353.15	0.7	0.52
	358.15	0.5	0.42

* Volume adjusted to 1 atm and 273 K.

COMPONENTS:	ORIGINAL MEASUREMENTS:
1. Nitric oxide; NO; [10102-43-9] 2. Water; H_2O; [7732-18-5] 3. Iron chloride; (ferrous chloride) $FeCl_2$; [7758-94-3]	Pozin, M.E.; Zubov, V.V.; Tereshchenko, L. Ya.; Tarat, E. Ya.; Ponomarev, Yu. L. *Izv. Vysshikh Ushebn. Zavedenii, Khim. i. Khim. Tecknol.* 1963, *6*, (4). 608-616.
VARIABLES:	PREPARED BY: W. Gerrard.

EXPERIMENTAL VALUES:

1 atm = 101.325 kPa.

Conc. of salt /mol dm^{-3}	T/K	Vol. No. abs. per mole of salt /dm^3*	p_{NO} /atm
1.771	283.15	13.8	0.82
	293.15	10.3	0.91
	303.15	7.15	0.93
	313.15	4.65	0.91
	323.15	2.95	0.87
	333.15	1.82	0.81
	348.15	0.734	0.63
0.885	283.15	13.98	0.81
	293.15	10.3	0.92
	303.15	6.95	0.94
	313.15	4.44	0.92
	323.15	2.69	0.88
	333.15	1.58	0.81
	348.15	0.657	0.63
0.435	283.15	14.4	0.88
	293.15	10.45	0.93
	303.15	6.8	0.92
	313.15	4.33	0.90
	323.15	2.52	0.86
	333.15	1.38	0.78
	348.15	0.468	0.62

* Volume adjusted to 1 atm, 273.15 K

AUXILIARY INFORMATION

METHOD/APPARATUS/PROCEDURE:

Volume of gas absorbed was measured by the change in volume in a measuring buret, water saturated with the gas being used as a liquid seal. Purified nitrogen was used to obtain lower partial pressure of nitric oxide. Vapor pressure of water was taken into account. Absorption was first effected at 363 K, and then the temperature was lowered. Desorption measurements were also mentioned.

SOURCE AND PURITY OF MATERIALS:

1. Dry gas stated to be 95-98%. Source not stated.
2. Water may be taken as pure
3. Ferrous chloride may be taken of satisfactory purity.

ESTIMATED ERROR:

REFERENCES:

COMPONENTS:

1. Nitric Oxide; NO; [10102-43-9]
2. Water; H_2O; [7732-18-5]
3. Iron (II) sulfate; $FeSO_4$; [7720-78-7]

ORIGINAL MEASUREMENTS:

Polovchenko, V.I.; Skvortsov, G.A.

Z. Prikl. Khim. 1974, *47*, 1917-1922

VARIABLES:

Pressure

PREPARED BY:

C.L. Young

EXPERIMENTAL VALUES:

T/°C	T/K	Partial pressure of NO p/atm	p/MPa	Nitric oxide concentration, C dm^3 (at 273.2 K and 101.3kPa) mol^{-1}
22.4	295.6	1.0	0.10	9.3
21.8	295.0	1.5	0.15	11.74
21.4	294.6	2.1	0.21	13.00
21.5	294.7	3.1	0.31	15.75
21.5	294.7	4.6	0.47	17.35
21.1	294.3	4.7	0.48	17.35
22.5	295.7	5.4	0.55	17.75
21.7	294.9	7.0	0.71	18.78
20.4	293.6	7.7	0.78	19.80
20.3	293.5	13.1	1.33	20.95
21.4	294.6	19.7	2.00	22.10

AUXILIARY INFORMATION

METHOD/APPARATUS/PROCEDURE:

High pressure static bomb. Known amount of gas dissolved in solution of known volume in a vessel of known volume. Pressure measured with Bourdon gauge. Solubilities determined for 20,10, 5 and 2.5 wt% iron sulfate solutions but results tabulated for 10 wt% only. Graphical results given for other concentrations.

SOURCE AND PURITY OF MATERIALS:

No details given.

ESTIMATED ERROR:

$\delta T/K = \pm 0.1$; $\delta C = \pm 0.1$

REFERENCES:

COMPONENTS:

1. Nitric oxide; NO; [10102-43-9]

2. Ethanol, C_2H_5OH; [64-17-5]

3. Iron bromide, (ferrous bromide); $FeBr_2$; [7789-46-0]

ORIGINAL MEASUREMENTS:

Thomas, V. *Bull Soc. Chem.* 1898, *19*, (3), 343.

Thomas, V. *Ann. Chim. Phys.* 1898, *13*, (7), 145.

VARIABLES:

Temperature, concentration

PREPARED BY:

W. Gerrard

EXPERIMENTAL VALUES:

Weight of ferrous Fe, g, in 100 cm^3 of solution	Volume of salt solution /cm^3	T/K	Volume of NO absorbed /cm^3	Mole of NO/56 of ferrous iron*
0.789	11.2	290.16	21	0.595
0.789	16.5	291.16	28	0.538
0.789	11.3	286.16	24.5	0.687
3.110	8	289.16	52	0.523
3.110	4.8	285.16	47	0.787
3.110	8	288.16	54	0.543
3.110	5	287.16	44	0.708

* Calculated by compiler.
p_{NO} assumed to be 101.325 kPa.

NOTE: The volume of pure ethanol equal to the volume of solution containing 56 g of ferrous iron would absorb approximately 0.09 mole of NO for the first group of data, and 0.08 mole NO for the second group.

In the calculation of the mole ratio, NO/Fe, the compiler has assumed that the volume of absorbed NO has been adjusted to 273.16 K, as appears to be done in the paper (1).

NOTE: In the second paper cited above, Thomas gave the weight of ferrous iron (Fe) as 8.89 g per 100 cm^3 instead of the above quoted 0.789 g per 100 cm^3, given in the first cited paper. The temperature was given as 17.0 °C in the second paper.

AUXILIARY INFORMATION

METHOD /APPARATUS/PROCEDURE:

The liquid was contained in a simple glass vessel furnished with an inlet tube which passed right into the liquid, and with an exit tube. Nitric oxide was slowly passed into the liquid, and the entrained water vapor was collected and weighed. It was simply stated that the quantity of absorbed gas was determined. The pressure was not mentioned.

SOURCE AND PURITY OF MATERIALS:

NO: Not mentioned; but presumably prepared from mercury and nitric acid as in ref. (1).

ESTIMATED ERROR:

REFERENCES:

(1) Thomas, V. *Bull. Soc. Chim.*, 1898, *19*, (3), 419.

COMPONENTS:	ORIGINAL MEASUREMENTS:
1. Nitric oxide; NO; [10102-43-9] 2. Ethanol; C_2H_5OH; [64-17-5] 3. Iron chloride (ferrous chloride); $FeCl_2$; [7758-94-3]	Thomas, V. *Bull.Soc. Chim.* 1898, *19*, (3), 419; *Ann. Chim. Phys.* 1898, *13*, (7), 145.
VARIABLES: Temperature.	PREPARED BY: W. Gerrard

EXPERIMENTAL VALUES:

The solution contained 4.6 g of ferrous iron in 1 dm^3. At 290.16 K 12.2 cm^3 of this solution absorbed 21.5 cm^3 of NO (adjusted to 273.16 K).

At 288.16 K 12.5 cm^3 of this solution absorbed 23.4 cm^3 of NO (adjusted to 273.16 K)

Although the pressure was stated to be 101.325 kPa, "augmentee de la pression de tension maxima du solvant", the actual pressure was not stated.

Assuming p_{NO} to be 101.325 kPa, the solution containing 56 g of ferrous iron absorbed 0.959 moles of NO at 290.16 K; and 1.02 moles of NO at 288.16 K. The alcohol contained in the solution would when pure absorb about 0.148 mole of NO.

AUXILIARY INFORMATION

METHOD /APPARATUS/PROCEDURE:

A measured volume of the alcoholic solution of ferrous salt was introduced into a graduated test-tube resting over mercury in a larger vessel. By means of a second graduated test-tube a known volume of nitric oxide was passed into the first tube.

There was no diagram, and the actual operation is difficult to visualize from this description.

SOURCE AND PURITY OF MATERIALS:

Nitric oxide was self prepared by the action of mercury on nitric acid.

ESTIMATED ERROR:

REFERENCES:

COMPONENTS:	ORIGINAL MEASUREMENTS:
1. Nitric oxide; NO; [10102-43-9] 2. Ethanol; C_2H_6O; [64-17-5] 3. Iron(II) iodide; FeI_2; [7783-86-0]	Thomas, V.; *Bull. Soc. Chim.* 1898, *19* (3), 343. Thomas, V.; *Ann. Chim. Phys.* 1898, *13* (7), 145.
VARIABLES:	PREPARED BY:
Temperature	W. Gerrard

EXPERIMENTAL VALUES:

Weight of ferrous iron, g, in 100 cm^3 of solution	Volume of salt soln. /cm^3	T/K	Volume of NO absorbed /cm^3	Mole NO/ for 56 g Fe*
1.925	8	288.16	23	0.374
1.925	9.1	294.16	26	0.371
1.925	9	285.16	27	0.390

* Calculated by the compiler. Based on the assumption that the observed volume of NO had been adjusted to 273.16 K, and that the pressure of NO, p_{NO}, was 101.325 kPa.

AUXILIARY INFORMATION

METHOD/APPARATUS/PROCEDURE:

The liquid was contained in simple glass vessel furnished with an inlet tube passing right into the liquid, and an exit tube. Nitric oxide was slowly passed into the liquid. It was simply stated that the quantity of absorbed gas was determined. The pressure was not stated.

SOURCE AND PURITY OF MATERIALS:

1. NO: not mentioned; but presumably it was prepared from mercury and nitric acid (1).

ESTIMATED ERROR:

REFERENCES:

1. Thomas, V.; *Bull. Soc. Chim.* 1898, *19* (3), 419.

COMPONENTS:	ORIGINAL MEASUREMENTS:
1. Nitric oxide; NO; [10102-43-9] 2. Ethanol; C_2H_5OH; [64-17-5] 3. Iron chloride (Ferrous chloride); $FeCl_2$; [7758-94-3]	Manchot, W.; Zechentmayer, K. *Annalen*, 1906, *350*, 368-389.
VARIABLES:	PREPARED BY:
Temperature, pressure	W. Gerrard

EXPERIMENTAL VALUES:

T/K	Weight of $FeCl_2$ added to 50 cm^3 ethanol + 2 cm^3 of hydrochloric acid, 14% HCl /g	Pressure of nitric oxide /kPa	Volume, V_1, of nitric oxide absorbed per mole of $FeCl_2$. /dm^3 (273.15 K, 101.325 kPa)
301.15	0.2464	89.164	21.2
284.75	0.2289	94.617	21.4
281.55	0.2766*	96.390	21.9
270.65	0.2210	98.257	21.9
270.95	0.2461	267.573	21.8
275.36	0.5716** +	183.448	21.7
270.95	0.2320 +	183.175	22.5
270.95	0.1369***	97.857	22.7

* 39.8% Fe.
** 43.88% Fe.
*** 20 cm^3 of ethanol, and no hydrochloric acid.
\+ Only 3 drops of hydrochloric acid (14% HCl).

AUXILIARY INFORMATION

METHOD: /APPARATUS/PROCEDURE:

Volumetric apparatus. A three-way tube was connected to a gas buret and an absorption pipet; the third member was used for exhausting the apparatus, and for passing in the gas. By means of a levelling tube containing mercury, the pressure and volume of gas in the buret could be regulated.

SOURCE AND PURITY OF MATERIALS:

1. Prepared from sodium nitrite and dilute sulfuric acid, and washed with alkali. Stated to be 100% pure.
2. Stated to be absolute.
3. Prepared from hydrogen chloride and hot sheet iron. Attested analytically.

ESTIMATED ERROR:

REFERENCES:

COMPONENTS:	ORIGINAL MEASUREMENTS:
1. Nitric oxide; NO; [10102-43-9] 2. Organic liquid 3. Iron(II) chloride; $FeCl_2$; [7758-94-3]	Manchot, W.; Huttner, F. *Annalen* 1910, *372*, 153-178.
VARIABLES:	PREPARED BY: W. Gerrard

EXPERIMENTAL VALUES: T/K = 273.15

Organic liquid	Conc. of Iron(II) /mol dm^{-3}	Pressure of NO /kPa	Volume, V, of NO per mole of Iron(II) salt /dm^3
Ethanol; C_2H_5OH; [64-17-5]	0.0102	100.257	22.2
Ethyl benzoate; $C_6H_5CO_2C_2H_5$; [93-89-0]	0.0292	100.257	22.5
	0.0667	100.790	22.6
Ethyl acetate; $CH_3CO_2C_2H_5$; [141-78-6]	0.0154	99.190	22.2
Diethyl propanedioate (Ethyl malonate); $CH_2(CO_2C_2H_5)_2$; [105-53-3]	0.0571	98.257	22.8
2-Propanone; CH_3COCH_3; [67-64-1]	0.0193	100.257	22.2
	0.0302	98.257	21.9
Pyridine; C_5H_5N; * [110-86-1]	0.0224	100.390	22.6
	0.0229	100.390	22.4

* not free from water.

V is the volume of NO after adjustment to 101.325 kPa.

AUXILIARY INFORMATION

METHOD/APPARATUS/PROCEDURE:

The volume of gas absorbed was measured by the Ostwald-type gas buret and absorption pipet. The apparatus and method were stated to be similar to those previously described (1), except that the technique for introducing the ferrous salt was modified.

SOURCE AND PURITY OF MATERIALS:

1. Nitric oside probably taken as 100% pure, see ref. (1).

2, 3. Appeared to be of satisfactory purity.

ESTIMATED ERROR:

REFERENCES:

1. Manchot, W.; Zechentmayer, K. *Annalen* 1906, *350*, 368-389.

COMPONENTS:

1. Nitric oxide; NO; [10102-43-9]
2. Water; H_2O; [7732-18-5]
3. Sulfuric acid; H_2SO_4; [7664-93-9]
4. Iron(III) sulfate; $Fe_2(SO_4)_3$; [10028-22-5]

ORIGINAL MEASUREMENTS:

Manchot, W.

Annalen

1910, *372*, 179-186.

VARIABLES:

Concentration of components 3 and 4

PREPARED BY:

W. Gerrard

EXPERIMENTAL VALUES:

T/K	Gram-atom of Fe in 1 dm^3 of solution	Conc. of H_2SO_4 /mol dm^{-3} (soln.)	Pressure of NO /kPa	Volume, V, of NO adjusted to 273.15 K and 101.325 kPa, absorbed per 1 mole Fe /dm^3
273.15	0.0096	18.32 (97.6%)	102.123	44.7
	0.0104	18.32 (97.6%)	99.857	44.4
	0.0106	0	99.857	0
	0.0108	16.92 (90.6%)	99.323	44.8
	0.0109	6.67 (47.6%)	99.323	0
	0.0109	11.73 (70.8%)	99.323	0
	0.0113	14.69 (82.0%)	99.323	0
	0.0500	16.92 (90.6%)	101.723	44.4
	0.1000	16.92 (90.6%)	101.723	42.2
	0.1960	16.92 (90.6%)	101.723	31.0
	1.4000	0	98.790	0
	0.0102	16.92 (90.6%)	166.783	44.9
	0.0117	16.92 (90.6%)	42.529	44.7
284.15	0.0100	16.92 (90.6%)	101.723	44.8

AUXILIARY INFORMATION

METHOD/APPARATUS/PROCEDURE:

Apparatus and method were similar to those described (1), see also (2). The volume of absorbed gas was measured by means of a gas buret and absorption pipet.

SOURCE AND PURITY OF MATERIALS:

1. Nitric oxide probably taken as being 100% pure as in refs. (1) and (2).

2, 3, 4. Appeared to be of satisfactory purity.

ESTIMATED ERROR:

REFERENCES:

1. Manchot, W.; Zechentmayer, K. *Annalen*, 1906, *350*, 368-389.
2. Manchot, W.; Huttner, F. *Annalen*, 1910, *372*, 153-178.

COMPONENTS:	ORIGINAL MEASUREMENTS:
1. Nitric oxide; NO; [10102-43-9] 2. Ethanol; C_2H_5OH; [64-17-5] 3. Iron chloride (Ferric chloride); $FeCl_3$; [7705-08-0]	Griffith, W. P.; Lewis, J.; Wilkinson, G. *J. Chem. Soc.* 1958, 3993-3998.
VARIABLES:	PREPARED BY: W. Gerrard

EXPERIMENTAL VALUES:

M = Molarity of $FeCl_3$/mol dm^{-3}. P = Pressure of NO/kPa. V_1 = Volume of NO/dm^3 (adjusted to 273.15 K, 101.325 kPa); V_1 is based on the authors' loose statement: "Vol of NO absorbed per mole at N.T.P.", the mole apparently referring to $FeCl_3$. K = Equilibrium constant, given by the authors as the mean value of P_{NO} $FeCl_3/FeCl_3NO$, in mmHg. K^* = The compiler's values based on $K = P_{NO}\ (22.4 - V_1)/V_1$, in mmHg. $\Delta H = 4.571(\log K_1 - \log K_2) \times T_1T_2/(T_1 - T_2)$, T = T/K.

NOTE: The authors simply stated: "heat of reaction 20.4 kcal/mole", but the basis of the calculation was not revealed.

$V_1(E)$ = Volume of NO (adjusted to 273.15 K, 101.325 kPa) which would be absorbed by that volume of ethanol equal to the volume of solution containing 1 mole of $FeCl_3$.

(cont.)

AUXILIARY INFORMATION

METHOD/APPARATUS/PROCEDURE:	SOURCE AND PURITY OF MATERIALS:
It was simply stated that the gas absorption was measured in a conventional Warburg-type apparatus.	1. Nitric oxide was prepared in a sodium nitrite-ferrous sulfate generator; nitrogen dioxide was removed by a concentrated aqueous solution of potassium hydroxide. 2, 3. Probably of satisfactory purity.
	ESTIMATED ERROR:
	REFERENCES:

1. Nitric oxide; NO; [10102-43-9] 2. Ethanol; C_2H_5OH; [64-17-5] 3. Iron chloride (Ferric chloride); $FeCl_3$; [7705-08-0]	Griffith, W. P.; Lewis, J.; Wilkinson, G. *J. Chem. Soc.* 1958, 3993-3998.

EXPERIMENTAL VALUES:

T/K	M	*P*/kPa	V_1/dm^3	*K*/(mmHg)	*K**/(mmHg)	V_1(E)	No.
282.4	0.034	98.657	21.3	-	38.2	8.4	1
	0.017		21.0	43.9	49.3	16.8	2
294.4	0.034	96.537	18.0	-	177.0	7.7	3
	0.017		17.4	196.0	208.1	15.4	4
298	0.034	93.99	15.2	-	333.9	7.6	5
	0.017		16.1	305.0	275.9	15.2	6

ΔH values, calculated by compiler: T/K range ( ): (1-3) 21.09; (1-4) 23.32; (1-5) 23.22; (1-6) 21.17; (2-3) 17.58; (2-4) 19.81; (2-5) 20.49; (2-6) 18.44; (3-5) 30.70; (3-6) 21.47; (4-5) 22.87; (4-6) 13.63 kcal mol^{-1}(1 kcal - 4.1840 kJ).

COMPONENTS:	ORIGINAL MEASUREMENTS:
1. Nitric oxide; NO; [10102-43-9] 2. Water; H_2O; [7732-18-5] 3. Hydrochloric acid; HCl; [7647-01-0] 4. Copper (II) chloride, (Cupric chloride); $CuCl_2$; [7447-39-4]	Kohlschutter, V.; Kutscheroff, M. *Ber*, 1904, *37*, 3044-3052
VARIABLES:	PREPARED BY:
Concentration	W. Gerrard.

EXPERIMENTAL VALUES:

Neither temperature nor pressure of nitric oxide were stated. T/K thought to be 293.16 K, and p_{NO} to be 101.325 kPa.

Volume, V_s, of hydrochloric acid (concentrated) containing one mole of cupric chloride /dm^3	Volume, V_1, absorbed by V_s /dm^3	Ostwald coefficient, *L*, calculated by compiler.
0.389	0.801	2.059
0.410	0.933	2.276
0.840	2.838	3.379
1.230	3.426	2.785
2.462	3.989	1.620
7.499	3.931	0.524
12.500	3.606	0.288
18.750	3.153	0.168
28.650	1.976	0.069
	Pure water	0.0505 (1)

AUXILIARY INFORMATION

METHOD/APPARATUS/PROCEDURE:

A gas buret and pipet of the Ostwald type were connected to a gasometer by a lead capillary tube. Two electrodes were fitted to the absorption tube, and the specific conductance was determined before and after the absorption. It was stated that the point of saturation could be sharply determined.

SOURCE AND PURITY OF MATERIALS:

No details given.

ESTIMATED ERROR:

REFERENCES:

1. Winkler, L.W. *Ber.* 1901, *34*, 1414

COMPONENTS:	ORIGINAL MEASUREMENTS:
1. Nitric oxide; NO; [10102-43-9] 2. Water; H_2O; [7732-18-5] 3. Copper chloride, (Cupric chloride); $CuCl_2$; [7447-39-4]	Kohlschutter, V.; Kutscheroff, M. *Ber.* 1904, *37*, 3044-3052.
VARIABLES: Concentration	PREPARED BY: W. Gerrard.

EXPERIMENTAL VALUES:

Neither temperature nor pressure of nitric oxide were stated. T/K thought to be 293.16 K, and p_{NO} to be 101.325 kPa.

Volume of solution, V_s containing 1 mole of cupric chloride $/dm^3$	Volume of nitric oxide, absorbed by the volume V_s, of solution $/dm^3$		Ostwald coefficient, L, calculated by compiler.
0.231	0.120		0.519
0.277	0.098		0.354
0.371	0.052		0.140
		Pure water (1)	0.0505

AUXILIARY INFORMATION

METHOD/APPARATUS/PROCEDURE:	SOURCE AND PURITY OF MATERIALS:
A gas buret and pipet of the Ostwald type were connected to a gasometer by a lead capillary tube. Two electrodes were fitted to the absorption tube, and the specific conductance was determined before and after the absorption. It was stated that the point of saturation could be sharply determined.	No details given. ESTIMATED ERROR: REFERENCES: 1. Winkler, L.W. *Ber.* 1901, *34*, 1414

COMPONENTS:

1. Nitric oxide; NO; [10102-43-9]
2. Water; H_2O; [7732-18-5]
3. Copper bromide (Cupric bromide); $CuBr_2$; [7789-45-9]

ORIGINAL MEASUREMENTS:

Kohlschutter, V.; Kutscheroff, M.

Ber. 1904, *37*, 3044-3052.

VARIABLES:

Concentration

PREPARED BY:

W. Gerrard

EXPERIMENTAL VALUES:

Neither temperature nor pressure of nitric oxide were stated. T/K thought to be 293.16 K, and p_{NO} to be 101.325 kPa.

Volume, V_s, of water containing one mole of cupric bromide /dm^3	Volume, V_1, of nitric oxide absorbed by V_s /dm^3	Ostwald coefficient, L, calculated by compiler
0.37	0.515	1.392
0.62	0.120	0.194
0.925	0.000	-
	Pure water (1)	0.0505

AUXILIARY INFORMATION

METHOD /APPARATUS/PROCEDURE:

A gas buret and pipet of the Ostwald type were connected to a gasometer by a lead capillary tube. Two electrodes were fitted to the absorption tube, and the specific conductance was determined before and after the absorption. It was stated that the point of saturation could be sharply determined.

SOURCE AND PURITY OF MATERIALS:

No details given.

ESTIMATED ERROR:

REFERENCES:

1. Winkler, L.W. *Ber.* 1901, *34*, 1414

COMPONENTS:

1. Nitric oxide; NO; [10102-43-9]
2. Water; H_2O; [7732-18-5]
3. Copper (II) sulfate; $CuSO_4$; [18939-61-2]
4. Sulfuric acid; H_2SO_4; [7664-93-9]

ORIGINAL MEASUREMENTS:

Manchot, W.

Annalen, 1910, *375*, 308-315

VARIABLES:

Temperature, concentration

PREPARED BY:

W. Gerrard

EXPERIMENTAL VALUES:

T/K	Conc. of cupric salt /mol dm^{-3} (soln).	Conc. of H_2SO_4 /mol dm^{-3} (soln)	Pressure of NO/kPa	Volume in dm^3 (adjusted to 101.325 kPa) of NO absorbed per 1 mole salt.
273.15	0.0053	18.32 (97.6%)	100.257	22.3
	0.0119	18.32 (97.6%)	100.390	22.4
	0.0159	16.92 (90.6%)	100.257	16.6
	0.0159	16.92 (90.6%)	100.257	22.4
	0.0236	0 0	100.390	0
	0.0241	16.92 (90.6%)	100.257	16.7
	0.0246	16.92 (90.6%)	100.257	16.9
	0.0251	6.67 (47.6%)	100.257	0
	0.0265	18.32 (97.6%)	100.257	22.4
	0.0265	11.73 (70.8%)	100.257	0
	0.0266	14.69 (82.0%)	100.257	1.4
	0.0586	0 0	100.390	0
	0.1183	18.32 (97.6%)	100.390	10.4
	0.1376	0 0	100.390	0
	7.833	0 0	98.790	0
	0.0179	18.32 (97.6%)	173.716	22.3
	0.0175	18.32 (97.6%)	42.929	22.4
284.15	0.0180	18.32 (97.6%)	101.723	22.7

AUXILIARY INFORMATION

METHOD/APPARATUS/PROCEDURE:

The volume of absorbed nitric oxide was observed presumably as previously stated (1).

SOURCE AND PURITY OF MATERIALS:

Presumably as previously stated (1).

ESTIMATED ERROR:

REFERENCES:

1. Manchot, W.; Zechentmayer, K. *Annalen*, 1906, *350*, 368-389.

COMPONENTS:	ORIGINAL MEASUREMENTS:
1. Nitric oxide; NO; [10102-43-9] 2. Water; H_2O; [7732-18-5] 3. Ethanol; C_2H_6O; [64-17-5] 4. Copper(II) chloride, (Cupric chloride); $CuCl_2$; [7447-39-4]	Manchot, W. *Annalen*, 1910, *375*, 308-315
VARIABLES:	PREPARED BY:
Temperature, concentration	W. Gerrard

EXPERIMENTAL VALUES:

T/K	Conc. of $CuCl_2$ /mol dm^{-3}(soln)	% (vol/vol) of ethanol in solution	Pressure of NO/kPa.	Volume, V_1, of NO (101.325 kPa) absorbed per 1 mole of CuCl /dm^3
273.15	0.005	Absolute	100.257	17.8
	0.0064	Absolute	100.257	18.0
	0.0123	Absolute	100.390	18.1
	0.0234	70	99.990	2.7
	0.0256	50	99.990	0.3
	0.0256	Absolute	100.390	17.5
	0.0271	90	99.990	10.4
	0.0277	0	100.390	0
	0.0353	Absolute	100.123	18.1
	0.0636	Absolute	100.390	17.4
	0.1268	0	100.390	0
	0.1352	Absolute	100.390	17.3
	0.2689	Absolute	98.257	15.9
	0.2806	Absolute	100.390	15.3
	0.6490	Absolute	100.390	14.1
	3.5650	0	98.657	0.3
	4.2800	Absolute	98.657	5.5
	11.7300	0	98.657	0.3
	0.0183	Absolute	173.983	19.3
	0.0176	Absolute	44.262	15.2
286.35	0.0494	Absolute	98.257	16.1

AUXILIARY INFORMATION

METHOD APPARATUS/PROCEDURE:

The volume of nitric oxide absorbed was observed as previously stated (1) using a volumetric apparatus. The volume of gas first absorbed by ethanol-water mixture first measured and then a small tube containing salt was broken and the absorption of solution measured.

SOURCE AND PURITY OF MATERIALS:

1. Presumably as previously stated (1). Prepared from sodium nitrate and dilute sulfuric acid.

ESTIMATED ERROR:

REFERENCES:

1. Manchot, W.; Zechentmayer, K. *Annalen*, 1906, *350*, 368-389.

COMPONENTS:

1. Nitric oxide; NO; [10102-43-9]
2. Methanol; CH_3OH; [67-56-1]
3. Copper chloride (Cupric chloride); $CuCl_2$; [7447-39-4]

ORIGINAL MEASUREMENTS:

Kohlschutter, V.; Kutscheroff, M.

Ber. 1904, *37*, 3044-3052

VARIABLES:

Concentration

PREPARED BY:

W. Gerrard

EXPERIMENTAL VALUES:

Neither temperature nor pressure of nitric oxide were stated. T/K thought to be 293.16 K, and p_{NO} to be 101.325 kPa.

Volume, V_s, containing 1 mole of cupric chloride in solution /dm^3	Volume V_1, of nitric oxide absorbed by V_s /dm^3	Ostwald coefficient, *L*, calculated by compiler.
1.60	3.3	2.062
8.22	5.6	0.681
20.50	6.15	0.300
82.25	4.9	0.0596
	Pure methanol (1)	0.350

82.25 dm^3 of pure methanol would absorb 28.79 dm^3 of nitric oxide.

AUXILIARY INFORMATION

METHOD/APPARATUS/PROCEDURE:

A gas buret and pipet of the Ostwald type were connected to a gasometer by a lead capillary tube. Two electrodes were fitted to the absorption tube, and the specific conductance was determined before and after the absorption. It was stated that the point of saturation could be sharply determined.

SOURCE AND PURITY OF MATERIALS:

No details given.

ESTIMATED ERROR:

REFERENCES:

1. Riccoboni, L. *Gazz. chim. ital.* 1941, *71*, 139

COMPONENTS:	ORIGINAL MEASUREMENTS:
1. Nitric oxide; NO; [10102-43-9] 2. Ethanol; C_2H_5OH; [64-17-5] 3. Copper chloride (Cupric chloride); $CuCl_2$; [7447-39-4]	Kohlschutter, V.; Kutscheroff, M. *Ber.* 1904, *37*, 3044-3052.
VARIABLES: Concentration	PREPARED BY: W. Gerrard.

EXPERIMENTAL VALUES:

Neither temperature nor pressure of nitric oxide were stated. T/K thought to be 293.16 K, and p_{NO} to be 101.325 kPa.

Volume, V_s, of ethanol containing one mole of cupric chloride /dm^3	Volume, V_1, of nitric oxide absorbed by V_s /dm^3	Ostwald coefficient, *L*, calculated by compiler.
1.50	8.7	5.80
3.84	12.38	3.22
12.8	15.43	1.205
38.41	18.15	0.473
76.83	18.05	0.235
192.1	15.92	0.0829
	Ethanol alone	0.285 (1)

AUXILIARY INFORMATION

METHOD/APPARATUS/PROCEDURE:

A gas buret and pipet of the Ostwald type were connected to a gasometer by a lead capillary tube. Two electrodes were fitted to the absorption tube, and the specific conductance was determined before and after the absorption. It was stated that the point of saturation could be sharply determined.

SOURCE AND PURITY OF MATERIALS:

No details given.

ESTIMATED ERROR:

REFERENCES:

1. Carius, L. *Annalen,* 1855, *94*, 129

COMPONENTS:	ORIGINAL MEASUREMENTS:
1. Nitric oxide; NO; [10102-43-9] 2. Ethanol; C_2H_5OH; [64-17-5] 3. Copper bromide, (Cupric bromide); $CuBr_2$; [7789-45-9]	Kohlschutter, V.; Kutscheroff, M. *Ber.* 1904, *37*, 3044-3052.
VARIABLES:	PREPARED BY:
Concentration	W. Gerrard.

EXPERIMENTAL VALUES:

Neither temperature nor pressure of nitric oxide were stated. T/K thought to be 293.16 K, and p_{NO} to be 101.325 kPa.

Volume, V_s, of ethanol containing one mole of cupric bromide /dm³	Volume, V_1, of nitric oxide absorbed by V_s /dm³	Ostwald coefficient, L, calculated by compiler
2.625	16.02	6.103
5.25	19.26	3.669
13.12	20.51	1.563
43.74	21.13	0.483
131.2	22.23	0.169
262.5	23.46	0.089
656.1	30.46	0.046
	Ethanol alone	0.285 (1)

AUXILIARY INFORMATION

METHOD/APPARATUS/PROCEDURE:

A gas buret and pipet of the Ostwald type were connected to a gasometer by a lead capillary tube. Two electrodes were fitted to the absorption tube, and the specific conductance was determined before and after the absorption. It was stated that the point of saturation could be sharply determined.

SOURCE AND PURITY OF MATERIALS:

No details given.

ESTIMATED ERROR:

REFERENCES:

1. Carius, L. *Annalen*, 1855, *94*, 129.

COMPONENTS:	ORIGINAL MEASUREMENTS:
1. Nitric oxide; NO; [10102-43-9] 2. Formic acid; CH_2O_2; [64-18-6] 3. Copper chloride, (Cupric chloride); $CuCl_2$; [7447-39-4]	Kohlschutter, V.; Kutscheroff, M. *Ber.* 1904, *37*, 3044-3052.
VARIABLES: Concentration	PREPARED BY: W. Gerrard

EXPERIMENTAL VALUES:

Neither temperature nor pressure of nitric oxide were stated. T/K thought to be 293.16 K, and p_{NO} to be 101.325 kPa.

Volume, V_s of formic acid (98%) containing one mole of cupric chloride /dm^3	Volume, V_1, of nitric oxide absorbed by V_s /dm^3	Ostwald coefficient, *L*, calculated by compiler.
27.9	12.76	0.457
56.0	13.17	0.235
140	14.34	0.102
280	18.68	0.0667
1400	27.29	0.0195

AUXILIARY INFORMATION

METHOD/APPARATUS/PROCEDURE:

A gas buret and pipet of the Ostwald type were connected to a gasometer by a lead capillary tube. Two electrodes were fitted to the absorption tube, and the specific conductance was determined before and after the absorption. It was stated that the point of saturation could be sharply determined.

SOURCE AND PURITY OF MATERIALS:

No details given.

ESTIMATED ERROR:

REFERENCES:

COMPONENTS:	ORIGINAL MEASUREMENTS:
1. Nitric oxide; NO; [10102-43-9] 2. Acetic acid; $C_2H_4O_2$; [64-19-7] 3. Copper chloride,(Cupric chloride); $CuCl_2$; [7447-39-4]	Kohlschutter, V.; Kutscheroff, M. *Ber.* 1904, *37*, 3044-3052.
VARIABLES: Concentration	**PREPARED BY:** W. Gerrard.

EXPERIMENTAL VALUES:

Neither temperature nor pressure of nitric oxide were stated. T/K thought to be 293.16 K, and p_{NO} to be 101.325 kPa.

Volume, V_s, of acetic acid containing one mole of cupric chloride $/dm^3$	Volume, V_1, of nitric oxide absorbed by V_s $/dm^3$	Ostwald coefficient, *L*, calculated by compiler.
252	51.77	0.205
504	39.67	0.079
1269	81.6	0.064

AUXILIARY INFORMATION

METHOD/APPARATUS/PROCEDURE:

A gas buret and pipet of the Ostwald type were connected to a gasometer by a lead capillary tube. Two electrodes were fitted to the absorption tube, and the specific conductance was determined before and after the absorption. It was stated that the point of saturation could be sharply determined.

SOURCE AND PURITY OF MATERIALS:

No details given.

ESTIMATED ERROR:

REFERENCES:

COMPONENTS:

1. Nitric oxide; NO; [10102-43-9]
2. 2-Propanone, (Acetone); C_3H_6O; [67-64-1]
2. Copper chloride, (Cupric chloride); $CuCl_2$; [7747-39-4]

ORIGINAL MEASUREMENTS:

Kohlschutter, V.; Kutscheroff, M.

Ber. 1904, *37*, 3044-3052

VARIABLES:

Concentration

PREPARED BY:

W. Gerrard.

EXPERIMENTAL VALUES:

Neither temperature nor pressure of nitric oxide were stated. T/K thought to be 293.16 K, and p_{NO} to be 101.325 kPa.

Volume, V_s, of acetone containing one mole of cupric chloride /dm^3	Volume, V_1, of nitric oxide absorbed by V_s /dm^3	Ostwald coefficient, *L*, calculated by compiler.
4.667	14.04	3.010
29.16	24.01	0.823
58.33	24.60	0.422
291.6	40.99	0.141
583.2	67.22	0.115
1166.4	81.96	0.070

AUXILIARY INFORMATION

METHOD/APPARATUS/PROCEDURE:

A gas buret and pipet of the Ostwald type were connected to a gasometer by a lead capillary tube. Two electrodes were fitted to the absorption tube, and the specific conductance was determined before and after the absorption. It was stated that the point of saturation could be sharply determined.

SOURCE AND PURITY OF MATERIALS:

No details given.

ESTIMATED ERROR:

REFERENCES:

COMPONENTS:

1. Nitric oxide; NO; [10102-43-9]
2. Ethanol; C_2H_6O; [64-17-5]
3. Copper (II) bromide, (Cupric bromide; $CuBr_2$; [7789-45-9]

ORIGINAL MEASUREMENTS:

Manchot, W.

Ber. 1914, *47*, 1601-1614.

VARIABLES:

Temperature, concentration

PREPARED BY:

W. Gerrard.

EXPERIMENTAL VALUES:

V_1 is the volume of NO (101.325 kPa) in dm^3 absorbed for 1 mole of $CuBr_2$.

T/K	Conc /mol dm^{-3} (soln)	Pressure of NO/kPa	V_1/dm^3
273.15	0.00498	99.190	20.02
	0.00548	99.857	20.32
	0.0107	99.457	21.36
	0.0599	99.590	20.57
	0.1187	99.590	21.48
	0.3833	99.857	20.90
	0.1126	149.185	19.98
293.65	0.05656	99.723	21.00
	0.0504	45.529	18.66
273.15+	0.0478	99.457	19.07

\+ "Spirit" was used instead of absolute ethanol.

AUXILIARY INFORMATION

METHOD/APPARATUS/PROCEDURE:

The volume of nitric oxide absorbed was observed as previously stated (1).

SOURCE AND PURITY OF MATERIALS:

(1) and (3) Presumably as previously described (1).

(2) Absolute ethanol; distilled from lime and sodium ethylate.

ESTIMATED ERROR:

REFERENCES:

1. Manchot, W.; Zechentmayer, K. *Annalen*, 1906, *350*, 368-389.

See also Manchot, W.; Huttner, F. *Annalen*, 1910, *372*, 153-178.

COMPONENTS:	ORIGINAL MEASUREMENTS:
1. Nitric oxide; NO; [10102-43-9] 2. Ethanol; C_2H_5OH; [64-17-5] 3. Copper chloride (Cupric chloride); $CuCl_2$; [7447-39-4]	Griffith, W. P.; Lewis, J.; Wilkinson, G. *J. Chem. Soc.* 1958, 3993-3998.
VARIABLES:	PREPARED BY:
	W. Gerrard

EXPERIMENTAL VALUES:

M = molarity of cupric chloride/mol dm^{-3}. *P* = Pressure of NO/kPa. V_1 = Volume of NO/dm^3 (adjusted to 273.15, 101.325 kPa); V_1 is based on the authors' loose statement: "Vol of NO absorbed per mole at N.T.P.", the mole apparently referring to $CuCl_2$. *K* = Equilibrium constant given by the authors as the mean value of P_{NO} $CuCl_2/CuCl_2NO$ in mmHg. K^* = Compiler's value based on P_{NO} $(22.4-V_1)/V_1$, in mmHg. The H values have been calculated by the compiler from the expression:

$$\Delta H = 4.571(\log K_1 - \log K_2) \times T_1T_2/(T_1 - T_2) \text{ in kcal mole}^{-1};$$

T = T/K. The *K* values are those of the original authors, and the temperature range is shown ( ). $V_1(E)$ = Volume of NO (adjusted to 273.15 K, 101.325 kPa) which would be absorbed by that volume of ethanol equal to the volume of solution containing 1 mole of $CuCl_2$.

(cont.)

AUXILIARY INFORMATION

METHOD/APPARATUS/PROCEDURE:	SOURCE AND PURITY OF MATERIALS:
It was simply stated that the gas absorption was measured in a conventional Warburg-type apparatus. NOTE: The authors did not show how the "heat of reaction 7.3 kcal/mole" was obtained.	1. Nitric oxide was obtained from a sodium nitrite-ferrous sulfate generator, and washed with a concentrated aqueous solution of potassium hydroxide. 2, 3. Probably of satisfactory purity.
	ESTIMATED ERROR:
	REFERENCES:

1. Nitric oxide; NO; [10102-43-9]	Griffith, W. P.; Lewis, J.;
2. Ethanol; C_2H_5OH; [64-17-5]	Wilkinson, G.
3. Copper chloride (Cupric chloride); $CuCl_2$; [7447-39-4]	*J. Chem. Soc.* 1958, 3993-3998.

EXPERIMENTAL VALUES:

No.	T/K	M	*P*/kPa	V_1/dm^3	*K*/(mmHg)	*K**/(mmHg)	$V_1(E)/dm^3$
1	273.4	0.05	99.043	19.1		128.35	6.3
2		0.025		20.1	115.0	85.05	12.6
3	283.0	0.05	97.577	17.3		215.76	5.7
4		0.025		17.0	226.4	232.46	11.4
5	297.6	0.05	92.911	15.6	340.8	303.7	5.2
6	304.4	0.025	91.458	12.2	568.0	573.54	10.2

ΔH values T/K range ( ): (2-4, 10.84; (2-5) 7.25; (2-6), 8.51; (4-6), 7.35; (5-6), 13.51. Mean is 9.5 kcal mol^{-1}. The authors gave ΔH as 7.3 kcal mol^{-1} (30.55 kJ mol^{-1}).

COMPONENTS:	ORIGINAL MEASUREMENTS:
1. Nitric oxide; NO; [10102-43-9] 2. Water; H_2O; [7732-18-5] 3. Nickel (II) sulfate; $NiSO_4$; [7786-81-4]	Hufner, G. *Z. Phys. Chem.* 1907, *59*, 416-423.
VARIABLES:	PREPARED BY:
Pressure	W. Gerrard.

EXPERIMENTAL VALUES:

V_1/cm^3, is the volume of nitric oxide absorbed by 205.69 cm of solution containing 0.0506 g of nickel as nickel sulfate, at the temperature and pressure, P/mmHg, of nitric oxide stated.

T/K	P/mmHg	V_1/cm^3
	760.0	25.4*
293.35	654.7	23.00
293.35	629.8	22.54
293.35	609.5	22.03
293.30	591.7	21.65
293.29	573.4	21.18

* Calculated by the original author from the equation: $V = a + bP$, where a = 8.3146, and b = 0.022493. The author referred to the product $760b/205.69$ as the "absorption coefficient α;" but this is not the quantity usually referred to as an absorption coefficient; α was stated to be 0.08311.

The following values were calculated by the compiler :

The volume of solution, V_3/dm^3, containing 58.71 g of nickel as sulfate = 238.7 dm³. The volume of nitric oxide absorbed by this volume of solution = 29.47 dm³ (760 mmHg, 101.325 kPa). The volume of nitric oxide which would be absorbed by 238.7 dm³ of water is 238.7 x 0.047 = 11.22 dm³ (101.325 kPa).

AUXILIARY INFORMATION

METHOD/APPARATUS/PROCEDURE:	SOURCE AND PURITY OF MATERIALS:
The volume of nitric oxide absorbed by 205.69 cm³ of solution was observed at the observed pressure of gas.	1. Nitric oxide was prepared from hydrogen iodide and nitric acid Winkler, (1). 3. Prepared from pure metal.
	ESTIMATED ERROR:
	REFERENCES:
	1. Winkler, L.W. *Ber*, 1901, *34*, 1408.

COMPONENTS:

1. Nitric oxide; NO; [10102-43-9]
2. Water; H_2O; [7732-18-5]
3. Cobalt (II) sulfate; $CoSO_4$; [10124-43-3]

ORIGINAL MEASUREMENTS:

Hufner, G.

Z. Phys. Chem. 1907, *59*, 416-423.

VARIABLES:

Pressure

PREPARED BY:

W. Gerrard

EXPERIMENTAL VALUES:

V_1/cm^3, is the volume of nitric oxide absorbed by 205.69 cm^3 of solution containing 0.0598 g of cobalt as cobalt sulfate, at the temperature and pressure, P/mmHg, of nitric oxide stated.

T/K	P/mmHg	V_1/cm^3
293.30	760	25.57*
293.30	678.3	23.47
293.31	653.5	23.01
293.35	636.6	22.55
293.45	615.9	21.99
293.55	600.0	21.56

* Calculated by the original author from the equation: $V = a + bP$, where $a = 6.7288$, and $b = 0.024791$. The author referred to the product 760 /205.69 (0.09146) as the "absorption coefficient, α;" but this is not equal to $V_1/205.69$.

The following values were calculated by the compiler: The volume, V_3/dm^3, of solution containing 58.93 g of cobalt as sulfate = 202.7 dm^3. The volume of nitric oxide absorbed by this volume of solution = 25.20 dm^3 (101.325 kPa). The volume of nitric oxide which would be absorbed by 202.7 dm^3 of water is 202.7 x 0.047 = 9.53 dm^3 (101.325 kPa).

AUXILIARY INFORMATION

METHOD/APPARATUS/PROCEDURE:

The volume of nitric oxide absorbed by 205.69 cm^3 of solution was observed at the observed pressure.

SOURCE AND PURITY OF MATERIALS:

1. Nitric oxide was prepared from hydrogen iodide and nitric acid (Winkler, (1)).

3. Prepared from pure metal.

ESTIMATED ERROR:

REFERENCES:

1. Winkler, L.W.

Ber. 1901, *34*, 1408.

COMPONENTS:	ORIGINAL MEASUREMENTS:
1. Nitric oxide; NO; [10102-43-9] 2. Water; H_2O; [7732-18-5] 3. Manganese chloride; $MnCl_2$; [7773-01-5]	Hufner, G. *Z. Phys. Chem.* 1907, *59*, 416-423.
VARIABLES:	PREPARED BY:
Pressure	W. Gerrard

EXPERIMENTAL VALUES: V_1/cm^3, is the volume of nitric oxide absorbed by 205.69 cm^3 of solution containing 0.0697 g of manganese as manganese chloride, at the temperature and pressure of NO, P/mmHg,

T/K	P/mmHg	V_1/cm^3
293.15	760.0	15.12*
293.15	711.96	14.25
293.20	686.5	13.99
293.35	657.4	13.49
293.45	638.9	13.05
293.60	621.0	12.81

* Calculated by the original author from the equation: $V_1 = a + bP$, where $a = 2.5518$, and $b = 0.016538$. The author referred to the product $760b/205.69$ (0.06111) as the "absorption coefficient, α"; but this is not equal to $V_1/205.69$.

The following values were calculated by the compiler:

The volume, V_3/dm^3, of solution containing 1 mole of manganese chloride = 162.1 dm^3.

The volume of nitric oxide absorbed by this volume of solution = 11.92 dm^3 (101.325 kPa).

The volume of nitric oxide which would be absorbed by 162.dm^3 of water is 162.1 x 0.047 = 7.62 dm^3 (101.325 kPa.)

AUXILIARY INFORMATION

METHOD/APPARATUS/PROCEDURE:

The volume of nitric oxide absorbed by 205.69 cm^3 of solution was observed at the observed pressure.

SOURCE AND PURITY OF MATERIALS:

(1) Nitric oxide was prepared from hydrogen iodide and nitric acid (Winkler, 1).

(3) Concentration attested as pyrophosphate.

ESTIMATED ERROR:

REFERENCES:

(1). Winkler, L.W. *Ber*, 1901, *34*, 1408.

COMPONENTS:	ORIGINAL MEASUREMENTS:
1. Nitric oxide; NO; [10102-43-9] 2. Water; H_2O; [7732-18-5] 3. Metal salts.	Usher, F.L. *Z. Phys. Chem.* 1908, *62*, 622-625.
VARIABLES:	PREPARED BY: W. Gerrard.

EXPERIMENTAL VALUES:

T/K = 293.15

The volume of nitric oxide absorbed was express as the Bunsen coefficient α (273.15 K, 101.325 kPa).

(3)	Sulfuric acid, nickel (2+) salt, (Nickel sulfate); $NiSO_4$; [7786-81-4]	"Hufner's concentration"* "Saturated solution"	α = 0.048 α = 0.0245
(3)	Sulfuric acid, cobalt (2+) salt, (Cobalt sulfate); $CoSO_4$; [10124-43-3]	"Saturated solution"	α = 0.0288
(3)	Manganese chloride; $MnCl_2$; [7773-01-5]	"Saturated solution"	α = 0.0082
(3)	Sulfuric acid, iron (2+) salt (Ferrous sulfate); $FeSO_4$; [7720-78-7]	"Hufner's concentration"** For water alone	α = 0.180 α = 0.049

* Hufner's concentration was 0.0506 g of nickel (as sulfate) in 203.69 cm^3 of solution.

** This concentration appears to refer to Hufner's solution that contained 0.099 g of iron (as sulfate) in 205.69 cm^3 of solution; the Bunsen α value was 0.170.

NOTE: Usher gave two sets of Bunsen α values for water alone. One set obtained by a large absorption pipet: 0.0451, 0.0449, 0.0448, were for NO washed with concentrated sulfuric acid. The other set: 0.0468,0.0471, 0.0473,0.0487 were for NO washed with aqueous potassium hydroxide. In the ferrous sulfate system he purported to show that discrepancies could occur by the reduction of NO to N_2O and N_2; but this contention was not confirmed by the detailed analysis of Tarte (2).

METHOD/APPARATUS/PROCEDURE:

Volume of nitric oxide absorbed was determined by the gas buret and absorption pipet technique of Ostwald. The maximum error was stated to be 3%.

SOURCE AND PURITY OF MATERIALS:

1. Nitric oxide was prepared from hydrogen iodide and nitric acid, Winkler (1). It was washed with aqueous potassium hydroxide.

ESTIMATED ERROR:

REFERENCES:

(1) Winkler, L.W. *Ber*, 1901, *34*, 1408.

(2) Tarte, P. *Ind. Chim. Belg.* 1952, *17*, 42.

COMPONENTS:	ORIGINAL MEASUREMENTS:
1. Nitrogen oxide; (Nitric oxide); NO; [10102-43-9] 2. Water; H_2O; [7732-18-5] 3. Sodium salts.	Kohlschutter, V.; Kutscheroff, M. *Ber*, 1907, *40*, 873-878
VARIABLES:	PREPARED BY: W. Gerrard

EXPERIMENTAL VALUES:

The pressure of nitric oxide, p_{NO}, was not stated. The compiler has taken p_{NO} to be 101.325 kPa. Temperature assumed to be 295.15K.

T/K	Volume of water, V_s, in which one mole of salt was dissolved / dm^3	Volume, V_1, of nitric oxide absorbed by one dm^3 of salt solution /dm^3
295.15	Sodium chloride; NaCl; [7647-14-5]	
	0.5	0.0580
	1.0	0.0359
	Sulfuric acid, disodium salt, (sodium sulfate); Na_2SO_4; [7757-82-6]	
	4.0	0.0397
	2.0	0.0277
	Water; H_2O; [7732-18-5]	0.0636

AUXILIARY INFORMATION

METHOD/APPARATUS/PROCEDURE:	SOURCE AND PURITY OF MATERIALS:
An Ostwald type gas buret and pipet were connected by a lead capillary.	No details given.
	ESTIMATED ERROR:
	REFERENCES:

COMPONENTS:	ORIGINAL MEASUREMENTS:
1. Nitric oxide; NO; [10102-43-9] 2. Water; H_2O; [7732-18-5] 3. Electrolytes.	Armor, J.N. *J. Chem. Engng. Data.* 1974, *19*, 82-84.
VARIABLES:	**PREPARED BY:**
Concentration	C.L. Young

EXPERIMENTAL VALUES:

T/K = 298.15 pH = 7.0

Electrolyte	Conc. of electrolyte /mol dm^{-3}	Solubility of nitric$^+$ oxide, S/mol dm^{-3}
H_2O	-	1.95
$H_2PO_4^-/OH^-$	0.1*	1.80
NaCl	0.1	1.62
	0.5	1.61
	1.0	1.38
LiCl	0.1	1.78
	1.0	1.52
$LiClO_4$	1.0	1.58
$NaClO_4$	1.0	1.28

\+ at a partial pressure of gas of 101.3 kPa

* Total ionic strength.

AUXILIARY INFORMATION

METHOD/APPARATUS/PROCEDURE:

Solution was saturated with gas for at least 30 minutes in a reaction vessel. A 5 cm^3 aliquot was removed and injected into 80 cm^3 of oxygen saturated water. This solution was analysed spectrophotometrically of product of reaction of NO_2 with sulfanilamide and N-(1-naphtyl)-ethylenediamine hydrochloride.

SOURCE AND PURITY OF MATERIALS:

1. Matheson sample. Nitrogen dioxide removed by "vigorous" scrubbing.
2. Distilled water, redistilled from alkaline permanaganate.
3. LiCl, recrystallised. $LiClO_4$ and $NaClO_4$ prepared by action of $HClO_4$ on corresponding carbonates, resulting solution concentrated and crystals obtained by cooling in ice bath

ESTIMATED ERROR:

REFERENCES:

COMPONENTS:	ORIGINAL MEASUREMENTS:
1. Nitric oxide; NO; [10102-43-9] 2. Water; H_2O; [7732-18-5] 3. Sodium chloride; NaCl; [7647-14-5] 4. Buffer solutions.	Armor, J.N. *J. Chem. Engng. Data.* 1974, *19*, 82-84.
VARIABLES:	PREPARED BY: C.L. Young

EXPERIMENTAL VALUES:

T/K = 298.15

Solution	Conc. of NaCl /mol dm^{-3}	pH	Solubility of nitric oxide,+ S/mol dm^{-3}
Buffer (1)	0.1	13.0	1.70
	0.10	12.9	1.52
	1.0	12.7	1.35
Buffer#	0.0	12.12	1.78
	0.10	11.92	1.77
	1.0	11.51	1.32
	0.10	10.8	1.73
	1.0	10.3	1.34
	0.0	9.87	1.78
	0.10	9.88	1.71
	1.0	9.56	1.29
	0	9.34	1.76
	0.1	9.27	1.68
	1.0	9.00	1.23
	1.0§	8.95	1.37
	0.0	8.15	1.85
	0.1	8.09	1.74
	1.0	7.82	1.31
	0.0	7.18	1.80
	0.1	7.06	1.69
	1.0	6.69	1.22
	0.0	4.24	1.73
	0.1	4.09	1.78

AUXILIARY INFORMATION

METHOD APPARATUS/PROCEDURE:	SOURCE AND PURITY OF MATERIALS:
Solution was saturated with gas for at least 30 minutes in a reaction vessel. A 5 cm^3 aliquot was removed and injected into a 80 cm^3 of oxygen saturated water. This solution was analysed spectrophotometrically of product of reaction of NO_2 with sulfanilamide and N-(1-naphtyl)-ethylenediamine hydrochloride.	1. Matheson sample-Nitrogen dioxide removed by "vigorous" scrubbing. 2. Distilled water, redistilled from alkaline permanagonate. 3. and 4. No details given.
	ESTIMATED ERROR: $\delta T/K = \pm 0.1$; $\delta S = \pm 4\%$
	REFERENCES:

COMPONENTS:	ORIGINAL MEASUREMENTS:
1. Nitric oxide; NO; [10102-43-9]	Armor, J.N.
2. Water; H_2O; [7732-18-5]	*J. Chem. Engng. Data.* 1974, *19*, 82-84.
3. Sodium chloride; NaCl; [7647-14-5]	
4. Buffer solutions.	

EXPERIMENTAL VALUES:

Solution	Conc. of NaCl /mol dm^{-3}	pH	Solubility of nitric oxide,+ S/mol dm^{-3}
Buffer#	1.0	3.91	1.36
Buffer (2)	0.0	2.0	1.69
Buffer (3)	0.1	1.0	1.47
Buffer (4)	1.0	0	1.02
Buffer (5)	0.0	2.0	1.71
Buffer (6)	1.0	0	1.37

\+ at a partial pressure of gas of 101.3kPa.

§ conc. of $NaClO_4$.

Buffer (1) 0.1 mol dm^{-3} NaOH

Buffer (2) 0.01 mol dm^{-3} $HClO_4$

Buffer (3) 0.1 mol dm^{-3} HCl

Buffer (4) 1.0 mol dm^{-3} HCl

Buffer (5) 0.01 mol dm^{-3} HCl

Buffer (6) 1.0 mol dm^{-3} $HClO_4$

\# Authors quoted "buffers were prepared from HOAc/NaOAc (pH4), $H_2PO_4^-/OH^-$ (pH 6-8), borax/HCl (pH 8-9.2) borax /OH^- (pH 9.2-10.5), HPO_4^{2-}/OH^- (pH 10.5-12.0)"

The total ionic strength of buffers appears to be about 0.1 mol dm^{-3}

COMPONENTS:	EVALUATOR:
1. Nitric oxide; NO; [10102-43-9] 2. Organic liquids	Colin L. Young, Department of Physical Chemistry, University of Melbourne, Parkville, Victoria 3052, Australia. October 1980.

CRITICAL EVALUATION:

The data of Vosper (1) are thought to be reliable since they were carried out with carefully purified materials and corrections for the vapor pressure of the liquid and non-ideality of the gas were applied. His data are classified as tentative.

The temperature-dependence of Klemenc *et al.*'s (2) mole fraction solubility of nitric oxide in tetrachloromethane is unusual and of opposite sign to that of Vosper (1). On the other hand Klemenc *et al.* (2) data for nitrobenzene show the normal increase in mole fraction solubility for a decrease in temperature but the opposite is true for benzene. Therefore the data of Klemenc *et al.* (2) on all three solvents are classified as doubtful. The data of Tsiklis and Svetlova (3) show a fairly large scatter and are classified as doubtful.

The data of Riccoboni (4) for methanol and Carius (5) for ethanol are included but in view of the year in which these data were determined must be regarded at best as doubtful. Garelli and Monath's data (6) based on the depression of freezing point are rejected.

The results of Trautz and Gerwig (7) for pentane and toluene are thought to be unreliable. The temperature-dependence of the solubility for pentane should be similar to that for hexane as measured by Vosper (1) but the pentane data show no such similarity. The Trautz and Gerwig (7) results for toluene are in poor agreement with those of Vosper (1) and are rejected.

References:

1. Shaw, A. W.; Vosper, A. J. *J. Chem. Soc., Faraday Trans. I* 1977, *73*, 1239.

2. Klemenc, A.; Spitzer-Neumann, E. *Monatsh.* 1929, *53*, 413.

3. Tsiklis, D. S.; Svetlova, G. M. *Zh. Fiz. Khim.* 1958, *32*, 1476.

4. Riccoboni, L. *Gazz. chim. ital.* 1841, *71*, 139.

5. Carius, L. *Annalen* 1855, *94*, 129.

6. Garelli, F.; Monath, E. *Att. Accad. Torino* 1926, *61*, 12.

7. Trautz, M.; Gerwig, W. *Z. anorg. Chem.* 1925, *146*, 1.

COMPONENTS:

1. Nitric oxide; NO; [10102-43-9]
2. Cyclohexane; C_6H_{12}; [110-82-7]

ORIGINAL MEASUREMENTS:

Tsiklis, D.S.; Svetlova, G.M.

Zh. Fiz. Khim. 1958, *32*, 1476-80.

VARIABLES:

Temperature, pressure

PREPARED BY:

C.L. Young

EXPERIMENTAL VALUES:

T/K	P/bar	Mole fraction of nitric oxide in liquid, x_{NO}
283.15	0.133	0.00030
	0.267	0.00059
	0.400	0.00089
	0.533	0.00119
	0.667	0.00148
	0.800	0.00178
	0.933	0.00208
	1.067	0.00237
293.15	0.133	0.00026
	0.267	0.00049
	0.400	0.00074
	0.533	0.00099
	0.667	0.00123
	0.800	0.00148
	0.933	0.00173
	1.067	0.00199
313.15	0.133	0.00025
	0.267	0.00049
	0.400	0.00074
	0.533	0.00099
	0.667	0.00123
	0.800	0.00148
	0.933	0.00173
	1.067	0.00199

AUXILIARY INFORMATION

METHOD/APPARATUS/PROCEDURE:

Volumetric apparatus. Gas admitted from a gas buret to absorption pipet. Mole fraction calculated from measurements of pressure and volume.

SOURCE AND PURITY OF MATERIALS:

1. Purified by passing through "caustic" at 233K to remove acidic oxides.
2. Commercial sample, twice distilled.

DATA CLASS:

ESTIMATED ERROR:

$\delta T/K = \pm 0.0.1$; $\delta P/\text{bar} = \pm 0.005$; $\delta x_{NO} = \pm 2 \times 10^{-5}$

REFERENCES:

COMPONENTS:	ORIGINAL MEASUREMENTS:
1. Nitric oxide; NO; [10102-43-9] 2. Hexane; C_6H_{14}; [110-54-3]	Shaw, A. W.; Vosper, A. J. *J. Chem. Soc. Faraday Trans. I* 1977, *73*, 1239-1244.
VARIABLES:	PREPARED BY:
Temperature	A. J. Vosper

EXPERIMENTAL VALUES:

Absorption measurements			Desorption measurements		
T/K	Mole fraction x_1*	Concentration* mol dm^{-3}	T/K	Mole fraction x_1*	Concentration* mol dm^{-3}
291.0	0.00240	0.0185	243.2	0.00289	0.0236
283.0	0.00248	0.0193	253.0	0.00277	0.0224
273.4	0.00256	0.0202	260.0	0.00270	0.0217
264.2	0.00264	0.0211	268.2	0.00262	0.0208
257.4	0.00271	0.0218			
248.5	0.00281	0.0229			
238.7	0.00294	0.0242			
233.2	0.00304	0.0252			

* The mole fractions follow directly from the authors' results whereas the concentrations were calculated using literature values (1) for the density of hexane, extrapolating where necessary.

Mean values for the partial molar enthalpy of solution and partial molar entropy of solution over the temperature range 233-291 K were given:

$$\Delta H^\circ = -2.23 \pm 0.03 \text{ kJ mol}^{-1} \qquad \Delta S^\circ = -57.8 \pm 0.01 \text{ J mol}^{-1} \text{ K}^{-1}$$

AUXILIARY INFORMATION

METHOD/APPARATUS/PROCEDURE:

A gas burette was used to determine the volume of gas absorbed by a known weight of the degassed liquid at a known temperature and at the prevailing barometric pressure. Measurements of desorption of gas from the saturated liquid with increasing temperature were also made. Corrections were applied for the vapor pressure of the liquid (1) and non-ideality of the gas (2). Results were calculated for a gas pressure of 101.325 kPa assuming Henry's Law which had previously been shown to be valid over the relevant pressure range.

SOURCE AND PURITY OF MATERIALS:

The gas (Matheson Co. Inc.) was passed through 90 per cent H_2SO_4 and a trap at *ca.* 190 K. It was solidified at 77 K and volatile impurities pumped away. The product contained less than 0.2 per cent N_2O as the only detectable impurity.

The liquid (BDH 99 per cent minimum) was dried over $CaCl_2$ then sodium wire and fractionated. The fraction boiling between 341.5-341.8K was used.

ESTIMATED ERROR:

REFERENCES:

1. Timmermans, J. *"Physico-chemical constants of pure organic compounds"*. Elsevier, Vol. 1, p.*42* and Vol. 2, p.*10*.
2. Johnson, H. L.; Weimer, H. W. *J. Amer. Chem. Soc.* 1934, *56*, 625.

COMPONENTS:	ORIGINAL MEASUREMENTS:
1. Nitric oxide; NO; [10102-43-9] 2. Benzene, C_6H_6, [71-43-2]	Klemenc, A.; Spitzer-Neumann, E. *Monatsh.* 1929, *53*, 413-419.
VARIABLES:	PREPARED BY:
Temperature, pressure	W. Gerrard

EXPERIMENTAL VALUES:

T/K	281.96	287.76	297.76	307.76
Pressure range, *				
from (kPa);	69.328	92.752	87.673	78.274
to (kPa):	132.53	90.673	86.260	80.513
Ostwald coefficient, *				
From :	0.284	0.279	0.300	0.316
To:	0.268	0.294	0.306	0.320
Mean: * *	0.275	0.284	0.300	0.318
Mole fraction, *** x_1:	0.00104	0.00106	0.00109	0.00114
Number of measurements:	11	5	5	5

* Irregular distribution.
** Given by authors.
*** Calculated by compiler.

T/K	281.96	287.6	287.76	307.76
ΔF^0, cal mole^{-1}	2480	2520	2600	2660

From ΔF^0 = -RT in (L/RT).
ΔH_{298} = 940 cal mole^{-1}.

* The Ostwald coefficient was given as :

$$L = \frac{\text{Concentration of gas in the liquid phase}}{\text{Concentration of gas in the gas phase.}}$$

AUXILIARY INFORMATION

METHOD/APPARATUS/PROCEDURE:

An absorption vessel, and a gas buret with a levelling tube of the usual form were used. The volume of NO absorbed was measured at a total pressure, $p_T = p_{NO} + p_S$, where p_{NO} is the partial pressure of the gas, and p_S is the vapour pressure of the liquid S over the solution.

SOURCE AND PURITY OF MATERIALS:

1. Prepared by the action of mercury on nitric acid in the presence of 90% sulfuric acid.

2. Rendered gas free.

ESTIMATED ERROR:

REFERENCES:

COMPONENTS:	ORIGINAL MEASUREMENTS:
1. Nitric oxide; NO; [10102-43-9] 2. Methylbenzene; C_7H_8; [108-88-3]	Shaw, A. W.; Vosper, A. J. *J. Chem. Soc. Faraday Trans. I* 1977, *73*, 1239-1244.
VARIABLES:	PREPARED BY:
Temperature	A. J. Vosper

EXPERIMENTAL VALUES:

Absorption measurements			Desorption measurements		
T/K	Mole fraction x_1*	Concentration mol dm^{-3}	T/K	Mole fraction x_1*	Concentration* mol dm^{-3}
			217.2	0.00171	0.0175
289.0	0.00120	0.0123	219.7	0.00169	0.0172
279.8	0.00125	0.0130	222.7	0.00167	0.0169
265.4	0.00137	0.0133	225.8	0.00164	0.0166
256.0	0.00141	0.0139	227.7	0.00162	0.0163
246.6	0.00146	0.0144	233.2	0.00158	0.0159
238.4	0.00151	0.0151	235.5	0.00156	0.0156
230.7	0.00157	0.0158	241.7	0.00152	0.0151
219.7	0.00168	0.0171	245.7	0.00149	0.0148
213.7	0.00175	0.0180	255.0	0.00142	0.0140
			272.1	0.00133	0.0128

* The mole fractions follow directly from the authors work whereas the concentrations were calculated using literature values for the density of toluene,[1] extrapolating where necessary.

Mean values for the partial molar enthalpy of solution and partial molar entropy of solution over the temperature range 214-289 K were given

$$\Delta H° = -2.28 \pm 0.04 \text{ kJ mol}^{-1}; \quad \Delta S° = -63.4 \pm 0.02 \text{ J mol}^{-1} \text{ K}^{-1}$$

AUXILIARY INFORMATION

METHOD/APPARATUS/PROCEDURE:

A gas burette was used to determine the volume of gas absorbed by a known weight of the degassed liquid at a known temperature and at the prevailing barometric pressure. Measurements of desorption of gas from the saturated liquid with increasing temperature were also made. Corrections were applied for the vapor pressure of the liquid[1] and nonideality of the gas[2]. Results were calculated for a gas pressure of 101.325 kPa assuming Henry's Law which had previously been shown to be valid over the relevant pressure range.

SOURCE AND PURITY OF MATERIALS:

The gas (Matheson Co. Inc.) was passed through 90 per cent H_2SO_4 and a trap at *ca.* 190 K. It was solidified at 77 K and volatile impurities pumped away. The product contained less than 0.2 per cent N_2O as the only detectable impurity. The liquid (BDH ANALAR GRADE) was dried over $CaCl_2$ then sodium wire and fractionated. The portion boiling between 383.8-383.9 K was used.

ESTIMATED ERROR:

REFERENCES:

1. Timmermans, J. *"Physico-chemical Constants of Pure Organic Compounds"*. Elsevier, Vol. 1, p.150 and Vol. 2, p.99.
2. Johnson, H. L.; Weimer, H. R. *J. Amer. Chem. Soc.* 1934, *56*, 625.

COMPONENTS:	ORIGINAL MEASUREMENTS:
1. Nitric oxide; NO; [10102-43-9] 2. 1,3-Dimethylbenzene; C_8H_{10}; [108-38-3]	Shaw, A. W.; Vosper, A. J. *J. Chem. Soc. Faraday Trans. I* 1977, *73*, 1239-1244.
VARIABLES:	PREPARED BY:
Temperature	A. J. Vosper

EXPERIMENTAL VALUES:

Absorption measurements			Desorption measurements		
T/K	Mole fraction x_1*	Concentration* mol dm^{-3}	T/K	Mole fraction x_1*	Concentration* mol dm^{-3}
294.1	0.00134	0.0111	243.2	0.00162	0.0139
284.1	0.00140	0.0115	251.3	0.00156	0.0133
274.1	0.00144	0.0120	259.2	0.00152	0.0128
263.2	0.00148	0.0125	267.4	0.00149	0.0124
255.2	0.00153	0.0129	271.7	0.00147	0.0122
249.0	0.00157	0.0134	279.4	0.00142	0.0118
245.4	0.00159	0.0136	285.4	0.00137	0.0113
239.4	0.00164	0.0141			
234.0	0.00169	0.0146			

* The mole fractions follow directly from the authors' work whereas the concentrations were calculated using literature values (1) for the density of m-xylene, extrapolating where necessary.

Mean values for the partial molar anthalpy of solution and partial molar entropy of solution over the temperature range 234-294 K were given:

$$\Delta H° = -2.17 \pm 0.04 \text{ kJ mol}^{-1} \quad \Delta S° = -62.4 \pm 0.2 \text{ J mol}^{-1} \text{ K}^{-1}$$

AUXILIARY INFORMATION

METHOD /APPARATUS/PROCEDURE:

A gas burette was used to determine the volume of gas absorbed by a known weight of the degassed liquid at a known temperature and at the prevailing barometric pressure. Measurements of desorption of gas from the saturated liquid with increasing temperature were also made. Corrections were applied for the vapor pressure of the liquid (1) and non-ideality of the gas (2). Results were calculated for a gas pressure of 101.325 kPa assuming Henry's Law which had previously been shown to be valid over the relevant pressure range.

SOURCE AND PURITY OF MATERIALS:

The gas (Matheson Co. Inc.) was passed through 90 per cent H_2SO_4 and a trap at *ca.* 190 K. It was solidified at 77 K and volatile impurities pumped away. The product contained less than 0.2 per cent N_2O as the only detectable impurity.

The liquid (BDH ANALAR GRADE) was dried over $CaCl_2$ then sodium wire and fractionated. The fraction boiling between 437.8-437.9 K was used.

ESTIMATED ERROR:

REFERENCES:

1. Timmermans, J. *"Physico-chemical constants of pure organic compounds"*. Elsevier, Vol. 1, p.*164* and Vol. 2, p.*112*.
2. Johnson, H. L.; Weimer, H. R. *J. Amer. Chem. Soc.* 1934, *56*, 625.

COMPONENTS:	ORIGINAL MEASUREMENTS:
1. Nitric oxide; NO; [10102-43-9] 2. Methanol; CH_3OH; [67-56-1]	Riccoboni, L. *Gazz. chim. ital.* 1841, *71*, 139-53.
VARIABLES:	PREPARED BY:
Temperature, pressure	W. Gerrard

EXPERIMENTAL VALUES:

T/K	273.16	283.16	293.16	303.16
Pressure, kPa, *				
from:	65.355	63.327	57.198	60.955
to:	150.821	174.767	188.730	181.396
Ostwald coefficient, L, *				
from:	0.367	0.366	0.355	0.358
to:	0.360	0.356	0.346	0.339
Mean: **	0.363	0.362	0.350	0.347
Mole fraction, : ***	0.000640	0.000623	0.000589	0.000571

* Irregular distribution
** Mean given by author
*** Calculated by compiler

The author assumed L to be independent of pressure.
The compiler used the gram-mole volume of NO (22388 cm^3 at 273.16, 101.325 kPa) based on the standard density, 1.3402 g/dm^3. The volume for other temperatures was obtained by 22388 × (T/273.16). The L value was assumed to be for 101.325 kPa.

AUXILIARY INFORMATION

METHOD/APPARATUS/PROCEDURE:

Determination of weight of gas absorbed by a known weight of liquid. Converted into the Ostwald coefficient, L. Solvent was freed from air and vacuum distilled into the ampoule. Solvent and solution were weighed in a removable ampoule which also served as a dilatometer. Diagram given by author.

SOURCE AND PURITY OF MATERIALS:

1. Nitric oxide. Self-prepared and rigorously purified and attested.
2. Methanol. Redistilled.

ESTIMATED ERROR:

REFERENCES:

COMPONENTS:	ORIGINAL MEASUREMENTS:
1. Nitric oxide; NO; [10102-43-9] 2. Ethanol, C_2H_6O; [64-17-5]	Carius, L. *Annalen*, 1855, *94*, 129-166.
VARIABLES:	PREPARED BY: W. Gerrard

EXPERIMENTAL VALUES:

T/K	Bunsen absortion coefficient, α	Mole fraction, x_{NO} (Calculated by compiler)*
275.15	0.30895	0.0007885
279.15	0.29684	0.0007615
284.95	0.28162	0.0007262
289.15	0.27250	0.0007062
293.15	0.26573	0.0006915
297.35	0.26014	0.0006800

Absorption coefficient, $\alpha = 0.31606 - 0.0034870t + 0.0000490\ t^2$ (From 273.15 to 298.15 K) where $t = T/K - 273.15$

Henrich (1) used Carius's data to give a modified smoothing equation :

$$\alpha = 0.31578 - 0.003469\ t + 0.00004827\ t^2$$

Note : Henrich did not give any experimental data.

* The gas molecular volume of NO at 273.15 K and 101.325 kPa was taken to be 22.385 dm^3, based on the standard density of 1.3402 g /dm^{-3}.

AUXILIARY INFORMATION

METHOD /APPARATUS/PROCEDURE:	SOURCE AND PURITY OF MATERIALS:
Measurement of volume by the Bunsen gas buret.	1. NO: Self prepared and purified. 2. Ethanol was distilled as "absolute", $d^{20} = 0.792$.
	ESTIMATED ERROR:
	REFERENCES: 1. Henrich, F. *Z. Phys. Chem.* 1892, *9*, 435.

COMPONENTS:	ORIGINAL MEASUREMENTS:
1. Nitric oxide; NO; [10102-43-9] 2. 1,1'-Oxybisethane, (diethyl ether); $C_4H_{10}O$; [60-29-7]	Shaw, A. W.; Vosper, A. J. *J. Chem. Soc. Faraday Trans. I* 1977, *73*, 1239-1244.
VARIABLES:	PREPARED BY:
Temperature	A. J. Vosper

EXPERIMENTAL VALUES:

Absorption measurements			Desorption measurements		
T/K	Mole fraction x_1*	Concentration* mol dm^{-3}	T/K	Mole fraction x_1*	Concentration* mol dm^{-3}
279.4	0.00264	0.0260	227.9	0.00355	0.0377
271.9	0.00271	0.0271	236.2	0.00335	0.0352
264.0	0.00283	0.0287	243.2	0.00321	0.0334
255.5	0.00294	0.0300	252.7	0.00306	0.0314
247.4	0.00306	0.0317	260.2	0.00296	0.0301
240.4	0.00320	0.0333	267.4	0.00287	0.0287
233.4	0.00337	0.0355			
230.7	0.00343	0.0364			
225.7	0.00360	0.0385			

* The mole fractions follow directly from the authors' work whereas the concentrations were calculated using literature values (1) for the density of diethyl ether.

Mean values for the partial molar enthalpy of solution and partial molar entropy of solution over the temperature range 226-279 K were given:

$$\Delta H° = -3.02 \pm 0.08 \text{ kJ mol}^{-1} \qquad \Delta S° = -60.2 \pm 0.3 \text{ J mol}^{-1} \text{ K}^{-1}$$

AUXILIARY INFORMATION

METHOD/APPARATUS/PROCEDURE:

A gas burette was used to determine the volume of gas absorbed by a known weight of the degassed liquid at a known temperature and at the prevailing barometric pressure. Measurements of desorption of gas from the saturated liquid with increasing temperature were also made. Corrections were applied for the vapor pressure of the liquid (1) and non-ideality of the gas (2). Results were calculated for a gas pressure of 101.325 kPa assuming Henry's Law which had previously been shown to be valid over the relevant pressure range.

SOURCE AND PURITY OF MATERIALS:

The gas (Matheson Co. Inc.) was passed through 90 per cent H_2SO_4 and a trap at *ca*. 190 K. It was solidified at 77 K and volatile impurities pumped away. The product contained less than 0.2 per cent N_2O as the only detectable impurity.

The liquid (BDH ANALAR GRADE) was dried over $CaCl_2$ then sodium wire and fractionated. The fraction boiling at 307.7 K was used.

ESTIMATED ERROR:

REFERENCES:

1. Timmermans, J. *"Physico-chemical constants of pure organic compounds"*. Elsevier, Vol. 1, p.*342*.
2. Johnson, H. L.; Weimer, H. R. *J. Amer. Chem. Soc.* 1934, *56*, 625.

COMPONENTS:	ORIGINAL MEASUREMENTS:
1. Nitric oxide; NO; [10102-43-9] 2. Ethyl acetate; $C_4H_8O_2$; [141-78-6]	Shaw, A. W.; Vosper, A. J. *J. Chem. Soc. Faraday Trans. I* 1977, *73*, 1239-1244.
VARIABLES:	PREPARED BY:
Temperature	A. J. Vosper

EXPERIMENTAL VALUES:

Absorption measurements			Desorption measurements		
T/K	Mole fraction x_1*	Concentration* mol dm^{-3}	T/K	Mole fraction x_1*	Concentration* mol dm^{-3}
286.6	0.00173	0.0179	226.2	0.00224	0.0249
277.8	0.00176	0.0184	236.0	0.00212	0.0234
266.2	0.00182	0.0193	242.0	0.00205	0.0224
257.8	0.00187	0.0201	249.2	0.00199	0.0216
247.2	0.00196	0.0213	253.7	0.00197	0.0211
239.8	0.00204	0.0224	260.7	0.00192	0.0205
231.7	0.00213	0.0236	271.2	0.00187	0.0197
224.7	0.00223	0.0249			
220.2	0.00230	0.0258			

* The mole fractions follow directly from the authors' results whereas the concentrations were calculated using literature values (1) for the density of ethyl acetate.

Mean values for the partial molar enthalpy of solution and partial molar entropy of solution over the temperature range 220-287 K were given:

$\Delta H° = -2.53 \pm 0.09$ kJ mol^{-1} $\Delta S° = -62.0 \pm 0.4$ J mol^{-1} K^{-1}

AUXILIARY INFORMATION

METHOD/APPARATUS/PROCEDURE:

A gas burette was used to determine the volume of gas absorbed by a known weight of the degassed liquid at a known temperature and at the prevailing barometric pressure. Measurements of desorption of gas from the saturated liquid with increasing temperature were also made. Corrections were applied for the vapor pressure of the liquid (1) and non-ideality of the gas (2). Results were calculated for a gas pressure of 101.325 kPa assuming Henry's Law which had previously been shown to be valid over the relevant pressure range.

SOURCE AND PURITY OF MATERIALS:

The gas (Matheson Co. Inc.) was passed through 90 per cent H_2SO_4 and a trap at *ca*. 190 K. It was solidified at 77 K and volatile impurities pumped away. The product contained less than 0.2 per cent N_2O as the only detectable impurity.

The liquid (BDH ANALAR GRADE) was dried over potassium carbonate then P_2O_5. It was fractionated and the portion boiling at 350.3 K was used.

ESTIMATED ERROR:

REFERENCES:

1. Timmermans, J. *"Physico-chemical constants of pure organic compounds"*. Elsevier, Vol. 1, p.415.
2. Johnson, H. L.; Weimer, H. R. *J. Amer. Chem. Soc.* 1934, *56*, 625.

COMPONENTS:	ORIGINAL MEASUREMENTS:
1. Nitric oxide; NO; [10102-43-9] 2. Tetrachloromethane; (Carbon tetrachloride); CCl_4; [56-23-5]	Klemenc, A.; Spitzer-Neumann, E.; *Monatsh.* 1929, *53*, 413-419.
VARIABLES:	**PREPARED BY:**
Temperature, pressure.	W. Gerrard

EXPERIMENTAL VALUES:

T/K	281.96	292.76	307.76
Pressure range,*			
from (kPa) :	59.462	108.391	75.754
to (kPa) :	133.456	58.823	100.192
Ostwald coefficient,*			
from :	0.336	0.338	0.368
to :	0.342	0.355	0.385
Mean :**	0.339	0.345	0.375
Mole fraction, x_1:***	0.00140	0.00139	0.00146
Number of measurements :	9	11	7

* Irregular distribution.
** Given by the authors.
*** Calculated by compiler

ΔF^0, cal $mole^{-1}$	2360	2460	2570

From $\Delta F^0 = -RT \ln (L/RT)$

$\Delta H_{298} = 700$ cal $mole^{-1}$

* The Ostwald coefficient was taken to be :

$$L = \frac{\text{Concentration of gas in the liquid phase}}{\text{Concentration of gas in the gas phase.}}$$

AUXILIARY INFORMATION

METHOD/APPARATUS/PROCEDURE:

An absorption vessel, and a gas buret with levelling tube of the usual form were used. The volume of NO absorbed at a total pressure $p_T = p_{NO} + p_S$, where p_{NO} is the partial pressure of the gas, and p_S is the vapor pressure of the solvent over the solution.

SOURCE AND PURITY OF MATERIALS:

1. Prepared by the action of mercury and nitric acid in the presence of 90% sulfuric acid.

2. The solvent was rendered gas free.

ESTIMATED ERROR:

REFERENCES:

COMPONENTS:	ORIGINAL MEASUREMENTS:
1. Nitric oxide; NO; [10102-43-9] 2. Tetrachloromethane, (carbon tetrachloride); CCl_4; [56-23-5]	Shaw, A. W.; Vosper, A. J. *J. Chem. Soc. Faraday Trans. I* 1977, *73*, 1239-1244.
VARIABLES:	**PREPARED BY:**
Temperature	A. J. Vosper

EXPERIMENTAL VALUES:

Absorption measurements			Desorption measurements		
T/K	Mole fraction x_1*	Concentration* mol dm^{-3}	T/K	Mole fraction x_1*	Concentration* mol dm^{-3}
292.2	0.00138	0.0144	256.2	0.00150	0.0163
288.0	0.00139	0.0145	261.2	0.00148	0.0160
283.5	0.00141	0.0149	266.4	0.00147	0.0157
278.4	0.00142	0.0150	270.7	0.00145	0.0154
273.7	0.00144	0.0153	275.7	0.00144	0.0153
268.0	0.00145	0.0155	285.0	0.00137	0.0143
264.0	0.00147	0.0158			
258.2	0.00149	0.0161			
253.7	0.00151	0.0165			

* The mole fractions follow directly from the authors' results whereas the concentrations were calculated using literature values (1) for the density of carbon tetrachloride extrapolating where necessary.

Mean values for the partial molar enthalpy of solution and partial molar entropy of solution over the temperature range 253-292 K were given:

$\Delta H° = -1.42 \pm 0.03$ kJ mol^{-1} $\Delta S° = -59.6 \pm 0.1$ J mol^{-1} K^{-1}

AUXILIARY INFORMATION

METHOD /APPARATUS/PROCEDURE:

A gas burette was used to determine the volume of gas absorbed by a known weight of the degassed liquid at a known temperature and at the prevailing barometric pressure. Measurements of desorption of gas from the saturated liquid with increasing temperature were also made. Corrections were applied for the vapor pressure of the liquid (1) and non-ideality of the gas (2). Results were calculated for a gas pressure of 101.325 kPa assuming Henry's Law which had previously been shown to be valid over the relevant pressure range.

SOURCE AND PURITY OF MATERIALS:

The gas (Matheson Co. Inc.) was passed through 90 per cent H_2SO_4 and a trap at *ca*. 190 K. It was solidified at 77 K and volatile impurities pumped away. The product contained less than 0.2 per cent N_2O as the only detectable impurity.

The liquid (BDH ANALAR GRADE) was refluxed with NaOH solution, dried over $CaCl_2$ and fractionated. The fraction boiling between 349.9-350.1 K was used.

ESTIMATED ERROR:

REFERENCES:

1. Timmermans, J. *"Physico-chemical constants of pure organic compounds"*. Elsevier, Vol. 1, p.*224*.

2. Johnson, H. L.; Weimer, H. R. *J. Amer. Chem. Soc.* 1934, *56*, 625.

COMPONENTS:	ORIGINAL MEASUREMENTS:
1. Nitric oxide; NO; [10102-43-9] 2. Acetonitrile; CH_3CN; [75-05-8]	Shaw, A. W.; Vosper, A. J. *J. Chem. Soc. Faraday Trans. I* 1977, *73*, 1239-1244.
VARIABLES:	PREPARED BY:
Temperature	A. J. Vosper

EXPERIMENTAL VALUES:

Absorption measurements			Desorption measurements		
T/K	Mole fraction x_1*	Concentration* mol dm^{-3}	T/K	Mole fraction x_1*	Concentration* mol dm^{-3}
282.9	0.000786	0.0152	235.7	0.000913	0.0188
274.7	0.000794	0.0155	242.0	0.000888	0.0181
264.7	0.000824	0.0163	248.6	0.000867	0.0175
252.4	0.000849	0.0171	256.7	0.000843	0.0169
244.9	0.000881	0.0179	260.7	0.000833	0.0166
237.6	0.000908	0.0186	269.3	0.000816	0.0161
232.5	0.000929	0.0192			

* The mole fractions follow directly from the authors' results whereas the concentrations were calculated using literature values (1) for the density of acetonitrile.

Mean values for the partial molar enthalpy and partial molar entropy over the temperature range 233-283 K were given:

$$\Delta H° = -1.83 \pm 0.05 \text{ kJ mol}^{-1} \qquad \Delta S° = -66.0 \pm 0.2 \text{ J mol}^{-1} \text{ K}^{-1}$$

AUXILIARY INFORMATION

METHOD/APPARATUS/PROCEDURE:

A gas burette was used to determine the volume of gas absorbed by a known weight of the degassed liquid at a known temperature and at the prevailing barometric pressure. Measurements of desorption of gas from the saturated liquid with increasing temperature were also made. Corrections were applied for the vapor pressure of the liquid (1) and non-ideality of the gas (2). Results were calculated for a gas pressure of 101.325 kPa assuming Henry's Law which had previously been shown to be valid over the relevant pressure range.

SOURCE AND PURITY OF MATERIALS:

The gas (Matheson Co. Inc.) was passed through 90 per cent H_2SO_4 and a trap at *ca.* 190 K. It was solidified at 77 K and volatile impurities pumped away. The product contained less than 0.2 per cent N_2O as the only detectable impurity.

The liquid (BDH ANALAR GRADE) was refluxed over P_2O_5 then fractionated. The fraction boiling at 354.8 K was used.

ESTIMATED ERROR:

REFERENCES:

1. Timmermans, J. *"Physico-chemical constants of pure organic compounds"*. Elsevier, Vol. 1, p.527 and Vol. 2, p.343.
2. Johnson, H. L.; Weimer, H. R. *J. Amer. Chem. Soc.* 1934, *56*, 625.

COMPONENTS:

1. Nitric oxide; NO; [10102-43-9]
2. Nitrobenzene; $C_6H_5NO_2$; [98-95-3]

ORIGINAL MEASUREMENTS:

Klemenc, A.; Spitzer-Neumann, E.

Monatsh. 1929, *53*, 413-419.

VARIABLES:

Temperature, pressure

PREPARED BY:

W. Gerrard

EXPERIMENTAL VALUES:

T/K	293.16	298.16	313.16	333.16	363.16
Pressure range,*					
from (kPa) :	62.128	133.056	136.522	144.921	142.388
to (kPa) :	140.225	-	65.995	67.728	61.195
Ostwald coefficient,*					
from :	0.176	-	0.182	0.184	0.196
to :	0.200	-	0.198	0.205	0.183
Mean: **	0.189	0.190	0.190	0.193	0.188
Mole fraction, x_1:***	0.000805	0.000799	0.000769	0.000747	0.000685
Number of measurements:	9	1	7	6	4

* Irregular distribution.
** Given by authors.
*** Calculated by compiler.

ΔF^0, cal mole^{-1}	2820	2870	3050	3276	3656

From $\Delta F^0 = -RT \ln (L/RT)$

$\Delta H_{298} = 0$

Henry's law was assumed.

AUXILIARY INFORMATION

METHOD/APPARATUS/PROCEDURE:

Gas buret and absorption vessel. Ostwald coefficient taken to be concentration of gas in the liquid phase/concentration of gas in the gas phase. Measurement of volume of gas absorbed at a measured pressure, p_{Total}. Partial pressure of gas =

$p_{Total} - p_S$, where p_S is the partial pressure of the solvent over the solution.

SOURCE AND PURITY OF MATERIALS:

1. Prepared by the action of mercury on nitric acid in the presence of 90% sulfuric acid.

2. Rendered gas free.

ESTIMATED ERROR:

REFERENCES:

COMPONENTS:	EVALUATOR:
1. Nitric oxide 2. Inorganic, nonaqueous liquids	W. Gerrard The Polytechnic of North London Holloway, London, N7 8DB UK March 1980

CRITICAL EVALUATION:

Sulfuric acid; H_2SO_4; [7664-93-9]

The data of Manchot, Konig, and Reimberger [1] on aqueous solutions of sulfuric acid of a range of concentrations up to 96% H_2SO_4 enables a rational, but speculative extrapolation to be made to 100% H_2SO_4. These data enable straightforward Ostwald coefficients to be recorded; and these are deemed acceptable. For 273 K and 1 atm the mole fraction x_{NO}, for the "pure" acid is calculated to be 0.000357, compared with 0.0000592 for water. Results by Lunge [2] and by Tower [3], and by Pinkus and Jacobi [4] fit approximately into the data of Manchot *et al.*

Nitrosyl chloride; NOCl; [2696-92-6]

The mole fraction, x_{NO}, at 220 K [1 atm], 0.0028, based on the observation of Trautz and Gerwig [5] appears to be approximately of the right magnitude.

Nitrose

Tseitlin [6] gave data on the solubility of nitric oxide in solutions of nitrosyl sulfuric acid which appear to be acceptable.

REFERENCES:

1. Manchot, W.; Konig, J.; Reimlinger, S. *Ber.* 1926, *59B*, 2672
2. Lunge, G. *Ber.* 1885, *18*, 1393
3. Tower, O.F. *Z. anorg. Chem.* 1906, *50*, 382
4. Pinkus, A.; Jacobi, J. *Bull. Soc. Chim. Belg.* 1927, *36*, 448
5. Trautz, M.; Gerwig, W. *Z. anorg. Chem.* 1925, *146*, 1
6. Tseitlin, A.N. *J. Applied Chem.* [USSR] 1946, *19*, 820

COMPONENTS:

1. Nitric oxide; NO; [10102-43-9]
2. Water; H_2O; [7732-18-5]
3. Sulfuric acid; H_2SO_4; [7664-93-9]

ORIGINAL MEASUREMENTS:

Manchot, W.; Konig, J.; Reimlinger, S.

Ber, 1926, *59B*, 2672-2681.

VARIABLES:

Composition

PREPARED BY:

W. Gerrard

EXPERIMENTAL VALUES:

T/K = 273.15 Pressure of nitric oxide appeared to be 1 atm.

Conc. of H_2SO_4 wt %	Volume of NO, cm^3, absorbed by 100 g of soln. a	100 g of soln. b	100 cm^3 of soln. a	100 cm^3 of soln. b
0	7.38		7.38	
8.8	6.5	6.1	6.9	6.5
18.2	5.2	4.9	5.9	5.6
28.0	4.5	4.3	5.4	5.2
38.6	3.8	3.6	4.9	4.6
48.0	2.9	2.7	4.0	3.8
52.6	2.4	2.3	3.4	3.2
58.7	2.2	2.1	3.3	3.1
66.5	1.9	1.8	3.0	2.8
70.8	1.9	1.8	3.0	2.8
76.7	1.8	1.7	3.1	2.9
78.0	1.9	1.8	3.2	3.1
88.3	2.0	1.9	3.5	3.4
89.1	2.1	1.9	3.7	3.6
90.0	2.3	2.1	4.1	3.8
90.4	2.4	2.2	4.3	4.0
91.9	2.4	2.2	4.3	4.0
92.4	2.5	2.4	4.6	4.4
95.0	3.8	3.7	7.1	6.9
95.9	4.2	3.9	7.7	7.2

a and b are duplicate measurements

AUXILIARY INFORMATION

METHOD /APPARATUS/PROCEDURE:

The apparatus and technique were stated to be those of Manchot (1).

SOURCE AND PURITY OF MATERIALS:

Appeared to be of satisfactory purity.

ESTIMATED ERROR:

REFERENCES:

1. Manchot, W. *Z. Anorg. Chem.* 1924, *141*, 38.

SYSTEM INDEX

Underlined page numbers refer to evaluation text and those not underlined to compiled tables. All compounds are listed as in Chemical Abstracts. For example, toluene is listed as benzene, methyl- and dimethylsulfoxide is listed as methane, sulfinylbis-.

G

H

I

REGISTRY NUMBER INDEX

Underlined page numbers refer to evaluation text

AUTHOR INDEX

AUTHOR INDEX

(cont.)